Matrices in Engineering Problems

Synthesis Lectures on Mathematics and Statistics

Editor
Steven G. Krantz, *Washington University, St. Louis*

Statistics is Easy!
Dennis Shasha and Manda Wilson
2008

A Gyrovector Space Approach to Hyperbolic Geometry
Abraham Albert Ungar
2008

Matrices in Engineering Problems

Marvin J. Tobias

ISBN: 978-3-031-01274-7   paperback
ISBN: 978-3-031-02402-3   ebook

DOI 10.1007/978-3-031-02402-3

A Publication in the Springer series
SYNTHESIS LECTURES ON MATHEMATICS AND STATISTICS

Lecture #10
Series Editor: Steven G. Krantz, *Washington University, St. Louis*
Series ISSN
Synthesis Lectures on Mathematics and Statistics
Print 1938-1743   Electronic 1938-1751

Matrices in Engineering Problems

Marvin J. Tobias

SYNTHESIS LECTURES ON MATHEMATICS AND STATISTICS #10

ABSTRACT

This book is intended as an undergraduate text introducing matrix methods as they relate to engineering problems. It begins with the fundamentals of mathematics of matrices and determinants. Matrix inversion is discussed, with an introduction of the well known reduction methods. Equation sets are viewed as vector transformations, and the conditions of their solvability are explored.

Orthogonal matrices are introduced with examples showing application to many problems requiring three dimensional thinking. The angular velocity matrix is shown to emerge from the differentiation of the 3-D orthogonal matrix, leading to the discussion of particle and rigid body dynamics.

The book continues with the eigenvalue problem and its application to multi-variable vibrations. Because the eigenvalue problem requires some operations with polynomials, a separate discussion of these is given in an appendix. The example of the vibrating string is given with a comparison of the matrix analysis to the continuous solution.

KEYWORDS

matrices , vector sets, determinants, determinant expansion, matrix inversion, Gauss reduction, LU decomposition, simultaneous equations, solvability, linear regression, orthogonal vectors & matrices, orthogonal transforms, coordinate rotation, Eulerian angles, angular velocity and momentum, dynamics, eigenvalues, eigenvalue analysis, characteristic polynomial, vibrating systems, non-conservative systems, Runge-Kutta integration

Contents

Preface

The primary objective of this book is to present matrices as they relate to engineering problems. It began as a set of notes used in lectures to "B" Course (applied mathematics) classes of the General Electric Advanced Engineering Program. Matrix analysis is a valuable tool used in nearly all the engineering sciences.

The approach is practical rather than strictly mathematical. Introductory mathematics is followed by example applications. Often, pseudo-programming ("Pascal-like") code is used in description of a method. In some parts of the book the emphasis is on the program. Matrix manipulations are fun to program and provide good learning/practice experience.

A working knowledge of matrix methods provides insight into coordinate transforms , rotations, dynamics, and vibrating systems, and many others problems. The fact that the subject matter is closely tied to programming makes it more interesting and more valuable to the engineer.

The first three chapters of the book introduce notation and basic matrix (and determinant) operations. It is well to study the notation, of course, but parts of Chapter 2 may already be known to the student. However, these chapters can be recommended for the programming exercise that they provide.

Chapter 3 is devoted to matrix inversion and its problems. The computer methods discussed are the Gauss reduction and LU decomposition.

Chapter 4 explores the solution to simultaneous equation sets. The equations of linear regression are developed as an example of a very "over-determined" set of linear equations.

Chapter 5 provides the reader with a matrix "framework" for visualizing in three dimensions, and extrapolating to n-dimensions. The equations of particle and rigid body dynamics are developed in matrix form.

Chapters 6 and 7 are largely concerned with the eigenvalue problem—especially as it relates to multi-dimensional vibration problems. The approach given for solving both conservative and non-conservative systems emphasizes the use of the computer.

Marvin J. Tobias
June 2011

CHAPTER 1

Matrix Fundamentals

1.1 DEFINITION OF A MATRIX

A matrix is defined to be a rectangular array of functional or numeric elements, arranged in row or column order. Most important in this definition is that (at most) two subscripts, or indices, are required to identify a given element: a row subscript, and a column subscript. That is, a matrix is a 2-dimensional array. Included within the definition are arrays in which the maximum value of one, or both subscripts is unity. For example, a single "list" of elements, arranged in a single row or column, is referred to as a "row" or "column" matrix. Even a single element may be referred to as a one-by-one (i.e., 1X1) matrix.

By way of illustration, the following matrix, "**A**," is diagrammed:

$$\mathbf{A} = \begin{bmatrix} a_{11} & a_{12} & a_{13} & \cdots & \cdots & a_{1n} \\ a_{21} & a_{22} & \cdots & \cdots & & a_{2n} \\ \cdots & \cdots & \cdots & \ddots & & \cdots \\ a_{m1} & a_{m2} & a_{m3} & \cdots & & a_{mn} \end{bmatrix} \qquad \begin{array}{c} \text{n columns} \\ \text{m rows} \end{array}$$

The above rectangular matrix has m rows, and n columns. The purpose of this book will be to discuss and define the arithmetic (and mathematics) of such arrays. Practical applications will be discussed, in which the array will often be viewed and manipulated as a ***single entity***. Once the notation of matrices is learned, there follows a very large advantage in being able to work with the array as an entity, without being encumbered with the arithmetic manipulation of the numeric values inside. That is, one of the big advantages is that of "bookkeeping."

Carrying this illustration further, we write an m-by-n set of linear algebraic equations as:

$$\begin{cases} a_{11}x_1 + a_{12}x_2 + a_{13}x_3 + \cdots + a_{1n}x_n & = c_1 \\ a_{21}x_1 + a_{22}x_2 + a_{23}x_3 + \cdots + a_{2n}x_n & = c_2 \\ \cdots \qquad \cdots \qquad \cdots \qquad \cdots \qquad \cdots & = \cdots \\ a_{m1}x_1 + a_{m2}x_2 + a_{m3}x_3 + \cdots + a_{mn}x_n & = c_m \,. \end{cases} \qquad (1.1)$$

The above defines a set of m-equations in n-unknowns, the solutions to which will be explored in a later chapter. Right now, the point is to compare the equation set (1.1) with the definition of the m row by n column matrix above. This chapter will concentrate on the basic rules of matrices, which will, among other things, allow us to write the set (1.1) as:

$$\mathbf{Ax} = \mathbf{c} \tag{1.2}$$

wherein the **A** matrix has the form diagrammed above. In (1.2), each of the literal symbols represents a matrix. The **A** matrix is a rectangular one, with m rows, and n columns. The **x** matrix has n rows and just one column. It is usually referred to as a "vector," as is the matrix, **c**, which has m rows and, again, just one column. As mentioned earlier, **x** and **c** can also be called column matrices (or column vectors).

It will be noted immediately that, although (1.2) is beautifully compact, it does not convey all the information of (1.1). That is, (1.2) does not make the "dimensionality" clear: It is not evident that **A** is m rows by n columns. This information must come from the context of the discussion—a fairly small price to pay.

If the set (1.2) is "square" (i.e., $m = n$), then associated with the matrix **A** will be a "***determinant***," written $|\mathbf{A}|$, or $|a_{ij}|$, whose elements are those of **A**, and in the same row, column relationship. Note the "absolute value" bars. This notation is not only convenient, but meaningful, since a determinant, though written as an array, does evaluate to a single functional, or numeric, value (but this $|\mathbf{A}|$ must not be assumed to be necessarily positive).

$$|\mathbf{A}| = \begin{vmatrix} a_{11} & a_{12} & \cdots & a_{1n} \\ a_{21} & a_{22} & \cdots & a_{2n} \\ \cdots & & \cdots & a_{kn} \\ a_{n1} & a_{n2} & \cdots & a_{nn} \end{vmatrix}.$$

Determinants are of great interest in this study of matrices. They "determine" the characteristics of the related matrix, and play a particularly important role in the solution to simultaneous equation sets. Some of the methods used to evaluate determinants will be discussed in the next chapter. At this point it is enough to simply establish that ***determinants are defined for square arrays only, and that they are scalar quantities***.

1.1.1   NOTATION

Matrices in which both indices are > 1, like the matrix **A** in (1.2), will be written using an upper case letter, boldfaced. Equivalently, we may denote such a matrix as $[a_{ij}]$. Since dimensionality must be set in the context of discussion, it will often be done as: **A**(mXn). The expression within these parentheses is read as: "m-by-n." ***The row index will always be stated first***. The vectors **x**, and **c** may be written as {x} and {c}, and when necessary, {x}(nX1), although it will be quite rare to have to write this in this way. In particular, once it is clear that **A** is (mXn), we will see that the dimensions of {x} and {c}, in (1.2), are determined.

The matrix or vector, itself (as an entity), is written in boldface type. However, the elements of the matrix are not bold, and may be written as $[a_{ij}]$, and {x}, for example (not bold.). However, it is sometimes necessary to refer to a row or column within a rectangular (or square) matrix. In such case it will be written in boldface; i.e., {$\mathbf{a}_1$} would refer to a column within **A**.

The {x} and {c} vectors in (1.2) are "column" vectors. There can also, be cases in which the row dimension is unity: a (1Xn) vector. Such a vector is called a "row" vector. It will be written within text as [v]. Please be careful to note the difference between [v] and {v}. For example, if we were to select vectors from the matrix $\mathbf{A}$, the row vectors would have n elements, but the column vectors would have m elements—a very significant difference. Notice also that [v] (a row vector) will not be confused with $[a_{ij}]$ (a rectangular matrix).

Within a text discussion, it would be very unwieldy to write the elements of a column vector vertically down the page. Therefore, if the elements of either a row or column vector must be delineated, it will be done across the page ("horizontally"). A three element column vector, {v}, would be written as: $\{v_1, v_2, v_3\}$.

$$\left\{ \begin{array}{c} v_1 \\ v_2 \\ v_3 \end{array} \right\} \quad \begin{array}{l} \text{written as } v \text{ or} \\ \{v_1, v_2, v_3\} \end{array}$$

A three element row vector would be written $[u_1, u_2, u_3]$, with square brackets.

Some notation examples, (numerical values chosen at random):

$$\mathbf{A} = \mathbf{A}(3X3) = \left[\begin{array}{ccc} a_{11} & a_{12} & a_{13} \\ a_{21} & a_{22} & a_{23} \\ a_{31} & a_{32} & a_{33} \end{array} \right] = \left\| \begin{array}{ccc} 3.1 & 0 & 1.6 \\ 2.2 & 5.2 & 1.1 \\ 1.0 & 3.2 & 4.4 \end{array} \right\| . \tag{1.3}$$

Note that a_{12} (for example) refers to the element in the first row and second column (in the example its value is 0). ***The row subscript is always given first***. Ordinarily, the square brace, $[..]$, is the notation for a matrix (while the single vertical bar denotes its determinant, $|\mathbf{A}|$), but, notice that ***the double vertical bar is sometimes used to denote a matrix.***

As will be seen in coming chapters, a matrix is often viewed as an assemblage of ***vectors.*** For example, $\mathbf{A}$ in (1.3), may be viewed as three row vectors, $[\mathbf{a}_k]$. Note that the entity within the square braces must be shown **bold,** because it refers to a vector, (i.e., $\mathbf{a}_k$), not an element. $\mathbf{A}$ could also be viewed as three column vectors, $\{\mathbf{a}_k\}$. Note that the type of braces used distinguishes between a row or a column vector. For example, with reference to (1.3):

$$[\ \mathbf{a}_2] = [\ 2.2 \quad 5.2 \quad 1.1 \ ]; \qquad \{\mathbf{a}_2\} = \{ \ 0 \quad 5.2 \quad 3.2 \ \}$$

and, also note that $\{\mathbf{a}_2\}$ is a ***column vector***, but, is written across the page (for convenience). Within text it would be written as $\{ \ \mathbf{a}_2 \ \} = \{ 0, 5.2, 3.2 \}$, with commas.

It is extremely difficult to strictly adhere to an unambiguous set of notation rules. Then, new rules, ***possibly contradictory***, may be found throughout the book. The most important 'rule' is to describe each topic clearly. Notation rules may sometimes be "bent" to fit the discussion.

1.2   ELEMETARY MATRIX ALGEBRA

In order to develop an elementary matrix algebra, the definitions of matrix equality, and the basic operations of addition, and multiplication, must be agreed upon. It will be found that there are some fundamental differences between matrix algebra and that of "ordinary" algebra, which deals with "scalar" entities—those ordinary numbers and functions whose dimension is 1X1. But, the rules of matrix algebra are logical, and will seem obvious rather than obtuse or complicated.

To begin, two matrices are equal iff (iff ≡"if and only if") the dimensions of each are the same, and their corresponding elements are equal. For example, A = B iff they both have the same dimensions, mXn, and $a_{ij} = b_{ij}$, for all i and j.

1.2.1   ADDITION (INCLUDING SUBTRACTION)

The sum of two (or more) matrices is formed by summing corresponding elements:

$$\mathbf{C} = \mathbf{B} \pm \mathbf{A} \text{ implies}$$

$$[c_{ij}] = [b_{ij}] \pm [a_{ij}] . \tag{1.4}$$

Note that if the two matrices are of different dimensionality then corresponding elements cannot be found, in which case addition is not defined. ***Matrix addition is defined only when B and A have the same numbers of rows and columns, respectively.*** When this is the case, the matrices **A** and **B** are said to be "conformable in addition." If all the elements of **A** are respectively the negatives of those of **B**, then the sum, **C,** will have all zero elements. In such case, **C** is known as a "null" matrix (the "zero" of matrix algebra). Also, if **A** happened to be null, then **C** would be equal to **B**, $c_{ij} = b_{ij}$ for all i and j.

Since addition is commutative for the elements of the matrix, then matrix addition itself is commutative. That is, $\mathbf{A} + \mathbf{B} = \mathbf{B} + \mathbf{A}$.

1.2.2   MULTIPLICATION BY A SCALAR

The matrix $(k)\mathbf{A}$ is formed by multiplying ***every element*** of **A** by the scalar (k). Note that the notation (k), with parentheses, is used here. However, the notation, $k\mathbf{A}$, will also be used. Neither $(k)\mathbf{A}$, nor $k\mathbf{A}$, will be confused with matrix multiplication, because row, or column, vectors (also expressed in lower case) must be written as {k}, or [k]. In passing, we note that if **A** is square (nXn), and is multiplied by the scalar, k, then the determinant of **A** will be multiplied by k^n. Conversely, then $(k)|\mathbf{A}|$ will mean the multiplication of a single row, or column, by k. More on this, later.

1.2.3   VECTOR MULTIPLICATION

Since rectangular matrices are composed of vectors, we will first discuss vector products, before defining the product of these "larger" matrices. The most important product of two vectors is their "dot product," or "scalar product." This product results in a scalar—just as does the vector dot product in vector analysis. Furthermore, the numerical result is the same also, since it is the sum of the products of the corresponding elements.

Vector dot product $\equiv \mathbf{u} \bullet \mathbf{v}$

$$\equiv [u_1 \ \cdots \ u_n] \left\{ \begin{array}{c} v_1 \\ \vdots \\ v_n \end{array} \right\} = (u_1 v_1 + u_2 v_2 + \ \cdots \ + u_n v_n) = \sum_{j=1}^{n} u_j v_j \ .$$

It may help to visualize the premultiplying row vector "swinging into the vertical," and then mul-tiplying element-by-element, as in the following diagram. Nevertheless, *the premultiplying vector must be a row vector*.

$$
\begin{array}{ll}
|u_1| \ |v_1| & \rightarrow u_1 v_1 \\
|u_2| \ |v_2| & \rightarrow u_2 v_2 \\
\vdots \ \ \vdots & \\
|u_n| \ |v_n| & \rightarrow u_n v_n \\
\hline
& \sum_{i=1}^{n} u_i v_i
\end{array}
$$

$$[u_1 \ u_2 \ ... u_n]$$

Note that both vectors must have the same number of terms (elements). That is, the two vectors must have the "same dimensions." If such were not the case, the two vectors would not be "conformable, in multiplication." *Most important is that the dot product is always seen as the product of a row vector times a column vector; and its result is a (1X1) matrix (i.e., a scalar).* In this regard, the most meaningful notation for the vector dot product is ⌊u⌋{v}, or [v]{u}.

In analytic geometry, two vectors are written: $u = u_1\mathbf{i} + u_2\mathbf{j} + u_3\mathbf{k}$, and $v = v_1\mathbf{i} + v_2\mathbf{j} + v_3\mathbf{k}$, where $\mathbf{i}, \mathbf{j}$, and $\mathbf{k}$, are "unit vectors" in the directions of an "xyz" coordinate set (for example, $\mathbf{i}$ may be the unit vector in the "x"-direction). The dot product of the two is:

$$\mathbf{u} \bullet \mathbf{v} = |\mathbf{u}| \ |\mathbf{v}| \cos \theta = (u_1\mathbf{i} + u_2\mathbf{j} + u_3\mathbf{k})(v_1\mathbf{i} + v_2\mathbf{j} + v_3\mathbf{k}) = \sum_{j=1}^{n} u_j v_j$$

where $|\mathbf{u}||\mathbf{v}|$ refers to the (scalar) product of their respective magnitudes, and θ is the angle between the two. In carrying out the multiplication, the following relationships are used:

$$\mathbf{i} \bullet \mathbf{j} = \mathbf{i} \bullet \mathbf{k} = \mathbf{j} \bullet \mathbf{k} = 0, \quad \text{Orthogonal axes;}$$
$$\mathbf{i} \bullet \mathbf{i} = \mathbf{j} \bullet \mathbf{j} = \mathbf{k} \bullet \mathbf{k} = 1, \quad \text{Unit length.}$$

In more than three dimensions, the idea is the same, but, we soon run out of $(\mathbf{i}, \mathbf{j}, \mathbf{k}, \ldots)$ unit vectors. When many dimensions are possible, the unit vectors might be denoted as $\mathbf{1}, \mathbf{2}, \mathbf{3}, \mathbf{4} \ldots$, and since there may be several coordinate sets in consideration, we might distinguish these by subscript. For example, $\mathbf{1}_x$ might be the unit vector along axis $\mathbf{1}$ of the x-set, while $\mathbf{1}_y$ would have the same meaning in the y-set. More often, the vector is simply written $\{v_1, v_2, v_3, \ldots\}$. Although, we may have trouble visualizing vectors in more than 3 dimensions, we simply draw the analogy to the 3 dimensional case.

Note that, just as in 3 dimensions, the n-dimensional dot product can produce a zero result even when neither of the vectors is zero. That is, $\cos\theta$ could be zero, in which case the vectors are perpendicular, or "orthogonal."

The product $\mathbf{v}\bullet\mathbf{v}$ is always *conformable*, and is the sum of the squared elements of $\mathbf{v}$. Again, by analogy with 3 dimensions, $\mathbf{v}\bullet\mathbf{v}$ is the "square of the length" of $\mathbf{v}$, and sqrt($\mathbf{v}\bullet\mathbf{v}$) is $|\mathbf{v}|$, the "length" of the n dimensional vector. Also $\mathbf{u}\bullet\mathbf{v}$ is the product $|\mathbf{u}||\mathbf{v}|$ multiplied by the cosine of the angle between $\mathbf{u}$ and $\mathbf{v}$ (as in vector analysis in 3 dimensions).

The product $\{v\}[u]$ (a column vector times a row vector) is also conformable, when $\mathbf{u}$ and $\mathbf{v}$ have the same dimensions. Given that both vectors are (nX1), the product is an (nXn) square matrix. This result will be reviewed again in the next paragraphs. See (1.21), Section 1.6.

1.2.4   MATRIX MULTIPLICATION

In (1.2), the product $\mathbf{A}x$ is set equal to the vector $\mathbf{c}$. Apparently, then, the product of a rectangular matrix and a vector is another vector. From (1.1), it will be seen that (in $\mathbf{A}x=\mathbf{c}$) *each (scalar) element of $\mathbf{c}$ is the sum of the element-by-element products of a <u>row vector</u> of $\mathbf{A}$ by the <u>column</u>* x: The first row vector of $\mathbf{A}$ is: $[\mathbf{a}_1] = [a_{11}, a_{12}, \ldots, a_{1n}]$. The product $[\mathbf{a}_1]\{x\}$ is c_1, the first element of the vector $\mathbf{c}$. That is (from (1.1)):

$$a_{11}x_1 + a_{12}x_2 + \cdots + a_{1n}x_n = [\mathbf{a}_1]\{x\} = \sum_{j=1}^{n} a_{1j}x_j = c_1 \,.$$

The above equation is nothing more than a rewrite of the first equation in (1.1). But, the important point to get here is that the left side of the above *is the dot product* $[\mathbf{a}_1]\{x\}$. The concept of matrix multiplication is simply the extension of this to the case where there are more columns in the "post-"multiplier.

In the general case, $\mathbf{C}=\mathbf{AB}$ (i.e., $\mathbf{C}=\mathbf{A}$ times $\mathbf{B}$), each element of $\mathbf{C}$ is the result of a dot product of a row from $\mathbf{A}$ and a column from $\mathbf{B}$. In particular, the general element $c_{ij} = \mathbf{a}_i \bullet \mathbf{b}_j$. The concept is shown diagrammatically in Figure 1.1.

$$\begin{bmatrix} a_{11} & a_{12} & \cdots & a_{1k} \\ a_{21} & \cdots & & a_{2k} \\ \cdots & \cdots & \cdots & \cdots \\ a_{m1} & & & a_{mk} \end{bmatrix} \begin{bmatrix} b_{11} & b_{12} & & b_{1n} \\ b_{21} & \cdots & & b_{2n} \\ \cdots & \cdots & \cdots & \cdots \\ b_{k1} & & & b_{kn} \end{bmatrix} = \begin{bmatrix} [\mathbf{a}_1]\{\mathbf{b}_1\} & [\mathbf{a}_1]\{\mathbf{b}_2\} & \cdots & [\mathbf{a}_1]\{\mathbf{b}_n\} \\ [\mathbf{a}_2]\{\mathbf{b}_1\} & [\mathbf{a}_2]\{\mathbf{b}_2\} & \cdots & [\mathbf{a}_2]\{\mathbf{b}_n\} \\ & \cdots & & \\ [\mathbf{a}_m]\{\mathbf{b}_1\} & & \cdots & \cdots & [\mathbf{a}_m]\{\mathbf{b}_n\} \end{bmatrix} \,.$$

Figure 1.1: The row-times-column dot product concept in matrix multiplicationn.

The figure is intended to emphasize the "row times column" dot product concept; so the **A** matrix is shown "partitioned" into rows (by the horizontal lines), and the **B** matrix is partitioned into columns. In the figure, the **A** matrix is shown with m rows, and k columns, i.e., **A**(mXk). The **B** matrix has k rows and n columns, **B**(kXn). The **C** matrix elements are all the results of a vector dot product. The following statements define matrix multiplication, and will clarify the dimensionality of **C**.

- Each element of the product matrix, c_{ij}, is the result of the dot product $[\mathbf{a}_i]\{\mathbf{b}_j\}$.

$$c_{ij} = [\mathbf{a}_i]\{\mathbf{b}_j\} = \sum_{s=1}^{k} a_{is}b_{sj} \ . \tag{1.5}$$

- If the dot product $[\mathbf{a}_i]\{\mathbf{b}_j\}$ is to be ***conformable***, the number of terms in $\mathbf{a}_i$ must be the same as the number of terms in $\mathbf{b}_j$. Then ***the number of columns in*** **A** ***must equal the number of rows in*** **B**. Thus, **B** must have k rows, conforming to the k columns in **A**.

- The conformability of **AB** does not depend on the number of rows of **A**, nor the number of columns of **B**.

- As each succeeding row in **A** is selected (to form the next dot product), a new row is created in the result, **C**. Then, **C** must have the same number of rows as **A**. The same reasoning shows that **C** must have the same number of columns as **B**. Therefore, **C** is (mXn).

Two (mXn) matrices are conformable in addition, but not in multiplication. For conformability in the multiplication, **AB**, we must have **A**(mX<u>k</u>)**B**(<u>k</u>Xn). That is, the underlined dimensions must be the same. At first, this may be confusing. But, there is a simple way to write down, and immediately determine conformability: Just write the two sets of dimensions within the same parentheses, and "cancel" the internal numbers, *IF they are the same*. Then if **A** is (mXk), and **B** is (kXn), we write (mX~~kXk~~Xn)→(mXn). In this case, since the columns of **A**, and the rows of **B** are equal in number; then the "k's cancel." This simple expression not only tells us that **A** and **B** are conformable, but also indicates that the resultant matrix will be (mXn). If the "k's don't cancel," i.e., the two inside dimensions in the expression are not the same, then **A** and **B** are not conformable in multiplication. When both matrices are (mXn), we have (mXnXmXn) in which case the internal subscripts do not match.

As a first example of matrix multiplication, consider the following product, (2X~~3X3~~X2)→(2X2):

$$\mathbf{AB} = \begin{bmatrix} 3 & 4 & -1 \\ 2 & 0 & 6 \end{bmatrix} \begin{bmatrix} 3 & 5 \\ 7 & -2 \\ 6 & 6 \end{bmatrix} = \begin{bmatrix} 31 & 1 \\ 42 & 46 \end{bmatrix}. \tag{1.6}$$

These matrices are also conformable in reverse order. The reader should calculate the product **BA**, and take special note that the result is (3X~~2X2~~X3) → (3X3). The very same matrices, **A** and **B,** but very different results—which illustrates that *matrix multiplication is <u>not commutative</u>* . That is, in general $\mathbf{AB} \neq \mathbf{BA}$. The product $\mathbf{BA}$ *may not even be conformable* in multiplication, even though $\mathbf{AB}$ is perfectly legal. For emphasis, however, please note that in general $\mathbf{AB} \neq \mathbf{BA}$ even if both products are conformable. Try a few simple matrix products to prove that this is the case (We will see, shortly, that in some cases, multiplication is commutative).

Because of the non-commutative nature of the matrix product the order of the product, must be stated explicitly. For example, $\mathbf{AB}$ can be described as "the *PRE* multiplication of $\mathbf{B}$, by $\mathbf{A}$," or alternatively, "the *POST* multiplication of $\mathbf{A}$, by $\mathbf{B}$."

Matrix multiplication is, however, associative. That is:

$$\mathbf{A(BC)} = \mathbf{(AB)C} = \mathbf{ABC} . \tag{1.7}$$

It does not matter whether we form the product $\mathbf{BC}$, first, then premultiply by $\mathbf{A}$, or form $\mathbf{AB}$, then postmultiply by $\mathbf{C}$. Further, it is distributive:

$$\mathbf{A(B + C)} = \mathbf{AB} + \mathbf{AC} . \tag{1.8}$$

From (1.7), we may draw the inference that the powers of a (*necessarily square*) matrix, say $\mathbf{A}$, are defined: $\mathbf{A}^2 = \mathbf{A(A)}$, $\mathbf{A}^3 = \mathbf{A(A)(A)}$, and so on. Then, it follows that matrix polynomials are also valid:

$$p(\mathbf{A}) = c_0 \mathbf{A}^n + c_1 \mathbf{A}^{n-1} + \cdots + c_{n-1} \mathbf{A} + c_n \mathbf{I} . \tag{1.9}$$

In (1.9), the coefficients, c_i, are scalar constants; c_n multiplies the "unit matrix," $\mathbf{I}$, defined in Section 1.3, below.

Because matrix multiplication is so fundamental to our study, the reader should try several examples, to become sure of the method. In each case, write the expressions like (2X~~3X3~~X2) to see how these indicate the conformability and the dimensions of the result.

1.2.5   TRANSPOSITION

The matrix transpose of $\mathbf{A}$ is written $\mathbf{A}'$ ("$\mathbf{A}$ prime"). $\mathbf{A}'$ is obtained by interchanging the rows and columns of $\mathbf{A}$. Then $\mathbf{A}$(mXn) becomes $\mathbf{A}'$(nXm) under transposition. Also, $\{v\}' => [v]$, that is, the transpose of a column is a row, and vice versa. The transpose operation is very important. For clarity, it may sometimes be necessary to write $\mathbf{A}$ transpose as $\mathbf{A}^t$, rather than $\mathbf{A}'$.

Transpose of a matrix product: Suppose $\mathbf{C} = \mathbf{AB}$, and we wish to express the transpose, $\mathbf{C}'$, in terms of $\mathbf{A}'$ and $\mathbf{B}'$. Remember that the c_{ij} element of $\mathbf{C}$ is the dot product of the ith row of $\mathbf{A}$ into the jth column of $\mathbf{B}$. Upon transposition, the columns of $\mathbf{B}$ become rows, and the rows of $\mathbf{A}$ become columns. It therefore follows that in order to preserve the correct dot products, we must take the $\mathbf{B}'$ and $\mathbf{A}'$ matrices in reverse order. That is:

$$\mathbf{C}' = (\mathbf{AB})' = \mathbf{B}'\mathbf{A}' \equiv \mathbf{B}^t \mathbf{A}^t . \tag{1.10}$$

As a check, consider the 2,3 element of $\mathbf{C}'$. It is the same as the 3,2 element of $\mathbf{C}$. From (1.6), it is clear that the element c_{32} is obtained by $[\mathbf{a}_3]\{\mathbf{b}_2\}$. The element c'_{23} is obtained by $[\mathbf{b}'_2]\{\mathbf{a}'_3\}$.

Apparently then, the reasoning of (1.10) is correct. This is known as the "reversal rule" of matrix transposition. By logical extension of this rule, to continued products:

$$\mathbf{D}' = (\mathbf{ABC})' = (\mathbf{C})'(\mathbf{AB})' = \mathbf{C}'\mathbf{B}'\mathbf{A}' . \tag{1.11}$$

Note that for any matrix, $\mathbf{A}(m\mathrm{X}n)$, that if $\mathbf{B} = \mathbf{A}'\mathbf{A}'$, then $\mathbf{B}' = \mathbf{B} = \mathbf{A}'\mathbf{A}$. That is, $\mathbf{B}$ is unchanged under transposition. Such a matrix is called "symmetric." See the next section, below.

1.3   BASIC TYPES OF MATRICES

1.3.1   THE UNIT MATRIX

A *square* ($n\mathrm{X}n$) matrix whose ij elements are zero for $i \neq j$, and whose elements ii are unity, is defined as the "unit matrix," $\mathbf{I}$. $\mathbf{I}$ corresponds to unity in scalar mathematics. For example, if they are conformable, $\mathbf{I}\{x\} = \{x\}$, or $\mathbf{AI} = \mathbf{A}$. Just as in scalar algebra, the multiplication of a matrix, $\mathbf{A}(n\mathrm{X}n)$, by the unit matrix, $\mathbf{I}(n\mathrm{X}n)$, leaves $\mathbf{A}$ unaltered. Further, $\mathbf{I}$ commutes with any square matrix of the same dimensions (i.e., $\mathbf{IA} = \mathbf{AI} = \mathbf{A}$). Note also that $\mathbf{I} = \mathbf{I}(\mathbf{I})$, and $\mathbf{I}' = \mathbf{I}$.

In the unit matrix, $\mathbf{I}$, the unity elements are said to lie in the "principal diagonal," or the "main diagonal." The "off-diagonal" elements are zero.

The unit matrix can also be written as $[\delta_{ij}]$. The symbol, "δ_{ij}," is known as the "**Kronecker Delta**." By definition, $\delta_{ij} = 1$, for $i = j$, and $\delta_{ij} = 0$, for $i \neq j$.

1.3.2   THE DIAGONAL MATRIX

If the main diagonal elements are not unity, but all elements off this diagonal are zero, then the matrix is a "diagonal matrix." The diagonal elements are not, in general, equal in value. In the cases in which the main diagonal elements are equal, the matrix is called a "scalar matrix." (In the matrix polynomial, written above, (1.9), $c_n\mathbf{I}$ is a scalar matrix.)

The product of two diagonal matrices is another diagonal matrix, whose main diagonal elements are the products of the corresponding elements of the two given matrices. Clearly, then, diagonal matrix products commute. However, if $\mathbf{A}$ is not diagonal, and $\mathbf{B}$ is diagonal, the product is not commutative. In $\mathbf{BA}$, the corresponding *rows* of $\mathbf{A}$ are multiplied by the diagonal elements of $\mathbf{B}$, while in $\mathbf{AB}$, the corresponding *columns* of $\mathbf{A}$ are multiplied by the diagonal elements of $\mathbf{B}$. Try both cases, to be assured that this is true.

1.3.3   ORTHOGONAL MATRICES

The rows (and/or) columns of an orthogonal matrix are perpendicular (orthogonal), in the very same sense meant in vector analysis. That is, the dot product of any row with another is zero. A simple example is:

$$\begin{bmatrix} \cos\theta & -\sin\theta \\ \sin\theta & \cos\theta \end{bmatrix} \quad \text{A (2X2) orthogonal matrix.}$$

Clearly, the rows and columns of the above 2X2 are orthogonal; their dot products are zero. In the case of this example, the matrix is also said to be "orthonormal," because the lengths of the rows/columns are normalized to 1.0 (i.e., the dot product of any row/column into itself is 1.0). The orthogonal matrix has frequent application in engineering problems.

Given an nXn orthonormal matrix, $\mathbf{A}$, it should be clear that $\mathbf{A'A} = \mathbf{I}$, the unit matrix, because $\mathbf{A'A}$ simply forms all the dot products of the columns of $\mathbf{A}$ with each other. Only when a column is dotted into itself is there a nonzero result, and that result will lie on the main diagonal, and will have the value unity. More generally, if $\mathbf{A}$ is just orthogonal (not normalized), then a diagonal matrix results from the $\mathbf{A'A}$ product.

1.3.4   TRIANGULAR MATRICES

If the matrix, $\mathbf{A}$, has all zero elements below the main diagonal, it is known as an "upper triangular" matrix. The transpose of an upper triangular matrix—one with all zero elements above the main diagonal—is called "lower triangular."

Such matrices are very important because (1) their determinant is easily calculated as the product of its main diagonal terms, and (2) its inverse is similarly easy to determine. The following example (though not a matrix inversion) indicates the ease of solution of a triagular set of equations:

$$\begin{bmatrix} 1 & 2 & -1 \\ 0 & 3 & 3 \\ 0 & 0 & 5 \end{bmatrix} \begin{Bmatrix} x_1 \\ x_2 \\ x_3 \end{Bmatrix} = \begin{Bmatrix} 7 \\ 9 \\ 5 \end{Bmatrix}.$$

Since the last equation is "uncoupled," $x_3 = 1$ by inspection. Once x_3 is known, x_2 can be solved, and then x_1 follows.

It is not surprising that many methods for solving determinants, equation sets, and matrix inversions incorporate matrix triangularization.

1.3.5   SYMMETRIC AND SKEW-SYMMETRIC MATRICES

A matrix which is unchanged under transposition is known as "symmetric." For example the matrix, $\mathbf{A}$, below, is symmetric ($\mathbf{A'} = \mathbf{A}$),

$$\mathbf{A} = \begin{bmatrix} a & e & f \\ e & b & g \\ f & g & c \end{bmatrix} \qquad \mathbf{W} = \begin{bmatrix} 0 & -w_3 & w_2 \\ w_3 & 0 & -w_1 \\ -w_2 & w_1 & 0 \end{bmatrix}$$

$$\text{Symmetric} \qquad\qquad \text{Skew-Symmetric}$$

(Note: a, b, c, e, f, g, and w_i, are scalar elements)

and we note that $\{\mathbf{a}_i\} = [\mathbf{a}_i]$, i.e., corresponding rows and columns are equal. For example, row 1: [a, e, f] equals column 1: {a, e, f}.

Symmetric matrices play a large part in engineering problems. For example, energy functions are usually symmetric. Later on, we will have use for the fact that, for any real matrix, $\mathbf{B}$, $\mathbf{B'B}$ is always a square, symmetric matrix. That is, in general, $\mathbf{B}$ is (mXn), and the product (nX~~m~~X~~m~~Xn) is (nXn), i.e., square. It is obvious that $(\mathbf{B'B})' = \mathbf{B'B}$, i.e., the product matrix is symmetric.

If $\mathbf{W'} = -\mathbf{W}$, then $\mathbf{W}$ is called a "Skew-symmetric matrix." Since the principal diagonal elements are unchanged under transposition, then necessarily, the main diagonal elements of a skew-symmetric matrix must be zero. The most prominent example of a skew-symmetric matrix is the angular velocity matrix (Chapter 5).

1.3.6   COMPLEX MATRICES

A matrix, $\mathbf{Z}$, whose elements are complex numbers can be written $[z_{ij}]$, where $z_{ij} = x_{ij} + jy_{ij}$, or $\mathbf{Z} = \mathbf{X} + j\mathbf{Y}$, (where "$j$" is the notation for $\sqrt{-1}$). The latter form shows a "separation" of the real and imaginary parts into separate matrices. In this notation, both $\mathbf{X}$ and $\mathbf{Y}$, are composed of real numbers. A matrix, $\mathbf{W} = \mathbf{X} - j\mathbf{Y}$, is called the "conjugate" of $\mathbf{Z}$. The transpose of $\mathbf{W}$ is referred to as the "associate" of $\mathbf{Z}$.

The sum, or product, of two complex matrices can be formed in the straightforward, element by element, way—using complex arithmetic—or using the second notation, $(\mathbf{Z} = \mathbf{X} + j\mathbf{Y})$, previously coded (real arithmetic) routines can be used, since $\mathbf{X}$ and $\mathbf{Y}$ are composed of real numbers. For example:

$$\mathbf{Z_1Z_2} = (\mathbf{X_1X_2} - \mathbf{Y_1Y_2}) + j(\mathbf{X_1Y_2} + \mathbf{Y_1X_2}) \ .$$

The Hermitian matrix: If the elements of the complex matrix, $\mathbf{Z} = \mathbf{X} + j\mathbf{Y}$, are such that $\mathbf{X}$ is symmetric, and $\mathbf{Y}$ is skew symmetric, then $\mathbf{Z}$ is known as an "Hermitian" matrix. The Hermitian matrix is equal to its "associate." That is, if $\mathbf{Z}$ is Hermitian, then $\mathbf{Z}$ is equal to the conjugate of its transpose. The Hermitian matrix (with its symmetrical real part) is similar in ways to the (entirely real) symmetric matrix.

1.3.7   THE INVERSE MATRIX

Thus, far, we have not defined matrix division. In the general case, no such operation as $\mathbf{A/B}$ exists. However, if $\mathbf{A}$ is a square matrix, then there may be a matrix, $\mathbf{B}$, such that $\mathbf{AB} = \mathbf{I}$. In this case, the matrix $\mathbf{B}$ is referred to as the "inverse" of $\mathbf{A}$, and is written with -1 in superscript as $\mathbf{B} = \mathbf{A}^{-1}$. Similarly, $\mathbf{A} = \mathbf{B}^{-1}$. The notation $\mathbf{A/B}$ or $\mathbf{A} = 1/\mathbf{B}$ is never used.

The matrices, $\mathbf{A}$ and $\mathbf{B}$, shown below, are examples:

$$\mathbf{A} = \begin{bmatrix} 1 & 2 & 2 \\ 1 & 0 & 1 \\ 1 & -3 & 0 \end{bmatrix} \qquad \mathbf{B} = \begin{bmatrix} -3 & 6 & -2 \\ -1 & 2 & -1 \\ 3 & -5 & 2 \end{bmatrix} \qquad \mathbf{AB} = \begin{bmatrix} 1 & 0 & 0 \\ 0 & 1 & 0 \\ 0 & 0 & 1 \end{bmatrix}$$

and, since $\mathbf{AB} = \mathbf{BA}$ then $\mathbf{B} = \mathbf{A}^{-1}$, and $\mathbf{A} = \mathbf{B}^{-1}$.

Note that inverse matrices commute (i.e., $\mathbf{AA}^{-1} = \mathbf{A}^{-1}\mathbf{A}$). Using the example, prove that this is true by multiplying $\mathbf{AB}$ and then $\mathbf{BA}$ to show that they are the same.

Finding the solution to a (square) set of linear algebraic equations (when the solution is unique) is equivalent to finding the inverse of the coefficient matrix:

Given
$$\mathbf{Ax} = \mathbf{c}, \text{ then}$$
$$(\mathbf{A}^{-1})\mathbf{Ax} = (\mathbf{A}^{-1})\mathbf{c}; \quad \text{assuming that } \mathbf{A}^{-1} \text{ exists.} \qquad (1.12)$$
$$\mathbf{x} = (\mathbf{A}^{-1})\mathbf{c}$$

Not every matrix has an inverse. For example, an nXm (non-square) matrix does not. Some square (nXn) matrices do not have an inverse. Those that do not are called "*singular matrices*."

The inverse of a diagonal matrix is another diagonal matrix, whose principal diagonal elements are the reciprocals of the corresponding elements of the given matrix. Clearly, then, a diagonal matrix with a zero element on the main diagonal, is "singular."

Also, the transpose of an inverse matrix is equal to the inverse of its transpose. That is, given a "non-singular" matrix, $\mathbf{A}$:

then
$$\mathbf{A}(\mathbf{A}^{-1}) = \mathbf{I}$$
$$(\mathbf{A}(\mathbf{A}^{-1}))' = (\mathbf{I})' = \mathbf{I}$$
$$(\mathbf{A}^{-1})'\mathbf{A}' = \mathbf{I}$$

and, by postmultiplying by the inverse of $\mathbf{A}$-transpose, $(\mathbf{A}')^{-1}$:

$$(\mathbf{A}^{-1})' = (\mathbf{A}')^{-1}.$$

The above equation shows not only the proof of the above statement, it also shows that the inverse of a symmetric matrix is also symmetric.

By similar reasoning, consider the matrix product, $\mathbf{C} = \mathbf{AB}$. Postmultiplying by $\mathbf{B}^{-1}$

$$\mathbf{CB}^{-1} = \mathbf{A}.$$

Now, postmultiply by $\mathbf{A}^{-1}$:

$$\mathbf{C}(\mathbf{B}^{-1}\mathbf{A}^{-1}) = \mathbf{I}.$$

Then, $\mathbf{C}^{-1}$ must be equal to the product $\mathbf{B}^{-1}\mathbf{A}^{-1}$. That is, the reversal rule applies to the product of matrices: *The inverse of the product of two matrices is equal to the product of their individual inverses, taken in the reverse order*. This fact is sometimes referred to as the "reversal rule" of matrix multiplication. It is worth reviewing that this reverse order phenomenon was also found in forming the transpose of the product of two matrices, (1.10).

1.4   TRANSFORMATION MATRICES

It is frequently necessary to manipulate rows, columns, elements within a matrix. Section 3.2 of Chapter 3 discusses three "Elementary Operations" that are useful in diagonalizing a matrix. These operations are briefly introduced here simply because they are good practice, and give excellent insight in the basic operations.

If a unit matrix row/column i is interchanged with row/column j, and that altered unit matrix is used as a premultiplier on **A**, the rows i and j of **A** are interchanged.

$$
\begin{bmatrix} 0 & 1 & 0 \\ 1 & 0 & 0 \\ 0 & 0 & 1 \end{bmatrix}
\begin{bmatrix} a_{11} & a_{12} & a_{13} \\ a_{21} & a_{22} & a_{23} \\ a_{31} & a_{32} & a_{33} \end{bmatrix}
=
\begin{bmatrix} a_{21} & a_{22} & a_{23} \\ a_{11} & a_{12} & a_{13} \\ a_{31} & a_{32} & a_{33} \end{bmatrix}. \tag{1.13}
$$

As a postmultiplier:

$$
\begin{bmatrix} a_{11} & a_{12} & a_{13} \\ a_{21} & a_{22} & a_{23} \\ a_{31} & a_{32} & a_{33} \end{bmatrix}
\begin{bmatrix} 0 & 1 & 0 \\ 1 & 0 & 0 \\ 0 & 0 & 1 \end{bmatrix}
=
\begin{bmatrix} a_{12} & a_{11} & a_{23} \\ a_{22} & a_{21} & a_{13} \\ a_{32} & a_{31} & a_{33} \end{bmatrix}. \tag{1.14}
$$

If the ith main diagonal element of the unit matrix is multiplied by a factor, k, and then that altered unit matrix is used as a premultiplier on **A**, the corresponding row of **A** is multiplied by $\underline{k}$:

$$
\begin{bmatrix} 1 & 0 & 0 \\ 0 & k & 0 \\ 0 & 0 & 1 \end{bmatrix}
\begin{bmatrix} a_{11} & a_{12} & a_{13} \\ a_{21} & a_{22} & a_{23} \\ a_{31} & a_{32} & a_{33} \end{bmatrix}
=
\begin{bmatrix} a_{21} & a_{22} & a_{23} \\ ka_{11} & ka_{12} & ka_{13} \\ a_{31} & a_{32} & a_{33} \end{bmatrix}. \tag{1.15}
$$

As a postmultiplier:

$$
\begin{bmatrix} a_{11} & a_{12} & a_{13} \\ a_{21} & a_{22} & a_{23} \\ a_{31} & a_{32} & a_{33} \end{bmatrix}
\begin{bmatrix} 1 & 0 & 0 \\ 0 & k & 0 \\ 0 & 0 & 1 \end{bmatrix}
=
\begin{bmatrix} a_{11} & ka_{12} & a_{13} \\ a_{21} & ka_{22} & a_{23} \\ a_{31} & ka_{32} & a_{33} \end{bmatrix}. \tag{1.16}
$$

Lastly, if the ijth ($i \neq j$) element of I is replaced by a factor k, and the altered unit matrix is used as a premultiplier, then to the elements of the ith row are added k times the elements of the jth row:

$$
\begin{bmatrix} 1 & 0 & 0 \\ k & 1 & 0 \\ 0 & 0 & 1 \end{bmatrix}
\begin{bmatrix} a_{11} & a_{12} & a_{13} \\ a_{21} & a_{22} & a_{23} \\ a_{31} & a_{32} & a_{33} \end{bmatrix}
=
\begin{bmatrix} a_{11} & a_{12} & a_{13} \\ ka_{11} + a_{21} & ka_{12} + a_{22} & ka_{13} + a_{23} \\ a_{31} & a_{32} & a_{33} \end{bmatrix}. \tag{1.17}
$$

Of the three operative matrices, this last one is the most important. It would be worthwhile for the reader to experiment with these operations—especially the last.

As an example use of such transformations, the following **A**(3X3) will be changed into triangle form (the original 3,1 element is already zero).

$$\text{Element 2, 1} \rightarrow 0 \begin{bmatrix} 1 & 0 & 0 \\ \frac{1}{3} & 1 & 0 \\ 0 & 0 & 1 \end{bmatrix} \begin{bmatrix} 3 & -1 & 0 \\ -1 & 5 & 2 \\ 0 & 2 & 1 \end{bmatrix} = \begin{bmatrix} 3 & -1 & 0 \\ 0 & \frac{14}{3} & 2 \\ 0 & 2 & 1 \end{bmatrix} \quad \mathbf{A} = \begin{bmatrix} 3 & -1 & 0 \\ -1 & 5 & 2 \\ 0 & 2 & 1 \end{bmatrix}$$

$$\text{Element 3, 2} \rightarrow 0 \begin{bmatrix} 3 & -1 & 0 \\ 0 & \frac{14}{3} & 2 \\ 0 & 2 & 1 \end{bmatrix} \begin{bmatrix} 1 & 0 & 0 \\ 0 & 1 & 0 \\ 0 & -2 & 1 \end{bmatrix} = \begin{bmatrix} 3 & -1 & 0 \\ 0 & \frac{2}{3} & 2 \\ 0 & 0 & 1 \end{bmatrix}.$$

1.5 MATRIX PARTITIONING

It is sometimes convenient to partition a given matrix into "submatrices," accomplished by drawing horizontal and vertical lines (the partitions) between the elements. Such partitions are often used in the multiplication of matrices. They are largely (but not completely) arbitrary. Consider the following matrix product, $\mathbf{C} = \mathbf{AB}$:

$$\mathbf{C} = \begin{bmatrix} a_{11} & a_{12} & a_{13} & a_{14} \\ a_{21} & a_{22} & a_{23} & a_{24} \\ a_{31} & a_{32} & a_{33} & a_{34} \\ a_{41} & a_{42} & a_{43} & a_{44} \end{bmatrix} \begin{bmatrix} b_{11} & b_{12} & b_{13} & b_{14} \\ b_{21} & b_{22} & b_{23} & b_{24} \\ b_{31} & b_{32} & b_{33} & b_{34} \\ b_{41} & b_{42} & b_{43} & b_{44} \end{bmatrix}. \tag{1.18}$$

Two lines (horizontal and vertical) partition the $\mathbf{A}$ matrix, while a single horizontal line partitions the $\mathbf{B}$ matrix, in (1.18). That is, $\mathbf{A}$ is partitioned into 4 submatrices, $\mathbf{B}$ into 2. The product, then, can be written:

$$\mathbf{C} = \begin{bmatrix} \mathbf{A}_1 & \mathbf{A}_2 \\ \mathbf{A}_3 & \mathbf{A}_4 \end{bmatrix} \begin{bmatrix} \mathbf{B}_1 \\ \mathbf{B}_2 \end{bmatrix} = \begin{bmatrix} \mathbf{A}_1\mathbf{B}_1 + \mathbf{A}_2\mathbf{B}_2 \\ \mathbf{A}_3\mathbf{B}_1 + \mathbf{A}_4\mathbf{B}_2 \end{bmatrix} \tag{1.19}$$

$$\mathbf{A}_1 = (3X3); \mathbf{A}_2 = (3X1); \mathbf{A}_3 = (1X3); \mathbf{A}_4 = (1X1)$$

$$\mathbf{B}_1 = (3X4); \mathbf{B}_2 = (1X4).$$

The check for conformable product matrices:

$$\mathbf{A}_1\mathbf{B}_1 = (3X\cancel{3}\cancel{3}X4) = (3X4) \qquad \mathbf{A}_2\mathbf{B}_2 = (3X\cancel{1}\cancel{1}X4) = (3X4)$$
$$\mathbf{A}_3\mathbf{B}_1 = (1X\cancel{3}\cancel{3}X4) = (1X4) \qquad \mathbf{A}_4\mathbf{B}_2 = (1X\cancel{1}\cancel{1}X4) = (1X4).$$

These checks show that the submatrices given as products in (1.19) are conformable in multiplication, and those shown as sums are conformable in addition. Note that the matrix $\mathbf{C}$ is 4X4. It is partitioned horizontally, into a 3X4 and a 1X4 (just like $\mathbf{B}$).

The submatrices of $\mathbf{A}$ and $\mathbf{B}$ are conformable because the vertical line in $\mathbf{A}$ divides the *columns* of $\mathbf{A}$ the same as the horizontal line in $\mathbf{B}$ divides its *rows*. *Note that if the vertical line in $\mathbf{A}$ changes* *position, it forces the line in* **B** *to change position. But, the horizontal line, in* **A** *is arbitrary.* It can be moved anywhere without destroying conformability.

Follow through the example below:

$$C = \begin{bmatrix} 3 & 1 & 2 & 0 \\ -1 & 4 & 2 & 2 \\ 0 & 1 & -1 & 3 \\ \hline 2 & -1 & 0 & 2 \end{bmatrix} \begin{bmatrix} 6 & -1 & 0 & 3 \\ 1 & 0 & -1 & 2 \\ 3 & 5 & -1 & 0 \\ 3 & 0 & 2 & 1 \end{bmatrix}.$$

Using the same definitions for the submatrices:

$$A_1 B_1 + A_2 B_2 = \begin{bmatrix} 25 & 7 & -3 & 11 \\ 4 & 11 & -6 & 5 \\ -2 & -5 & 0 & 2 \end{bmatrix} + \begin{bmatrix} 0 & 0 & 0 & 0 \\ 6 & 0 & 4 & 2 \\ 9 & 0 & 6 & 3 \end{bmatrix}$$

$$A_3 B_1 + A_4 B_2 = [11 \ -2 \ 1 \ 4] + [6 \ 0 \ 4 \ 2].$$

Then, the product **AB** is:

$$C = \begin{bmatrix} 25 & 7 & -3 & 11 \\ 10 & 11 & -2 & 7 \\ 7 & -5 & 6 & 5 \\ 17 & -2 & 5 & 6 \end{bmatrix}$$

Now, move the horizontal partitioning line in **A** up one row. Note that the check of matrix conformability is:

$$A_1 B_1 = (2X\cancel{3}X\cancel{3}X4) = (2X4)$$
$$A_2 B_2 = (2X\cancel{1}X\cancel{1}X4) = (2X4)$$
$$A_3 B_1 = (2X\cancel{3}X\cancel{3}X4) = (2X4)$$
$$A_4 B_1 = (2X\cancel{1}X\cancel{1}X4) = (2X4).$$

The important point is that these remain conformable ***no matter where the horizontal line is moved in matrix*** **A**. It may be worthwhile to continue, by finding the **AB** product, as done above, but with the new partitioning.

In order to introduce partitioning, a simple 4X4 example was used. Such an example fails to show the value of partitioning (it would be simpler to just multiply **AB**). Partitioning is of value in cases of very large matrices. For example, partitioning can be used in the inversion process for large matrices as a method for controlling roundoff error. Also, partitioning is sometimes used conceptually—where the submatrices are actually the given matrices of the problem. Both of these uses will be seen later in this book.

Matrix multiplication, itself, is done (conceptually) by first partitioning the premultiplier by rows, and the postmultiplier by columns. Then, each element of the product matrix is the "dot product" of these partitions. Yet, this rather basic conception can be changed. For example, try to visualize the premultiplier partitioned into *columns* and the postmultiplier in *rows*—in, say, an nXn product. Now, each (of the *n*) column times row products yields an nXn matrix; the sum of these *n* products produces the end result.

Finally, please note that partitioning is here referred to product matrices. It should be clear that partitioning for addition (somewhat trivial) would be quite different. For example, none of the matrices above are partitioned to be conformable in addition (i.e., for $A + B = C$).

1.6   INTERESTING VECTOR PRODUCTS

1.6.1   AN INTERPRETATION OF $Ax = c$

In the previous discussion of matrix multiplication, the equation set $Ax = c$ was used to show that each element c_i is the dot product $[a_i]\{x\}$. But, there is another, very interesting, interpretation of the equation set $Ax = c$. A review of Equation (1.1) shows that each x_i multiplies only the terms in the *column* $\{a_i\}$. Then, Equation (1.1) can be written:

$$\{a_1\}x_1 + \{a_2\}x_2 + \cdots + \{a_n\}x_n = \{c\} . \tag{1.20}$$

The vector c is therefore seen to be formed from the weighted sum of the column vectors of A, the weighting factors being the variables, x_i. It is this interpretation of the equation set that leads to the terminology of "transform" when referring to the set.

As an example of this interpretation, we return to an earlier example

$$\begin{bmatrix} 3 & 4 & -1 \\ 2 & 0 & 6 \end{bmatrix} \begin{bmatrix} 3 & 5 \\ 7 & -2 \\ 6 & 6 \end{bmatrix} = \begin{bmatrix} 31 & 1 \\ 42 & 46 \end{bmatrix} .$$

The columns $\{31, 42\}$ and $\{1, 46\}$ are found as weighted sums of the premultiplier columns:

$$\begin{Bmatrix} 31 \\ 42 \end{Bmatrix} = 3 \begin{Bmatrix} 3 \\ 2 \end{Bmatrix} + 7 \begin{Bmatrix} 4 \\ 0 \end{Bmatrix} + 6 \begin{Bmatrix} -1 \\ 6 \end{Bmatrix}, \text{ and } \begin{Bmatrix} 1 \\ 46 \end{Bmatrix} = 5 \begin{Bmatrix} 3 \\ 2 \end{Bmatrix} - 2 \begin{Bmatrix} 4 \\ 0 \end{Bmatrix} + 6 \begin{Bmatrix} -1 \\ 6 \end{Bmatrix} .$$

1.6.2   THE (nX1X1Xn) VECTOR PRODUCT

In the paragraph on vector products, it was mentioned that two vectors could be multiplied to form a rectangular (very much non-vector) matrix. In three dimensions, consider $v(3X1)$ times $u(1X3)$. Note that they are conformable, (3X1X1X3), and this particular result is (3X3):

$$\begin{Bmatrix} v_1 \\ v_2 \\ v_3 \end{Bmatrix} \begin{bmatrix} u_1 & u_2 & u_3 \end{bmatrix} = \begin{bmatrix} v_1 u_1 & v_1 u_2 & v_1 u_3 \\ v_2 u_1 & v_2 u_2 & v_2 u_3 \\ v_3 u_1 & v_3 u_2 & v_3 u_3 \end{bmatrix} . \tag{1.21}$$

Each row of v consists of just one element, and each column of u has one element. Then, each dot product elements of the product matrix has just the one vu term.

This is an unusual product of two vectors and is not at all the same result as $\mathbf{v} \bullet \mathbf{u}$, (the dot product). The result shows again that matrix multiplication is *non-commutative*. This particular product is very important and useful when $\mathbf{u}$ and $\mathbf{v}$ are "eigenvectors"—(Chapters 6 and 7).

1.6.3 VECTOR CROSS PRODUCT

There is no direct operation between two vectors (written as nX1 matrices) that results in the vector product (or cross-product) of the two. However, by expressing the first vector, say {u} as a (3X3) matrix, we can obtain a vector that is the equivalent of the vector analysis product, $(\mathbf{u} \times \mathbf{v})$ (*this is only defined in three dimensions, of course*):

$$\begin{bmatrix} 0 & -u_3 & u_2 \\ u_3 & 0 & -u_1 \\ -u_2 & u_1 & 0 \end{bmatrix} \begin{Bmatrix} v_1 \\ v_2 \\ v_3 \end{Bmatrix} = \begin{Bmatrix} u_2 v_3 - u_3 v_2 \\ u_3 v_1 - u_1 v_3 \\ u_1 v_2 - u_2 v_1 \end{Bmatrix}. \tag{1.22}$$

The above may seem to be a very contrived construction of the vector product—and it is. However, this kind of matrix will be seen to come up in just this way, in problems in kinetics, where the $\mathbf{U}$ matrix contains the elements of an angular velocity vector. The (3X3) $\mathbf{U}$ matrix is "skew-symmetric" (see Section 1.3).

1.7   EXAMPLES

1.7.1 AN EXAMPLE MATRIX MULTIPLICATION

Given $\mathbf{A} = \begin{bmatrix} 2 & 1 \\ -1 & 3 \\ 1 & 2 \end{bmatrix}$; $\mathbf{B} = \begin{bmatrix} 5 & 0 & 4 \\ 1 & -1 & 0 \end{bmatrix}$. The check for conformability is (3X2̶X̶2X3). Then the result will be $\mathbf{C}(3X3) = \mathbf{AB}$. When the elements are written out to show the operations involved, the result is:

$$\mathbf{C} = \mathbf{AB} = \begin{bmatrix} 2(5) + 1(1) & 2(0) + 1(-1) & 2(4) + 1(0) \\ -1(5) + 3(1) & -1(0) + 3(-1) & -1(4) + 3(0) \\ 1(5) + 2(1) & 1(0) + 2(-1) & 1(4) + 2(0) \end{bmatrix}.$$

That is, all the column vectors of the product $\mathbf{C}$ are linear combinations of the two column vectors in $\mathbf{A}$. For example: $\mathbf{c}_1 = 5\mathbf{a}_1 + 1\mathbf{a}_2$ (note the bold, lower case "a", subscripted, denotes a vector in $\mathbf{A}$—usually a column vector). Thus, all three of the column vectors of $\mathbf{C}$ lie in the same plane—the plane defined by the intersecting column vectors of $\mathbf{A}$.

The same points can be made concerning the row vectors of $\mathbf{C}$. These are all linear combinations of the rows of $\mathbf{B}$, and they lie in the plane defined by the intersection of the $\mathbf{B}$ row vectors.

In later chapters, matrices like $\mathbf{C}$, above, will be discussed in some length. It will be shown that they are "singular" matrices, whose determinant is zero.

1.7.2   AN EXAMPLE MATRIX TRIPLE PRODUCT

In the study of vibrating systems, a particular triple product is important—one in which the middle term is a diagonal matrix. $\mathbf{P} = \mathbf{ADC}$, where $\mathbf{D} = [\delta_{ii}d_{ii}]$, with nonzero elements on the main diagonal only. The matrices involved are square, nXn. When a matrix is postmultiplied by a diagonal matrix (as $\mathbf{A}$ is here), the effect is that the diagonal elements multiply onto the respective columns of the premultiplier ($\mathbf{A}$, in this case). The $\mathbf{AD}$ product is shown as

$$\mathbf{AD} = [\{\mathbf{a}_1\}d_{11}, \{\mathbf{a}_2\}d_{22}, \cdots \{\mathbf{a}_n\}d_{nn}],$$

where $\mathbf{A}$ *is partitioned into columns,* and then those columns are multiplied by their respective d_{ij} elements. That is, d_{11} multiplies $\{\mathbf{a}_1\}$, and so forth. Now postmultiply by $\mathbf{C}$, *having first partitioned it into rows.* Note that $\{\mathbf{a}_i\}[\mathbf{c}_j]$ is conformable: (nX1X1Xn) = (nXn). So the product

$$\mathbf{ADC} = \sum_j d_{jj}\{\mathbf{a}_j\}[\mathbf{c}_j] \text{ (the sum of } n \text{ matrices, each nXn .)}$$

Although this result appears cumbersome, it can be a delight, because the d_{jj} factor and the two corresponding vectors, all are related, in an "eigenvalue" analysis (Chapters 6 and 7).

1.7.3   MULTIPLICATION OF COMPLEX MATRICES

This example shows the product of two complex matrices, $\mathbf{A}$ and $\mathbf{C}$. The $\mathbf{A}$ matrix can be written:

$$\mathbf{A} = \begin{bmatrix} 1.021 & 1.503 & 2.001 \\ 1.000 & 0.002 & -5.247 \\ 1.002 & 0.002 & -8.055 \end{bmatrix} + j \begin{bmatrix} 0.010 & 2.330 & 10.258 \\ 1.123 & 3.884 & 14.055 \\ 1.222 & 5.566 & 20.103 \end{bmatrix}.$$

However, the intent here is to emphasize another way in which a matrix may be shown in this book—as a tabulation of values (usually with double bars at left and right). In the case of complex matrices, the imaginary parts will be *shown immediately under the real parts*, as in $\mathbf{A}$ and $\mathbf{C}$, below. For example, $a_{12} = 1.503 + j2.330$.

Matrix $\mathbf{A}$				Matrix $\mathbf{C}$		
1.021	1.503	2.001		−4.120	3.259	3.124
0.010	2.330	10.258		6.110	−3.589	0.011
1.000	0.002	−5.247		5.225	−2.661	2.125
1.123	3.884	14.055		−3.840	6.005	−3.010
1.002	0.002	−8.055		0.000	6.120	1.751
1.222	5.566	20.103		0.000	−3.580	4.777

Being square, $\mathbf{A}$ and $\mathbf{C}$ are conformable in either order. This example finds the product $\mathbf{CA}$, obtained in the usual way—using complex arithmetic. Complex matrices are manipulated just like real ones—but, with the considerable increase in operations required, simply because of the complex numbers involved. It is recommended that the reader calculate a few terms of **CA** just "for practice."

Matrix **CA**

$$\begin{Vmatrix} 6.157 & -6.537 & -63.107 \\ 10.119 & 29.723 & 97.675 \\ \\ 1.776 & 10.230 & 22.801 \\ -1.271 & 7.901 & 43.970 \\ \\ 6.057 & -12.668 & -91.931 \\ 10.219 & 33.519 & 101.522 \end{Vmatrix}$$

A sample calculation for the element 2,3 of **CA** is shown here:

$$ca_{23} = [c_2]\{a_3\} = c_{21}a_{13} + c_{22}a_{23} + c_{23}a_{33}$$

$c_{21}a_{13}$	2.001	$+j10.258$	$\times$	5.225	$-j3.840$	$=$	49.846	$+j45.914$
$c_{22}a_{23}$	-5.247	$+j14.055$	$\times$	-2.661	$+j6.005$	$=$	-70.438	$-j68.909$
$c_{23}a_{33}$	-8.055	$+j20.103$	$\times$	2.125	$-j3.010$	$=$	43.393	$+j66.964$

$$ca_{23} \quad = \quad 22.801 \quad +j43.970$$

CA could also be calculated by separating real and imaginary parts into separate matrices, as discussed earlier. In this case, the product would be found as

$$\mathbf{CA} = (\mathbf{C}_R\mathbf{A}_R - \mathbf{C}_I\mathbf{A}_I) + j(\mathbf{C}_R\mathbf{A}_I + \mathbf{C}_I\mathbf{A}_R)$$

where, for example, $\mathbf{C}_R$ = the matrix formed from just the real part of **C**. All the arithmetic in this way becomes real. Note that the order of the product matrices is important.

1.8   EXERCISES

1.1.   Show that $\mathbf{A} + \mathbf{B} = \mathbf{B} + \mathbf{A}$; that is, matrix addition is always "associative."

1.2.   How many vectors can be selected from the (mXn) matrix, **A**?

1.3.   Given two matrices, **A** and **B**, that are conformable (i.e., the product **AB** is conformable), is the product **BA** ever conformable? Is $\mathbf{A}'\mathbf{B}'$ conformable? Is $\mathbf{B}'\mathbf{A}'$ conformable?

1.4.   Given **U**(4X4) whose first row is $\mathbf{u}_1 = [5.11\ 2.46\ 0.567\ 6.91]$, and **V**(4X4) whose first column is $\mathbf{v}_1 = \{3.03\ -0.821\ 1.44\ -2.02\}$, find the value of w_{11} in the product $\mathbf{W} = \mathbf{UV}$. Time yourself in this calculation and use it to estimate how long it would take to manually determine **W**, given all the terms of **U** and **V**.

1.5. Determine the product $\{v_1\}[u_1]$, using the definitions of u_1 and v_1 from problem 1.4.

1.6. Given the matrix equation $A(nXn)x = c$, express the vector c as a weighted sum of the column vectors of A.

1.7. Find the vectors u and v,

$$v = \begin{bmatrix} -4 & 1 & -2 \\ 1 & 2 & -1 \\ 5 & 1 & 1 \end{bmatrix} \begin{bmatrix} -1 \\ 2 \\ 3 \end{bmatrix}$$

and $\quad u = \begin{bmatrix} -1 & 2 & 3 \end{bmatrix} \begin{bmatrix} -4 & 1 & -2 \\ 1 & 2 & -1 \\ 5 & 1 & 1 \end{bmatrix}.$

1.8. Given the matrix definitions at right: Find the most efficient way to calculate

a) $ABCv$
b) $v_1 u_1 v_2 u_2$
c) $u_1 AB$

$A, B, C = (nXn)$
$v_1, v_2 = (nX1)$

1.9. For $A = \frac{1}{\sqrt{7}} \begin{bmatrix} 2 & 3 \\ 1 & -2 \end{bmatrix}$ find A^2, A^3, and A^{10}.

1.10. Solve the following equations for A: $C = A + B$ and $C = AB$.

1.11. Given that $P(x) = x^2 - 2x - 2$, find $P(A)$, for $A = \begin{bmatrix} 1 & -2 \\ 2 & -1 \end{bmatrix}$.

1.12. If $AB = k[\delta_{ij}]$ find A^{-1}.

1.13. Given the (2X2) orthogonal matrix, $T(\theta)$, find T^2 and compare the result with $T(2\theta)$. Using this information, find $T^6(\frac{\pi}{36})$.

$$T(\theta) = \begin{bmatrix} \cos\theta & -\sin\theta \\ \sin\theta & \cos\theta \end{bmatrix}.$$

1.14. Show that the matrices T_1 and T_2 are orthogonal, i.e., that their rows/columns are mutually perpendicular. The notation T'_1 indicates the transpose of T_1. Find the product, $T_1 T_2$ and show that this product is orthogonal.

$$T_1 = \begin{bmatrix} \cos\theta & -\sin\theta & 0 \\ \sin\theta & \cos\theta & 0 \\ 0 & 0 & 1 \end{bmatrix}; \quad T_2 = \begin{bmatrix} \cos\varphi & 0 & \sin\varphi \\ 0 & 1 & 0 \\ -\sin\varphi' & 0 & \cos\varphi \end{bmatrix}.$$

1.15. Find the product $\begin{bmatrix} 1 & 1 & 1 \\ 1 & 1 & 1 \\ 1 & 1 & 1 \end{bmatrix} \begin{bmatrix} 3 & 6 & 4 \\ -2 & 3 & -5 \\ 0 & -8 & 1 \end{bmatrix}$.

1.16. In the matrix product $\mathbf{P} = \mathbf{ABC}$ show that $p_{ij} = [\mathbf{a}_i]\mathbf{B}\{\mathbf{c}_j\}$.

1.17. Given the (4X4) matrix $[a_{ij}]$, determine an elementary transformation matrix that will cause element a_{31} to vanish (go to zero).

CHAPTER 2

Determinants

2.1 INTRODUCTION

The definition of a determinant is derived from the solution of linear algebraic equations. Since the single variable case is trivial, we will begin with the (2X2):

$$a_{11}x_1 + a_{12}x_2 = c_1$$
$$a_{21}x_1 + a_{22}x_2 = c_2 . \tag{2.1}$$

To eliminate x_2, we multiply the first equation by a_{22}, and the second by a_{12}, then subtract the second from the first:

$$(a_{11}a_{22} - a_{12}a_{21})x_1 = (a_{22}c_1 - a_{12}c_2) . \tag{2.2}$$

Equivalently, we may eliminate x_1 (by the same methods):

$$(a_{11}a_{22} - a_{12}a_{21})x_2 = (a_{11}c_2 - a_{21}c_1) . \tag{2.3}$$

The coefficients on both sides of (2.2) and (2.3) can be viewed as **expansions** via "**cross-multiplication**" of determinant arrays, as follows:

$$\begin{cases} \begin{vmatrix} a_{11} & a_{12} \\ a_{21} & a_{22} \end{vmatrix} = (a_{11}a_{22} - a_{12}a_{21}) \\[12pt] \begin{vmatrix} a_{11} & c_1 \\ a_{21} & c_2 \end{vmatrix} = (a_{11}c_2 - c_1a_{21}) \\[12pt] \begin{vmatrix} c_1 & a_{12} \\ c_2 & a_{22} \end{vmatrix} = (c_1a_{22} - a_{12}c_2) \end{cases} \tag{2.4}$$

The square arrays in (2.4) "*expand*," by the cross—multiplication indicated, to the scalars shown on the right sides. And, we *define the determinant in terms of its expansion*. Expansions are defined only for **square** arrays. The result of the expansion is a scalar expression, or numeric value. That is, the determinant is a scalar value. Further, from (2.2) and (2.3), the values of the variables are found as the ratio of these expanded determinants—all of which are known, given in the problem.

Three Variables:

$$a_{11}x_1 + a_{12}x_2 + a_{13}x_3 = c_1$$
$$a_{21}x_1 + a_{22}x_2 + a_{23}x_3 = c_2 \qquad\qquad (2.5)$$
$$a_{31}x_1 + a_{32}x_2 + a_{33}x_3 = c_3 \ .$$

By the very same processes of elimination used above, we find that x_j is expressed as the ratio of two determinants:

$$x_j = \frac{\mathbf{D}_j}{\mathbf{D}}; \quad \text{where } \mathbf{D} = \begin{vmatrix} a_{11} & a_{12} & a_{13} \\ a_{21} & a_{22} & a_{23} \\ a_{31} & a_{32} & a_{33} \end{vmatrix} \qquad\qquad (2.6)$$

known as Cramer's Rule, where the expansion of $\mathbf{D}$ is given by:

$$\mathbf{D} = a_{11}a_{22}a_{33} - a_{11}a_{23}a_{32} + a_{12}a_{23}a_{31}$$
$$- a_{12}a_{21}a_{33} + a_{13}a_{21}a_{32} - a_{13}a_{22}a_{31} \qquad\qquad (2.7)$$

and the expansion of $\mathbf{D}_j$ follows the same rules, after replacing the ith column of $\mathbf{D}$ with the vector, $\mathbf{c} = \{c_i\}$.

The "cross-multiplication" algorithm for the (3X3) is much more complex than for the (2X2). It is shown diagrammatically below. Comparison of the diagram to Equation (2.7) shows them to be the same.

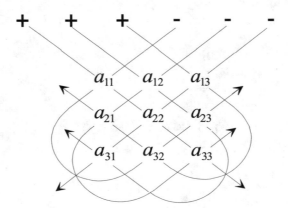

It will be noted that there are six terms in the expansion, rather than two, and further, in both expansions, *half the terms are negative, half positive.*

Note also, that each term in the expansion of the (2X2) has two factors, and each term in the expansion of the (3X3) has three factors.

Now examine the row and column subscripts within each term: *Every row (and column) subscript is represented once—and only once*. This fact is extremely important to the developments that follow.

Continuing with the expansions of the (4X4), then (5X5), and so on, it would be found that there are 24 terms in the expansion of the (4X4), and 120 terms in the (5X5) expansion. In these cases, also, exactly half of the terms are positive, half negative. (Note: The statement that a term is "positive" or "negative" does not refer to the signs of the factors within the term. Half the terms in the expansion of a determinant *with all positive elements* will be negative.) In these cases we would also find that within every term, each row, and each column, subscript is represented exactly once. Then, each term in the expansion of a (4X4) has four factors, and each term in the expansion of the (5X5) contains five factors.

2.2   GENERAL DEFINITION OF A DETERMINANT

The general (nXn), determinant, $|\mathbf{A}|$, can be represented as a *square*, two-dimensional array of elements, each with two subscripts—the first indicating the element row position, the second indicating column position. *Notwithstanding the two-dimensional representation, the determinant is a scalar. That is, it "expands" (as discussed above) to a scalar value—either a numeric, or a literal (function)*.

$$\begin{vmatrix} a_{11} & a_{12} & a_{13} & \cdots & a_{1n} \\ a_{21} & a_{22} & a_{23} & \cdots & a_{2n} \\ a_{31} & a_{32} & a_{33} & \cdots & a_{3n} \\ \cdots & \cdots & \cdots & \ddots & \cdots \\ a_{n1} & a_{n2} & a_{n3} & \cdots & a_{nn} \end{vmatrix}.$$

The expansion of $|\mathbf{A}|$ can be written as shown here:

$$|\mathbf{A}| = \sum (-1)^s a_{1i} a_{2j} a_{3k} \cdots \tag{2.8}$$

Most of the remaining discussion in Sections 2.2 and 2.3 will refer to, and clarify, this equation. With respect to it, we note the following definitive statements:

1. The determinant, $|\mathbf{A}|$ expands into a sum of product terms.

2. Each term is the product of n elements from the array, where (nXn) is the dimension of the array.

3. No two elements in any one term can come from the same row, or column. For example, Equation (2.8) implies that the elements were selected "in row order." The first element from row 1 (any column, the ith), the second from row 2 (and any column *except* the ith), and so on. This process continues until all possible terms have been selected. As an example, every term in the expansion of a (4X4) will look like:

$$a_{1i} a_{2j} a_{3k} a_{4l}$$

where i can be any one of 4 columns, j can be any one of three columns (but not the ith), k can be one of two, and l must be the one remaining. In the (4X4) expansion, then, there will be 4X3X2X1, or 24 terms.

4. In general, there will be $n!$ terms in the expansion (2.8).

5. The sign to be affixed to each term will depend on the value of s, the superscript of the (-1) factor, in (2.8). In every expansion, exactly half the terms will have a positive leading sign, half a negative one.

The 5 statements given above define the general determinant, except for the term, $(-1)^s$, in Equation (2.8), whose value depends on whether the superscript, s, is odd or even. This will be the subject of the next section.

Also, Equation (2.8) and its discussion imply that elements are to be selected "in row order" (the first element coming from the first row, etc.). But, of course, these factors could obviously be reordered within the terms. That is, they could have been chosen "in column order," or indeed, in any order—just so long as no row or column index appears more than once in every term. But, it is important to note that every term in a determinant expansion *can* be arranged such that either row or column index appears in numeric order. The other index will then be in some permutation of this order. Since there are $n!$ permutations of n things, there are $n!$ terms in the expansion.

2.3   PERMUTATIONS AND INVERSIONS OF INDICES

The subject of permutations is concerned with the arrangements of given things, or objects, in which the order of the arrangements is important. As an example, we ask: In how many ways can the digits 1,2,3,4 be arranged into a four digit number? (Equivalently: How many permutations are there of 4 things taken four at a time?). All the possible permutations of this example are:

1234	2134	3124	4123
1243	2143	3142	4132
1324	2314	3214	4213
1342	2341	3241	4231
1423	2413	3412	4312
1432	2431	3421	4321

There are clearly 24 permutations. This was pretty obvious from the beginning, since in choosing an arrangement, we can choose any of the digits in the first place, (4), then we have one less to choose from (3) for the second, two for the third place, and just one for the fourth. There are, then, 4! total choices.

In general, there are $n!$ permutations of n things taken n at a time.

In the expansion of a (4X4) determinant, every term contains four elements, chosen (as described previously) from the rows and columns of the array, such that each row/column index appears just once in each term. The four (product) elements in every term can be arranged such that either row (or column) indices are in numeric order. When this is done, the other index will appear in all possible ($n!$) permutations. These permutations, for a (4X4) case, are given in the table, above.

The above table represents only the column indices. Adding the row indices (in 1234 order), the following lists all term indices.

11 22 33 44	12 21 33 44	13 21 32 44	14 21 32 43
11 22 34 43	12 21 34 43	13 21 34 42	14 21 33 42
11 23 32 44	12 23 31 44	13 22 31 44	14 22 31 43
11 23 34 42	12 23 34 41 ←	13 22 34 41	14 22 33 41
11 24 32 43	12 24 31 43	13 24 31 42	14 23 31 42
11 24 33 42	12 24 33 41	13 24 32 41	14 23 32 41

In the table, only the indices are shown. For example, (12 23 34 41), see arrow, represents the term $a_{12}a_{23}a_{34}a_{41}$. The column indices are in the same order as those given in the earlier table. In fact, the earlier table is clearer. There is no information contained in the repetitive numeric order of row indices.

The main point is: ***Given a method to write down all the permutations of n things taken n at a time, we can directly write down all the terms in the expansion of an (nXn) determinant***. Developing the method is non-trivial, and more important, it is still unclear how to affix leading signs to these terms.

2.3.1   INVERSIONS

Given a permutation of n indices: $ijklm\cdots$, an "inversion" is defined as a transposition of adjacent indices. For example, $ijklm$ undergoes one inversion by interchanging, say k with j (notice: the inversion is the interchange of **adjacent** indices). More specifically, ***we define the "inversions" in the permutation ijklm·····, to be equal to the number of such transpositions of adjacent indices to arrive at the numeric order 12345····***. For example, the permutation 3241 has four inversions:

$$\begin{matrix} (1) & & (2) & & (3) & & (4) \\ 3241 & \rightarrow & 3214 & \rightarrow & 3124 & \rightarrow & 1324 & \rightarrow & 1234 \end{matrix}$$

With these four inversions, "natural numeric order" (1234) is restored to the permutation (3241). In Equation (2.8), defining the expansion of a determinant, ***the exponent, "s," on the (−1) factor is***

defined as the number of inversions in the indices of that term. Given that row indices are (arbitrarily taken) in numeric order, then s would be the inversions in the column indices. In the example (3241) given here, since s=4, the term (13 22 34 41) would be given a positive leading sign. Note that the numeric value of s is not important—just whether it is odd, or even.

*There is an easy way to determine the inversions, **s***. Given the permutation, take each digit in turn, and determine the number of digits to its right which are numerically less. Example: (3241): 3 has two inversions (it is larger than both 2 and 1, which are to its right). 2 has one inversion, and 4 has one. The total is 4.

Just for practice, consider (45312). There are 8 total inversions. Now, interchange the "5" with the "2," and determine inversions of (42315). There are 5. Note that the inversions changed by an *odd* number. It may be a good exercise to write the inversions down in each case—showing that numeric order is restored.

We now have the capability to expand any (nXn) determinant via (2.8). Without a computer, it would be a lengthy, and arduous process if $n > 4$. Even with a computer, there are very few programs available which use (2.8) directly to evaluate determinants.

Up to this point there is an implication that the value of s depends upon just how the elements are ordered within the terms in the expansion. But, the sign of terms in the expansion must not depend on an arbitrary order of the products. We will now show that the sign will not change—however, it is true that the numeric value of s depends on the ordering.

Given a permutation *ijklm*· · ·, if any two adjacent indices are transposed, the change in s will be either +1, or −1. For example, if k and j are transposed, giving *ikjlm*· · ·, the change will be +1 if $k > j$, or it will be −1 if $k < j$. Obviously, the contributing inversions from the other indices will be unchanged. For example, (45312) → (43512), changes by −1, since $3 < 5$. By inspection, (43512) does indeed have 7 inversions. (45312) has 8.

If the elements in a term in (2.8) are reordered, both row and column indices are reordered. Then, if two adjacent elements in a term are transposed, the row inversions will change by +1, or −1. The column inversions will also change by +1, or −1. Therefore, as the term is reordered by a series of such transpositions, *the total inversions, considering both row and column, must change by an **even** number*. The numeric value of s will, in general, change, but it will remain either even or odd. Thus, the sign of the term does not change by some arbitrary reordering of terms. An example: in a (4X4) the term:

$$13 \quad 21 \quad 34 \quad 42 \quad s=3 \text{ (inversions of column indices)}$$

will have a leading negative sign (s odd). If we "scramble" the elements to:

$$34 \quad 42 \quad 13 \quad 21 \quad s=9 \text{ (inversions of both row and column) .}$$

Note that s has changed significantly, but the leading sign of the term still is determined to be negative. The numeric value of s is always minimized if the elements are given in row order, or column order. Furthermore, s is the same in either case. The same term with column indices in order gives:

$$21 \quad 42 \quad 13 \quad 34 \quad s=3 \text{ (inversions of row indices)} .$$

Using this fact, an important conclusion can be drawn: A square matrix has the same determinant as its transpose. That is, $|\mathbf{A}| = |\mathbf{A}'|$. By definition, the matrix $[a_{ij}]$ is transposed by interchanging rows and columns. One term in $|\mathbf{A}|$ (in a (4X4) case) would be (21 42 13 34). Its corresponding term in $|\mathbf{A}'|$ will be (12 24 31 43). Clearly, these both have the same values for s, and the numeric values of the elements are identical. This is true for all corresponding terms; and the argument obviously holds for the (nXn) case. Thus, the assertion is proved.

   Before leaving the subject of inversions, we will show that if any two indices in a permutation are transposed, the inversions change by an odd number. Given $ijklm\cdots$, let p equal the number of indices between the two which are to be interchanged. For example, if j is to be interchanged with m then $p=2$, since there are two indices between m and j. Choosing j first, it is moved to the right, over k, and then over l, and reinserted. This amounts to p (2) transpositions of adjacent indices. Now, m is removed, and moved to the left over $p+1$ (3) indices, and inserted into the place vacated by j. In the whole operation, there are $2p+1$ transpositions. Since $2p+1$ is necessarily odd, then s will have changed by an odd number. This fact proves that if two rows, or columns of a determinant are interchanged, the sign of the determinant is reversed. This will be discussed later as one property of a determinant.

2.3.2   AN EXAMPLE DETERMINANT EXPANSION

The following (4X4) determinant expansion is shown as an example of the method discussed in this article. That is, each of the 24 (4!) terms is found by determining all possible permutations of 1,2,3,4, and using these as the column subscripts. The row subscripts are taken in numeric order. The leading sign of each term is determined by the method of inversions. Note: By coincidence, the products in every term turned out to be positive.

$$|\mathbf{A}| = \begin{vmatrix} -2 & 3 & 2 & -5 \\ 3 & -4 & -5 & 6 \\ 4 & -7 & -6 & 9 \\ -3 & 5 & 4 & -10 \end{vmatrix} .$$

The two tables, below, show each *Term* of the expansion followed by its *Value*. The s($\pm$) column gives the value of the inversions ("s") and the leading sign. The "sum" is the running accumulated value of the signed terms. The accumulation runs from top to bottom of the 1st (left) table, then continues in the second. The final value of "sum" is $|\mathbf{A}|$.

Term	Value	$s\,(\pm)$	Sum	Term	Value	$s\,(\pm)$	Sum
$a_{11}a_{22}a_{33}a_{44}$	480	0(+)	+480	$a_{13}a_{21}a_{32}a_{44}$	420	2(+)	+353
$a_{11}a_{22}a_{34}a_{43}$	288	1(−)	+192	$a_{13}a_{21}a_{34}a_{42}$	270	3(−)	+83
$a_{11}a_{23}a_{32}a_{44}$	700	1(−)	−508	$a_{13}a_{22}a_{31}a_{44}$	320	3(−)	−237
$a_{11}a_{23}a_{34}a_{42}$	450	2(+)	−58	$a_{13}a_{22}a_{34}a_{41}$	216	4(+)	−21
$a_{11}a_{24}a_{32}a_{43}$	336	2(+)	+278	$a_{13}a_{24}a_{31}a_{42}$	240	4(+)	+219
$a_{11}a_{24}a_{33}a_{42}$	360	3(−)	−82	$a_{13}a_{24}a_{32}a_{41}$	252	5(−)	−33
$a_{12}a_{21}a_{33}a_{44}$	540	1(−)	−622	$a_{14}a_{21}a_{32}a_{43}$	420	3(−)	−453
$a_{12}a_{21}a_{34}a_{43}$	324	2(+)	−298	$a_{14}a_{21}a_{33}a_{42}$	450	4(+)	−3
$a_{12}a_{23}a_{31}a_{44}$	600	2(+)	+302	$a_{14}a_{22}a_{31}a_{43}$	320	4(+)	+317
$a_{12}a_{23}a_{34}a_{41}$	405	3(−)	−103	$a_{14}a_{22}a_{33}a_{41}$	360	5(−)	−43
$a_{12}a_{24}a_{31}a_{43}$	288	3(−)	−391	$a_{14}a_{23}a_{31}a_{42}$	500	5(−)	−54
$a_{12}a_{24}a_{33}a_{41}$	324	3(+)	−67	$a_{14}a_{23}a_{32}a_{41}$	525	6(+)	−18

$$|\mathbf{A}| = -18$$

2.4   PROPERTIES OF DETERMINANTS

1. *A square matrix, $\mathbf{A}$, and its transpose, $\mathbf{A}'$, have the same determinant.* This property is proven in Section 2.3.1, top of page 29.

2. *If any row, or column, of a determinant contains all zero elements, that determinant equals zero.* Every term in the expansion of $|\mathbf{A}|$ must contain exactly one element from every row (column) of $|\mathbf{A}|$. Then, every term in the expansion contains a zero factor. Thus, $|\mathbf{A}|= 0$.

3. *The determinant of a diagonal matrix is equal to the product of its diagonal elements.* Clearly, in the expansion of any determinant, one term is $(11\ 22\ 33 \cdots)$; and this term will have a leading + sign (since there are no inversions in either index). Every other term in the expansion will contain a zero factor.

4. *If any row, or column, of a determinant is multiplied by a constant value, the result is that the determinant is multiplied by this amount.* Each term in the expansion must contain a factor that is multiplied by the constant.

   It is interesting to note this difference between matrices and determinants. If a matrix is multiplied by a scalar, k, every element is multiplied by that scalar. Then, if the matrix is (nXn), the effect is that its determinant is multiplied by k^n.

5. *If two rows, or columns, of a determinant are interchanged, the sign of the determinant is reversed.* When any two rows of $|\mathbf{A}|$ are interchanged, the order of the column indices in the general term will not have changed, but two of the row indices will have been exchanged. Since the exchange of two indices changes the inversions by an odd number, the sign affixed to this

term must be reversed. Because every term in the expansion must contain elements from these two rows, the signs of all terms in the expansion change; the sign of $|\mathbf{A}|$ must be reversed. If instead of two rows, two columns are interchanged, the columns indices in the general term are exchanged causing the same sign reversal.

Very similar reasoning is used in the proof of the next property.

6. *If two rows, or columns, of a determinant are identical, its expansion is zero.* To start, consider the example of a (4X4), whose 2nd and 4th rows are the same.

$$
\begin{array}{l}
a_{12}a_{24}a_{33}a_{41} \\
a_{12}a_{21}a_{33}a_{44}
\end{array}
\qquad
\begin{vmatrix}
a_{11} & a_{12} & a_{13} & a_{14} \\
a_{21} & a_{22} & a_{23} & a_{24} \\
a_{31} & a_{32} & a_{33} & a_{34} \\
a_{41} & a_{42} & a_{43} & a_{44}
\end{vmatrix}
$$

Two terms in the expansion of $|\mathbf{A}|$ are also shown—and, note that these terms are equal in value, because $a_{21} = a_{41}$ and $a_{24} = a_{44}$. But, these terms will have opposite leading signs (the column subscripts are 2431 in the first term and 2134 in the other). That is, 2134 is derived from 2431 by interchanging the second and fourth subscripts. Interchanging these two subscripts changes the inversions by an odd number.

This argument holds in the general (nXn) case. Every term in the expansion has a corresponding identical term, the one whose column subscripts are reversed in the elements whose rows are identical (note that in the example, the column subscripts interchanged are the second and fourth—in the rows that are the same). Thus, the leading signs are always opposite, and all corresponding terms cancel—giving a zero result.

7. *If some amounts are added to the elements of a row, or column, then the effect is the same as the sum of the original determinant, plus a new determinant with the row (column) in question replaced by the adders.*

For example, a (2X2), with the amounts d_1 and d_2 added to the first column:

$$
\begin{vmatrix}
a_{11} + d_1 & a_{12} \\
a_{21} + d_2 & a_{22}
\end{vmatrix}
=
\begin{vmatrix}
a_{11} & a_{12} \\
a_{21} & a_{22}
\end{vmatrix}
+
\begin{vmatrix}
d_1 & a_{12} \\
d_2 & a_{22}
\end{vmatrix}.
\tag{2.9}
$$

After addition of the d_i factors, we will refer to the resulting determinant as $|\mathbf{B}|$. Since every term in the expansion of $|\mathbf{A}|$ contains a factor a_{i1}, then every term in the $|\mathbf{B}|$ expansion will have a factor:

$$
(a_{i1} + d_i).
$$

Then, every term breaks into two, the first being the same as that in the expansion of $|\mathbf{A}|$, and the other, from the expansion of a determinant $|\mathbf{A}|$ but, whose first column (in this case) is replaced by the additive factors. Nothing in this argument depends upon the adders necessarily being added to the first column. And the argument holds if the adders are on a row rather than a column.

Now, if the d_i adders happen to be a constant, k, times the elements of some other column, then (after factoring the k) the second determinant in (2.9) is one in which two rows are identical. In that case, the second determinant is zero, by property 6. Then:

8. ***If to any row, or column, there is added a constant factor multiplied by the corresponding elements of any other row, or column, the value of the determinant is unchanged***. This is an extremely important property. It is almost always utilized in the expansion of determinants.

It is important to generalize property 2. Consider an (nXn) determinant, $|\mathbf{A}|$, one of whose rows (or columns), say the jth, is initially zero—i.e., all the jth elements are zero. Then, by property 2, $|\mathbf{A}|=0$. Now, by repetitive use of property 8, add to the jth (row/column) arbitrary multiples of other rows (columns). The result is (assuming a row):

$$\text{row } j = c_1(\text{row } 1) + c_2(\text{row } 2) + \cdots + c_n(\text{row } n) \tag{2.10}$$

where the row-sums in (2.10) are to be viewed as element-by-element additions. For example, "row1+row2" would be viewed just like two vectors would be added:

$$|(a_{11} + a_{21})(a_{12} + a_{22}) \cdots (a_{1n} + a_{2n})| .$$

Also in (2.10), some, but not all, of the c_k values could be zero.

Now, the jth row (originally all zero), is no longer zero, and its elements are not equal to those of any other single row. Yet, property 8 insists that the value of the determinant has not changed by these additions (or subtractions; note that some, or all of the c_k could be negative). Then, the value of $|\mathbf{A}|$ must still be zero. We, therefore, conclude the property:

9. ***If any row, or column, of a determinant is a "linear combination" of the other rows, or columns, then that determinant is zero***. By definition, the "linear combination" is the summation given in (2.10).

The reader may remember this property from the algebraic solution to sets of linear equations. In order to achieve a unique solution of n equations in n unknowns, n *independent* equations are needed. If one (or more) of these equations is a "linear combination" of the others, then a unique solution does not exist. This will be a subject in a later chapter.

2.5   THE RANK OF A DETERMINANT

Continuing the discussion of the last article, consider an (nXn) determinant whose value is zero by virtue of property 9. If just one of its rows (or columns) is a linear combination of the others, then its "rank" is said to be $(n-1)$. If two rows (or columns) are linear combinations of the others, then its rank is $(n-2)$. And so on. On the other hand, if all n of the rows (or columns) of $|\mathbf{A}|$ are "linearly independent" (i.e., none of the rows (columns) is a linear combination of the others), then the rank of $|\mathbf{A}|$ is n (and its determinant is nonzero).

A more accurate way to say the above is: If the rows (columns) of a determinant are linear combinations of $(n - k)$ independent rows (columns), then the rank of the determinant is $(n - k)$.

In summary, an (nXn) determinant (or square matrix) may have a maximum rank of n, if the determinant does not vanish (not zero). Its minimum rank would be zero, if all its elements are zero (a trivial case).

If $|\mathbf{A}|$ is not zero, its rank is n. If its rank is $(n - 1)$ then there exists at least one (n−1Xn−1) determinant made up of the rows (columns) of $|\mathbf{A}|$ that is not zero. If its rank is $(n - 2)$, then at least one (n−2Xn−2) non-zero determinant can be found. And so on. The subject of "rank of a matrix" will come up in a future chapter.

2.6   MINORS AND COFACTORS

If one, or more, rows and columns are deleted from a determinant, the result is a determinant of lower order, and is called a "*minor*" of the original. If just one row and one column are deleted, the resulting "*first minor*" is of order $(n - 1)$. Clearly, within an $|nXn|$, there exist n^2 first minors. The "*second minor*" is of order $(n - 2)$, and is the result of deleting 2 rows and 2 columns. In this same way, minors of various orders can be defined. A minor is a determinant, and must, therefore, always have the same number of rows as columns.

The elements which lie at the intersections of the deleted rows and columns also form a determinant, which is called the "*complement*" of the minor. Note that the complement of a first minor is a single 1X1 element.

Of particular interest are the first minors. These are of order $n - 1$, and result from the deletion of the ith row, and jth column. The complement is the element a_{ij}, and the minor will be denoted M_{ij}.

2.6.1   EXPANSIONS BY MINORS—LAPLACE EXPANSIONS

The LaPlace expansion is defined as follows. Select any number, say r, rows (or columns) from $|\mathbf{A}|$. Then, the value of $|\mathbf{A}|$ is equal to the sum of products of all the rth order minors contained in these r rows (columns) each multiplied by its corresponding algebraic complement (the complement with the correct leading sign attached). Of greatest importance are the first minors.

Expansion by First Minors

The table below has been taken from Section 2.3. It lists all the indices in the expansion of a (4X4). In this table, leading signs have been added, according to the inversions rules already discussed.

Inspection of the first column in Table 2.1 shows that, when a_{11} is factored, the terms represented in this column can be written:

$$a_{11}(a_{22}a_{33}a_{44} - a_{22}a_{34}a_{43} - a_{23}a_{32}a_{44} + a_{23}a_{34}a_{42} + a_{24}a_{32}a_{42} - a_{24}a_{33}a_{42})$$

Table 2.1:			
+ 11 22 33 44	− 12 21 33 44	+ 13 21 32 44	− 14 21 32 43
− 11 22 34 43	+ 12 21 34 43	− 13 21 34 42	+ 14 21 33 42
− 11 23 32 44	+ 12 23 31 44	− 13 22 31 44	− 14 22 31 43
+ 11 23 34 42	− 12 23 34 41	+ 13 22 34 41	− 14 22 33 41
+ 11 24 32 43	− 12 24 31 43	+ 13 24 31 42	− 14 23 31 42
− 11 24 33 42	+ 12 24 33 41	− 13 24 32 41	+ 14 23 32 41

The terms within parentheses are the expansion of the determinant below.

$$\begin{vmatrix} a_{22} & a_{23} & a_{24} \\ a_{32} & a_{33} & a_{34} \\ a_{42} & a_{43} & a_{44} \end{vmatrix}$$

But, this is M_{11}, the minor of a_{11}. This column, then, can be expressed as $a_{11}M_{11}$.

Using the same reasoning on the second column of Table 2.1, the result is $-a_{12}M_{12}$. Note that the sign is negative, because the sign of all the terms in the second column are reversed from those in the first (the "algebraic complement" then is $-a_{12}$). Continuing for all four columns:

$$|\mathbf{A}| = a_{11}M_{11} - a_{12}M_{12} + a_{13}M_{13} - a_{14}M_{14} .\tag{2.11}$$

In the general (nXn) case, with row number 1:

$$|\mathbf{A}| = \sum_{j=1}^{n} (-1)^{j-1} a_{1j}M_{1j} .\tag{2.12}$$

This is proven in the following manner:

There are n $a_{1j}M_{1j}$ terms in (2.12), and each of these terms contain $(n-1)!$ product terms of the original determinant. All of the product terms are unique, and fall within the definition of terms in $|\mathbf{A}|$. That is, they are all terms in $|\mathbf{A}|$. Since there are $n(n-1)!$ total terms, and all from $|\mathbf{A}|$, then all $n!$ terms in $|\mathbf{A}|$ are represented. Note, again, that all are unique. None of the terms containing a_{1k} contain a_{1m}, and vice versa. It must be concluded that (2.12) contains all the terms in $|\mathbf{A}|$.

Inversions and the Leading Signs in (2.12)

Any minor, M_{ij}, expansion has the same leading term signs as the expansion of any (n−1Xn−1) determinant. That is, the deletion of the ith row and jth column does not introduce any inversions; an obvious, but important point. Then, the main diagonal term in M_{ij} will always be positive, in $|M_{ij}|$. Now, considering $a_{1j}M_{1j}$, we will choose a leading sign by considering the product of a_{1j} times this main diagonal term in $|M_{1j}|$.

Then, this sign is determined only by the inversions of the j subscript in the a_{1j} factor (remember, there are no inversions in the diagonal term in M_{1j}). The number of inversions is $j - 1$. Therefore, the superscript on the (-1) factor is $j - 1$.

The General LaPlace Expansion of $|\mathbf{A}|$ in First Minors

In general, $|\mathbf{A}|$ can be expanded in terms of any row or column:

$$\begin{cases} |\mathbf{A}| &= \sum_{j=1}^{n} (-1)^{(i+j)} a_{ij} M_{ij} \ i\text{th row minors} \\ |\mathbf{A}| &= \sum_{i=1}^{n} (-1)^{(i+j)} a_{ij} M_{ij} \ j\text{th column minors} \end{cases} \tag{2.13}$$

The generalization to any (ith) row (the above proof concerned the first row), follows directly, after first reversing the ith with the $(i - 1)$th, then with the $(i - 2)$nd, and so on, until the ith row appears in the first row position, followed by row1, then row2, etc. Note that this is *not* the same as just interchanging the first and ith rows. With the ith row in the first position, the same arguments as above lead to the result. The row reversing operation, described above occurs $i - 1$ times, and each one introduces a change in sign (by property 5 of Section 2.4). Then, when we combine these sign changes with those in Equation (2.12), the exponent on the (-1) term becomes $i - 1 + j - 1$, or $i + j - 2$, or $i + j$.

The argument which shows that $|\mathbf{A}|$ can be expanded in terms of column minors as well as row minors is simply based on the property that $|\mathbf{A}| = |\mathbf{A}'|$, i.e., property 1. After transposition of $|\mathbf{A}|$, all the above arguments hold.

Note that the (first) minor of an element is the coefficient of that element in the general expansion—i.e., the element a_{ij} occurs in exactly $(n - 1)!$ terms in $|\mathbf{A}|$, and those terms are given by M_{ij}.

In summary, the LaPlace expansion provides a concise and clear picture of the expansion of a determinant—easier to visualize than the term by term expansion defined in Equation (2.8). However, expansion by minors is no more, or less, than the term by term expansion.

The ideas of the present section are illustrated in the example, below:

$$|\mathbf{A}| = \begin{vmatrix} 3 & 1 & 4 \\ 0 & -2 & 1 \\ -5 & -1 & 1 \end{vmatrix}$$

Determinant
to be Expanded

term-by-term expansion
Sign

		Sign	
$a_{11}a_{22}a_{33}$	$=$	$+$	(-6)
$a_{11}a_{23}a_{32}$	$=$	$-$	(-3)
$a_{12}a_{21}a_{33}$	$=$	$-$	$(\ 0)$
$a_{12}a_{23}a_{31}$	$=$	$+$	(-5)
$a_{13}a_{21}a_{32}$	$=$	$+$	$(\ 0)$
$a_{13}a_{22}a_{31}$	$=$	$-$	(40)

$$\overline{|\mathbf{A}| = -48}$$

Expansion of $|\mathbf{A}|$ *by minors of the first row*

$$|A| = +a_{11}M_{11} - a_{12}M_{12} + a_{13}M_{13}$$
$$= +3(-2 + 1) - 1(0 + 5) + 4(0 - 10) = -48 .$$

(Note that the 6 terms of the above equation correspond to those in the term by term expansion).

Expansion of $|\mathbf{A}|$ *by minors of the second row*

$$|\mathbf{A}| = 0 - 2(3 + 20) - 1(-3 + 5) = -48 .$$

Expansion of $|\mathbf{A}|$ *by minors of column one*

$$|\mathbf{A}| = +a_{11}M_{11} - a_{21}M_{21} + a_{31}M_{31}$$
$$= +3(-2 + 1) - 0(1 + 4) + (-5)(1 + 8) = -48 .$$

Expansion by minors of any row or column would yield the same result. The reader should prove this, for practice.

Other sums of products of elements times their minors can be formed. For example, consider the main diagonal elements times their minors:

$$\sum_i a_{ii} M_{ii} .$$

This summation at first appears to represent n! terms. But, this summation contains non-unique terms, and is NOT the expansion of $|\mathbf{A}|$. For example, both $a_{11}M_{11}$ and $a_{22}M_{22}$ contain the main diagonal product term $a_{11}a_{22}a_{33} \cdots$ *The only expansions of first minors that result in* $|\mathbf{A}|$ *are given by* (2.13).

Cofactors

The leading signs in Equations (2.13) produce an alternating pattern of signs, as shown in the diagram below (and also evident in Equation (2.11)). If we associate these signs with their corresponding first minors, the results are defined as "*cofactors.*"

$$\begin{vmatrix} + & - & + & - & + & \cdots \\ - & + & - & + & - & \cdots \\ + & - & + & - & + & \cdots \\ \cdots & & & & & \cdots \end{vmatrix}$$

The cofactor of the ijth element will be denoted as A_{ij}:

$$A_{ij} = (-1)^{i+j} M_{ij} \tag{2.14}$$

and Equations (2.13) can be rewritten as:

$$|\mathbf{A}| = \sum_j a_{ij} A_{ij} \quad \text{(row cofactors)}$$

$$|\mathbf{A}| = \sum_i a_{ij} A_{ij} \quad \text{(column cofactors)} . \tag{2.15}$$

If the ith row of $|\mathbf{A}|$ is replaced by some new elements, d_j, then the new determinant so defined is:

$$|\mathbf{D}| = \sum_j d_j A_{ij} . \tag{2.16}$$

Note, especially, that the ith row cofactors of $|\mathbf{D}|$ are the same as those of $|\mathbf{A}|$, and this fact is reflected in (2.16). Now, if the new elements d_j are the elements from some other row (say, the kth), then the expansion of $|\mathbf{D}|$ is that of a determinant with two identical rows; and $|\mathbf{D}|$ must be zero, by property 6 of Section 2.4.

Then, the sum of products of any row (or column) elements times the cofactors of any other row (or column) is identically zero.

$$\sum_j a_{ij} |\mathbf{A}_{kj}| = |\mathbf{A}|, \quad \text{if } i = k$$

$$\sum_j a_{if} |\mathbf{A}_{kj}| = 0, \quad \text{if } i \neq k . \tag{2.17}$$

The above is an important and informative result, as is illustrated by a continuation of the previous (3X3) example:

The original determinant is: $|\mathbf{A}| = \begin{vmatrix} 3 & 1 & 4 \\ 0 & -2 & 1 \\ -5 & -1 & 1 \end{vmatrix}$.

Now arrange all of the signed M_{ij} minors (i.e., cofactors) into a matrix, as follows:

$$\begin{bmatrix} M_{11} & -M_{12} & M_{13} \\ -M_{21} & M_{22} & -M_{23} \\ M_{31} & -M_{32} & M_{33} \end{bmatrix} = \begin{bmatrix} -1 & -5 & -10 \\ -5 & 23 & -2 \\ 9 & -3 & -6 \end{bmatrix} = \text{Matrix of cofactors}$$

where, for example, $A_{22} = M_{22} = (3 + 20) = 23$, and $A_{31} = M_{31} = (1-4(-2)) = 9$.

Now, postmultiply the **A** matrix by the ***transpose of the matrix of cofactors***:

$$\underset{[A]}{\begin{bmatrix} 3 & 1 & 4 \\ 0 & -2 & 1 \\ -5 & -1 & 1 \end{bmatrix}} \underset{[\mathbf{A}_{adj}]}{\begin{bmatrix} -1 & -5 & 9 \\ -5 & 23 & -3 \\ -10 & -2 & -6 \end{bmatrix}} = \underset{|\mathbf{A}|\,\mathbf{I}}{\begin{bmatrix} -48 & 0 & 0 \\ 0 & -48 & 0 \\ 0 & 0 & -48 \end{bmatrix}} \qquad (2.18)$$

Equation (2.18) is a direct illustration of Equations (2.17). The *transposed* matrix of cofactors is defined as the "***adjoint***" of the original **A** matrix. It is written as $\mathbf{A}^{\mathbf{a}}$, or $\mathbf{A}_{adj}$. The product of the first row of **A** times the adjoint columns gives a nonzero result only when the column contains the row 1 cofactors—i.e., the 1st column of $\mathbf{A}^{\mathbf{a}}$. Section 2.8, below continues the discussion of the adjoint matrix, and its relation to the inverse matrix.

2.6.2   EXPANSION BY LOWER ORDER MINORS

The LaPlace expansion is simply a systematic method of deriving all the terms in the term-by-term expansion, Equation (2.8). Although expansion by first minors is probably the most important, it is of interest to note that $|\mathbf{A}|$ can be expanded by other minors, as well.

Starting again from the definition of the LaPlace expansion, we can select any number, say r, rows (columns) within which to form complements. ***Each of these complements will be rXr determinants.*** The (n-rXn-r) minors of these complements will then be "lower order minors." Both the complement and its minor are minors of the original determinant, a source of confusion. In this discussion, the complements formed within the chosen r rows will be called "complementary minors." Each of these will have a "minor" (and a signed minor, or cofactor).

Within the r rows of an nXn determinant we can form $\frac{n!}{r!(n-r)!}$ complements (i.e., combinations of n things taken r at a time). Each complement will have $r!$ terms, while its minor will have $(n - r)!$ terms. Then the sum of products of all complements by their minors will produce

$$\text{Total number of terms} = \frac{n!}{r!(n - r)!} \times r!(n - r)! = n! \, .$$

Since complement and minor are formed from different columns and rows, then each of the terms so formed are truly from $|\mathbf{A}|$. Therefore, the $n!$ totality of them are the expansion of $|\mathbf{A}|$.

In determining the cofactor leading sign we look at the term which arises from the main diagonal of the complement and multiplies the main diagonal terms of its minor. Since both of these factors are main diagonal, there are no inversions within them. However, when they are multiplied together, the number of inversions determines the leading sign.

The method will be numerically illustrated by using the (4X4) example given in Section 2.3.2, page 29, shown again here.

$$
|\mathbf{A}| = \begin{vmatrix} -2 & 3 & 2 & -5 \\ 3 & -4 & -5 & 6 \\ 4 & -7 & -6 & 9 \\ -3 & 5 & 4 & -10 \end{vmatrix}.
$$

There are 4(4–1)/2 (=6) complement 2X2s that can be formed in the first two rows of $|\mathbf{A}|$. These are from columns: (1&2), (1&3), (1&4), (2&3), (2&4), (3&4). Each of these has a 2X2 minor. Their products are summed to expand $|\mathbf{A}|$:

Col's	Compl't	Minor	Product	Result

(1&2)

$$
\begin{vmatrix} a_{11} & a_{12} \\ a_{21} & a_{22} \end{vmatrix} \begin{vmatrix} a_{33} & a_{34} \\ a_{43} & a_{44} \end{vmatrix} = \begin{vmatrix} -2 & 3 \\ 3 & -4 \end{vmatrix} \begin{vmatrix} -6 & 9 \\ 4 & -10 \end{vmatrix} = -24
$$

Leading sign = inv$\{a_{11}a_{22}a_{33}a_{44}\}$ = + ; signed result = −24

(1&3)

$$
\begin{vmatrix} a_{11} & a_{13} \\ a_{21} & a_{23} \end{vmatrix} \begin{vmatrix} a_{32} & a_{34} \\ a_{42} & a_{44} \end{vmatrix} = \begin{vmatrix} -2 & 2 \\ 3 & -5 \end{vmatrix} \begin{vmatrix} -7 & 9 \\ 5 & -10 \end{vmatrix} = 100
$$

Leading sign = inv$\{a_{11}a_{23}a_{32}a_{44}\}$ = − ; signed result = − 100

(1&4)

$$
\begin{vmatrix} a_{11} & a_{14} \\ a_{21} & a_{24} \end{vmatrix} \begin{vmatrix} a_{32} & a_{33} \\ a_{42} & a_{43} \end{vmatrix} = \begin{vmatrix} -2 & -5 \\ 3 & 6 \end{vmatrix} \begin{vmatrix} -7 & -6 \\ 5 & 4 \end{vmatrix} = 6
$$

Leading sign = inv $\{a_{11}a_{24}a_{32}a_{43}\}$ = − ; signed result = + 6

Col's	Compl't	Minor	Product	Result

(2&3)
$$\begin{vmatrix} a_{12} & a_{13} \\ a_{22} & a_{23} \end{vmatrix}\begin{vmatrix} a_{31} & a_{34} \\ a_{41} & a_{44} \end{vmatrix} = \begin{vmatrix} 3 & 2 \\ -4 & -5 \end{vmatrix}\begin{vmatrix} 4 & 9 \\ -3 & -10 \end{vmatrix} = 91$$

Leading sign = inv$\{a_{12}a_{23}a_{31}a_{44}\}$ = + ; signed result = + 91

(2&4)
$$\begin{vmatrix} a_{12} & a_{14} \\ a_{22} & a_{24} \end{vmatrix}\begin{vmatrix} a_{31} & a_{33} \\ a_{41} & a_{43} \end{vmatrix} = \begin{vmatrix} 3 & -5 \\ -4 & 6 \end{vmatrix}\begin{vmatrix} 4 & -6 \\ -3 & 4 \end{vmatrix} = 4$$

Leading sign = inv$\{a_{12}a_{24}a_{31}a_{43}\}$ = − ; signed result = − 4

(3&4)
$$\begin{vmatrix} a_{13} & a_{14} \\ a_{23} & a_{24} \end{vmatrix}\begin{vmatrix} a_{31} & a_{32} \\ a_{41} & a_{42} \end{vmatrix} = \begin{vmatrix} -2 & -5 \\ -5 & 6 \end{vmatrix}\begin{vmatrix} 4 & -7 \\ -3 & 5 \end{vmatrix} = 13$$

Leading sign = inv$\{a_{13}a_{24}a_{31}a_{42}\}$ = + ; signed result = + 13 .

Adding the signed results, above, yields the value of $|\mathbf{A}| = -18$, (the same as in Section 2.3.2).

If $|\mathbf{A}|$ were (7X7), and its first 3 rows are chosen in which to form the complements (which will be 3X3's), the number of complements will be equal to 35, the number of combinations of 7 things taken 3 at a time:

$$\text{Number of complements} = \frac{n!}{r!(n-r)!} = \frac{7!}{3!4!} = \frac{7 \bullet 6 \bullet 5}{3 \bullet 2} = 35 .$$

Each of the 3X3 complements will have a 4X4 cofactor, formed within the lower 4 rows, and using the 4 columns that are not used in the complement. Each complement expands to 3! = 6 terms, and its cofactor has 4! = 24 terms. Then the total number of terms will be 35•6•24 = 7!. This is the correct number of terms needed in the expansion of a 7X7, and note that every term is taken from elements of separate rows and columns, as required.

In determining the cofactor leading sign we look at the term which arises from the main diagonal of the 3X3 and multiplies the main diagonal terms of its minor. For example, one of the complements will be formed using the first 3 rows and columns 1, 4, and 6. Its main diagonal term is

$a_{11}a_{24}a_{36}$ and the cofactor main diagonal is $a_{42}a_{53}a_{65}a_{77}$.

Of course, there are no inversions within these. However, when these terms are multiplied:

$a_{11}a_{24}a_{36}a_{42}a_{53}a_{65}a_{77} \Rightarrow$ Inversions in 1462357 = 5 (odd) .

Therefore, the leading sign of the product of this complement times its minor must be negative.

2.6.3   THE DETERMINANT OF A MATRIX PRODUCT

The LaPlace expansion methods are not convenient for use in expanding determinants, but they give valuable insight into the problem. For example, consider:

$$\mathbf{C} = \begin{bmatrix} \mathbf{0} & \mathbf{X} \\ -\mathbf{I} & \mathbf{B} \end{bmatrix} \text{ where } \mathbf{C} = (2n\mathrm{X}2n) \text{ and the partitioned matrices are } (n\mathrm{X}n) .$$

To find $|\mathbf{C}|$, expansion by complements in the first n rows is the obvious choice—there will only be one such complement since all others will have a zero column. The complement will be $|\mathbf{X}|$, but it might appear that the negative sign on $\mathbf{I}$ may alter the sign of the result depending upon n-odd or even. But, look at the inversions in the column indices. They will be:

$$[(n + 1)(n + 2) \cdots (2n)][(1 \cdot 2 \cdot 3 \cdot, , \cdot n)] \text{ Example: If } n = 2: \text{ column indices are } 3412 .$$

Clearly, there will always be n^2 inversions. How nice. Whenever n is odd, the leading sign is negative, just "canceling" the negative value of $|-\mathbf{I}|$. Thus, $|\mathbf{C}| = |\mathbf{X}|$ for any n.

This result is prominent in the proof that the determinant of a matrix product is the product of their determinants. Consider the matrix equation

$$\mathbf{C} = \begin{bmatrix} \mathbf{I} & \mathbf{A} \\ \mathbf{0} & \mathbf{I} \end{bmatrix} \begin{bmatrix} \mathbf{A} & \mathbf{0} \\ -\mathbf{I} & \mathbf{B} \end{bmatrix} = \begin{bmatrix} \mathbf{0} & \mathbf{AB} \\ -\mathbf{I} & \mathbf{B} \end{bmatrix} . \tag{2.19}$$

Since the matrices on each side of the equality are identical, they must have the same determinant. So, we may "take determinants of both sides." In so doing, note that the first matrix on the left is a "Fundamental Operations" matrix which causes sums and/or differences of rows to be combined with the original rows in the second matrix. These operations do not affect the value of the determinant (Property 8). So:

$$|\mathbf{C}| = \begin{vmatrix} \mathbf{A} & \mathbf{0} \\ -\mathbf{I} & \mathbf{B} \end{vmatrix} = \begin{vmatrix} \mathbf{0} & \mathbf{AB} \\ -\mathbf{I} & \mathbf{B} \end{vmatrix} . \tag{2.20}$$

The determinant on the left, expanded by minors of the first n rows, is clearly equal to $|\mathbf{A}||\mathbf{B}|$. The determinant on the right has just been shown to be $|\mathbf{AB}|$. Then:

$$|\mathbf{C}| = |\mathbf{A}| \, |\mathbf{B}| = |\mathbf{AB}| . \tag{2.21}$$

The extension of this to multiple matrices in the product is obvious.

2.7   GEOMETRY: LINES, AREAS, AND VOLUMES

The "Two-point form" of the equation of a line can be written as the following determinant

$$\begin{vmatrix} x & y & 1 \\ x_1 & y_1 & 1 \\ x_2 & y_2 & 1 \end{vmatrix} = 0 \Rightarrow y = \frac{y_2 - y_1}{x_2 - x_1}x + \frac{y_1 x_2 - y_2 x_1}{x_2 - x_1} . \tag{2.22}$$

Note that the equation is satisfied at both x_1, y_1 and x_2, y_2, from determinant property 6.

   The equation of a parabola passing through points (x_1, y_1), (x_2, y_2), (x_3, y_3) is given by

$$\begin{vmatrix} y & x & x^2 & 1 \\ y_1 & x_1 & x_1^2 & 1 \\ y_2 & x_2 & x_2^2 & 1 \\ y_3 & x_3 & x_3^2 & 1 \end{vmatrix} = 0 \Rightarrow y = ax^2 + bx + c \tag{2.23}$$

where the coefficients a, b, and c are the ratios of the Minors of the determinant. The equation is often used in parabolic interpolation, wherein given data is locally fitted to a parabola. In that case, $(x_1, x_2, \text{ and } x_3)$ are taken to be (-1, 0, and +1), and the resulting equation becomes

$$y = y_2 + \tfrac{1}{2}(y_3 - y_1)x + (y_1 - 2y_2 + y_3)x^2; \text{ for } y \text{ in the local interval .} \tag{2.24}$$

   The area of a triangle $(\triangle)$, one of whose vertices at the origin is given by

$$\tfrac{1}{2}(x_1 y_2 - x_2 y_1) = \tfrac{1}{2} \begin{vmatrix} x_1 & y_1 \\ x_2 & y_2 \end{vmatrix}. \tag{2.25}$$

For example, area $\triangle OAB$, in the diagram.

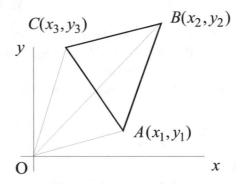

To find the area of $\triangle ABC$, whose vertices are not at the origin, use (2.25)

$$\triangle ABC = \triangle OAB + \triangle OBC - \triangle OAC$$

$$\triangle ABC = \tfrac{1}{2} \begin{vmatrix} x_1 & y_1 \\ x_2 & y_2 \end{vmatrix} + \tfrac{1}{2} \begin{vmatrix} x_2 & y_2 \\ x_3 & y_3 \end{vmatrix} - \tfrac{1}{2} \begin{vmatrix} x_1 & y_1 \\ x_3 & y_3 \end{vmatrix}.$$

But, this is just the expansion of a 3X3 determinant:

$$\triangle ABC = \tfrac{1}{2} \begin{vmatrix} x_1 & y_1 & 1 \\ x_2 & y_2 & 1 \\ x_3 & y_3 & 1 \end{vmatrix}. \tag{2.26}$$

The determinant value interpreted as a volume

A point of greater interest and importance is made by considering the equation of a plane defined by three points, P_1, P_2, and P_3, in space. Equation 2.27, below, first shows the general equation of a plane, and, second, the expansion of the determinant $F(x, y, z)$ by its first row complements:

$$F(x, y, z) = \begin{vmatrix} x & y & z & 1 \\ x_1 & y_1 & z_1 & 1 \\ x_2 & y_2 & z_2 & 1 \\ x_3 & y_3 & z_3 & 1 \end{vmatrix} = 0$$

$$\left. \begin{aligned} Ax + By + Cz + D = 0 \\ \begin{vmatrix} y_1 & z_1 & 1 \\ y_2 & z_2 & 1 \\ y_3 & z_3 & 1 \end{vmatrix} x + \begin{vmatrix} z_1 & x_1 & 1 \\ z_2 & x_2 & 1 \\ z_3 & x_3 & 1 \end{vmatrix} y + \begin{vmatrix} x_1 & y_1 & 1 \\ x_2 & y_2 & 1 \\ x_3 & y_3 & 1 \end{vmatrix} + \begin{vmatrix} x_1 & y_1 & z_1 \\ x_2 & y_2 & z_2 \\ x_3 & y_3 & z_3 \end{vmatrix} = 0 \end{aligned} \right\} \quad (2.27)$$

Comparison of the two equations shows that F is the equation of a plane defined by the three points, P_j. Note the following diagram of a (three dimensional) tetrahedron with the triangle $P_1 P_2 P_3$, as its base (shown shaded):

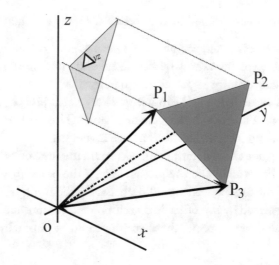

The coefficients of the variable in (2.27) are triangular areas, as shown in the previous paragraphs (see equation 2.26). These triangles are the projections of triangle $P_1 P_2 P_3$ onto the coordinate planes. Let Δ represent the are of triangle $P_1 P_2 P_3$. Then $\Delta_{yz} = \Delta \cos \alpha$, $\Delta_{xz} = \Delta \cos \beta$, $\Delta_{xy} = \Delta \cos \chi$, where the angles α, β, and γ are the direction cosines of the normal from O, perpendicular to the plane (e.g., α is the angle between the x-axis the normal)[1].

Division of the first equation (2.27) by $\sqrt{A^2 + B^2 + C^2}$ converts it to the "normal form" of the equation of a plane in which the coefficients of the variables become the direction cosines of the

[1]This "cosine effect" will be seen again in Chapter 5, in the Section "Solar Angles," on page 116.

normal, and the constant term the distance from O to the plane—the length, p, of the normal. That is:

$$p = \frac{-D}{\sqrt{A^2 + B^2 + C^2}} . \tag{2.28}$$

The volume of this tetrahedron is given as $\frac{1}{3} \times$ area of the base(triangle $P_1 P_2 P_3$) $\times$ the length of the normal to the plane.

Noting that A, B, and C are related to the areas of the projected triangle (e.g., $A = 2\Delta_{yz}$), then

$$\sqrt{A^2 + B^2 + C^2} = 2\sqrt{\Delta_{yz}^2 + \Delta_{zx}^2 + \Delta_{xy}^2} = 2\Delta\sqrt{\cos^2 \alpha + \cos^2 \beta + \cos^2 \gamma} = 2\Delta .$$

The term $\sqrt{\cos^2 \alpha + \cos^2 \beta + \cos^2 \gamma} = 1$, the direction cosines are the coordinates of a "unit vector." Therefore,

$$\text{volume} = \frac{1}{3} p \times \Delta = \frac{1}{3} \frac{D}{\sqrt{A^2 + B^2 + C^2}} \times \frac{\sqrt{A^2 + B^2 + C^2}}{2} = \frac{1}{6} \begin{vmatrix} x_1 & y_1 & z_1 \\ x_2 & y_2 & z_2 \\ x_3 & y_3 & z_3 \end{vmatrix}$$

D is the value of the determinant defined by the vectors OP_i, the constant term in $F(x, y, z)$.

This important result shows that the value of a determinant can be equated to a "volume." In more than 3 dimensions, the volume cannot be visualized—but, just envision the 3 dimensional case and let the mathematics take over for larger dimensionality.

A determinant (its expanded value) can become very small just because its vectors (rows or columns) are themselves small, or a subset of them is small. This will be easy to see, and can be changed by re-scaling, making its rows balanced in size numerically.

After rescaling, the value of the determinant becomes a measure of its "skew"—the orientation of vectors within the set. For example, in the present case, if the point P_1 were to move toward the line $P_2 P_3$, the volume of the tetrahedron would decrease. At the limit, if P_1 reaches this line then OP_1 falls into the plane defined by the other two vectors—the volume, and hence the determinant value, will be zero. At the other extreme, these vectors could be mutually orthogonal, minimum "skew."

2.8   THE ADJOINT AND INVERSE MATRICES

The adjoint matrix—defined in Section 2.6, page 38 as the transpose of the matrix of cofactors—is denoted as $\mathbf{A^a}$ or $\mathbf{A}_{adj}$. Equation (2.18) leads directly to the statement of (2.29):

$$\mathbf{AA^a} = |\mathbf{A}| \, \mathbf{I}, \text{ for any square matrix.} \tag{2.29}$$

$$[\mathbf{A}] \frac{\mathbf{A^a}}{|\mathbf{A}|} = \mathbf{I}, \text{ when } |\mathbf{A}| \neq 0 . \tag{2.30}$$

In Chapter 1, the "inverse matrix" (of $\mathbf{A}$) was defined as a matrix which, when pre- or postmultiplied by $\mathbf{A}$, produces the unit matrix, $\mathbf{I}$. (2.30) shows just such a case. The adjoint matrix with each of its elements divided by $|\mathbf{A}|$, as shown in (2.30) is clearly the inverse of $\mathbf{A}$. The adjoint and inverse matrices are defined only for square matrices. If $|\mathbf{A}| = 0$ the inverse of $\mathbf{A}$, written $\mathbf{A}^{-1}$, is not defined—the matrix is "singular."

The inverse matrix, defined in (2.30), also commutes with $\mathbf{A}$. That is:

$$\mathbf{A}[\mathbf{A}^{\mathbf{a}}] = [\mathbf{A}^{\mathbf{a}}]\mathbf{A} = |\mathbf{A}|\mathbf{I} \qquad (\mathbf{A}^{\mathbf{a}} \equiv \mathbf{A}_{\text{adj}}) . \tag{2.31}$$

The column cofactors of $|\mathbf{A}|$ are in the rows of $\mathbf{A}_{\text{adj}}$ ($\mathbf{A}_{\text{adj}}$ is the transpose of the matrix of cofactors). So, the product $[\mathbf{A}_{\text{adj}}]\mathbf{A}$ forms the products of these column cofactors by the columns of $\mathbf{A}$. The arguments already given show that this result is $|\mathbf{A}|\mathbf{I}$.

$$\mathbf{A}^{-1} = [\mathbf{A}_{\text{adj}}]/|\mathbf{A}| \tag{2.32}$$
$$\mathbf{A}^{-1}\mathbf{A} = \mathbf{A}\mathbf{A}^{-1} = \mathbf{I} \tag{2.33}$$

Equations (2.32) and (2.33) define a unique inverse. Suppose, to the contrary, that a matrix, $\mathbf{B}$, exists such that $\mathbf{BA} = \mathbf{I}$. By simply postmultiplying by $\mathbf{A}^{-1}$, the result is $\mathbf{B} = \mathbf{A}^{-1}$. By starting with $\mathbf{AB} = \mathbf{I}$, it is similarly shown that $\mathbf{A}^{-1}$ is unique.

2.8.1 RANK OF THE ADJOINT MATRIX

In Section 2.5 the "rank" of a determinant was discussed. The rank of a square matrix is the same as that of its determinant. If a matrix is non-singular then its rank is the same as its order (i.e., an nXn matrix is of order "n," and its rank is "n"). In this case, the rank of its adjoint matrix is also n. Conversely, the rank of a singular matrix is necessarily less than n. If that rank is $n - 1$, then, from Section 2.5, at least one determinant of order $n - 1$ can be found that is nonzero.

The adjoint matrix is made up of these $n - 1$ determinant values. Therefore, the adjoint matrix cannot be null, yet its product with the original $\mathbf{A}$ matrix has to be null, from Equation (2.29), above. If the rank of $\mathbf{A}$ is less than $n - 1$, then every $n - 1$ minor of $\mathbf{A}$ is null, and the adjoint therefore is null (its rank is zero).

The interesting case is when $\mathbf{A}$ has rank $n - 1$. In this case, the rank of $\mathbf{A}_{\text{adj}}$ is unity. All of its rows (columns) are linear combinations of a single row (column). This important result will be discussed in some detail in the chapter on solutions to linear simultaneous equations. For now, consider an example 4X4:

$$\mathbf{A} = \begin{bmatrix} 11 & 12 & 27 & 17 \\ 1 & -1 & -3 & 0 \\ 5 & 8 & 13 & 7 \\ 26 & 37 & 42 & 27 \end{bmatrix}; \ \mathbf{A}^{\mathbf{a}} = \begin{bmatrix} -186 & 620 & 930 & -124 \\ 84 & -280 & -420 & 56 \\ -90 & 300 & 450 & -60 \\ 204 & -680 & -1020 & 136 \end{bmatrix}.$$

$\mathbf{A}$ is singular, $|\mathbf{A}| = 0$, with rank 3. Its adjoint has the rank of one. All columns of $\mathbf{A}^{\mathbf{a}}$ are a multiple of $\{-31, 14, -15, 34\}$. Also, note that any column of the adjoint and, in fact, any multiple of $\{-31, 14, -15, 34\}$ is a solution to $\mathbf{Ax} = \mathbf{0}$.

2.9   DETERMINANT EVALUATION

The foregoing lays out the characteristics and properties of determinants, but implies very laborious work in actually calculating their values. Fortunately, this is not the case. Modern methods of expansion are straightforward, and easy to program. They do involve a lot of calculation but far less than the direct methods already discussed.

   Practical evaluation of determinants involves some method of condensation (i.e., reduction to a lower order of determinant). Repeated applications of the method eventually lead to the scalar result, $|\mathbf{A}|$. These methods are equivalent to the "elimination of x_j," as discussed in the very beginning of this chapter. The array concept of the determinant lends itself to the definition of arrays in popular programming languages, and the "repeated applications" mentioned above lead to program looping.

2.9.1   PIVOTAL CONDENSATION

"Pivotal condensation" is a name more difficult than the method. The idea is easily described, easily understood, easily done—and fun to program. The description here will be via example using a 4X4:

$$\begin{vmatrix} a_{11} & a_{12} & a_{13} & a_{14} \\ a_{21} & a_{22} & a_{23} & a_{24} \\ a_{31} & a_{32} & a_{33} & a_{34} \\ a_{41} & a_{42} & a_{43} & a_{44} \end{vmatrix} = \pm d \begin{vmatrix} b_{11} & b_{12} & b_{13} & b_{14} \\ 0 & 0 & 1 & 0 \\ b_{31} & b_{32} & b_{33} & b_{34} \\ b_{41} & b_{42} & b_{43} & b_{44} \end{vmatrix} = -d \begin{vmatrix} b_{11} & b_{12} & b_{14} \\ b_{31} & b_{32} & b_{34} \\ b_{41} & b_{42} & b_{44} \end{vmatrix}. \qquad (2.34)$$

The determinant $|\mathbf{A}|$ is manipulated to produce an equivalent $d|\mathbf{B}|$. The determinant $|\mathbf{B}|$ in (2.34) can be expanded by the elements of its second row. The result will clearly be a determinant of 3rd order as shown. A minus sign is chosen, in this case, because of the factor $(-1)^{(2+3)}$ resulting from having chosen the 2,3 element as the "pivot," The resulting "condensed" determinant is then operated upon in the same manner to produce further condensations until the product string of d factors multiply to the final result.

   In this example, the second row is arbitrarily chosen for the first "pivot row." Rather than make an arbitrary choice, the largest element (absolute value) is chosen as the pivotal element. The pivot does not have to be the largest, but that's a good choice—and avoids the possible choice of a pivot equal to zero. If the matrix is complex, just choose the element, $x + jy$, with the largest sum of absolute values of real and imaginary parts ($|x| + |y|$).

   At the beginning step in each cycle, let the pivot (largest) element be a_{pq}. If the pth row is divided by this value, the new pqth element will have the value 1.0. In the equivalent $d|\mathbf{B}|$, the factor d is set equal to a_{pq}, (determinant property 4, in Section 2.4). Now, from each column the proper multiple of column q is subtracted such that all elements of row p become zero. These operations do not change the value of the determinant (by property 8). In the column subtractions it is not necessary to actually calculate the values in row p, since it is already known that they will be zeros. Also, column q remains the same—is skipped from calculations. The new, condensed determinant does not take any of its elements from the pth row, or qth column of $|\mathbf{B}|$

Condensing the determinant uses Equation (2.13), expansion by first minors. The minus sign is taken if the cofactor has the opposite sign from the minor, as in (2.14), i.e., $(-1)^{(p+q)}$.

2.9.2 GAUSSIAN REDUCTION

The foregoing paragraphs indicate that determinant evaluation amounts to the repeated application of a simple algorithm. Gaussian reduction is one of these simple algorithms, and it is not very different than pivotal condensation. As it progresses, the determinant is condensed to a smaller and smaller array until the determinant value becomes the product of n factors (and these factors are the "pivots," just as before).

The objective of Gaussian Reduction is to reduce the given determinant to an equivalent triangular one like the following:

$$\begin{vmatrix} a_{11} & a_{12} & a_{13} & a_{14} \\ 0 & a_{22} & a_{23} & a_{24} \\ 0 & 0 & a_{33} & a_{34} \\ 0 & 0 & 0 & a_{44} \end{vmatrix}$$

4x4 upper triangular determinant

Its value is easily seen to be equal to the product of its main diagonal elements. All other terms contain the zero as a factor. The pivots, then, are these a_{ii} elements.

As before, at each stage, the pivots chosen are the largest elements in the condensed determinant. In general, of course, these are not found on the main diagonal. They must be moved there by row and column exchanges. If both a row and a column exchange occur, the determinant value is not changed. As an example: at the first stage the largest element is found to be a_{34}. To bring this element to the a_{11} position, row three is exchanged with row one and column four is exchanged with column one. Since two sign changes are made, they cancel. If only one (either column or row) exchange occurs, the value of the determinant changes sign. Thus, it is necessary to keep track of these exchanges.

After the exchange(s) occur the method is very like pivotal condensation, except it is not desired to divide the pivotal row by the pivot (to reduce the pivot position to unity). But, the rest of the pivotal column is reduced to zero, just as the pivotal rows were reduced in the previous method.

$$|\mathbf{A}| = \begin{vmatrix} a_{11} & a_{12} & \cdots & a_{1n} \\ 0 & c_{22} & \cdots & \cdots \\ 0 & c_{k2} & c_{kk} & \\ 0 & \cdots & & c_{nn} \end{vmatrix} \qquad \begin{aligned} c_{ij} &= a_{ij} - a_{1j}\frac{a_{i1}}{a_{11}} \\ ij &= 2, 3 \cdots n \end{aligned}$$

The above display shows the first stage, after the largest element has been moved to the a_{11} position. If we make the calculations indicated, the elements in the first column become zero:

$$\text{At } j = 1; \quad c_{ij} = a_{ij} - \frac{a_{i1}}{a_{11}}a_{1j} = a_{i1} - \frac{a_{i1}}{a_{11}}a_{11} = 0$$

and the condensed determinant is $|c_{ij}|$. In (row) vector terms, the elements c_{ij} are formed by subtracting from each ith row the proper multiple of the pivot row (row 1, in stage 1):

$$[\mathbf{c}_i] = [\mathbf{a}_i] - \frac{a_{i1}}{a_{11}}[\mathbf{a}_1] . \tag{2.35}$$

In (2.35) the boldface type identifies vectors and the square brackets indicate [row] vectors, not {column} vectors.

The method is:

(1) Set $p = 1$; p is defined as the pivot row index. As such, it will take values $1..n$ (n is the order of the determinant). The pivots, then will have the subscripts pp.

(2) Find the element with the largest absolute value (if the determinant is complex, the absolute value could be used, or just the largest sum of abs values of real and imaginary parts).

Exchange rows and columns to move the largest element to the a'_{pp} position. The "prime" is used here to indicate that the values of these elements change as the procedure continues.

(3) Now, for all i rows below the pth, subtract $\dfrac{a'_{ip}}{a'_{pp}}$ times the pth row, as in (2.35). ***Note that it is unnecessary to operate on elements in, or to the left of, the pth column***. They will all be zero.

(4) Now, increment p. If this new value is less than n, then repeat steps 2 and 3. If $p = n$, the procedure is complete. The determinant can now be evaluated as the product of the diagonal elements.

The method can become confusing with the exchange of rows and columns. Otherwise, it is quite straightforward. The labor is alleviated by the use of a computer; and the programming is enjoyable, tricky only in keeping track of row/column exchanges.

In this regard, ***it is unnecessary to actually exchange data rows/columns***. Lists can be kept, indicating where they are. For example, evaluating a 4X4, the row list:

row list = rlist(i) = 1,2,3,4. If rows 2 and 4 are exchanged, rlist(i) = 1,4,3,2 .

The same thing can be said for the column list.

Of course, this leads to the complication that the elements must be accessed through these lists. That is, an element $a(i, j)$, in $|\mathbf{A}|$, now must be referred to as $a(\text{rlist}(i), \text{clist}(j))$.

The Gaussian Reduction method can be done using "partial pivoting," in which the pivot elements are always chosen from the pivot column. This reduces the exchanges down to just row interchanges. Within the subject at hand, full pivoting is just as easy. The advantage (reduced complexity) of partial pivoting is noticeable in solving simultaneous equations, and/or calculating the inverse of a matrix. There will be more about this in the following chapter.

The reduction of a 5X5 will provide an example. At each stage, the new pivot is shown within a box. In each case the pivot is the largest element within the condensed matrix. At stage 1, the

2,3 element is the largest in the entire 5X5; at stage 2, 3.4 is the largest within the 4X4 (row and column 1 excluded). One by one, these pivots are brought to the main diagonal, and the elements below them are zeroed by subtraction of $\dfrac{a'_{ip}}{a'_{pp}}$ times the pivot row, as discussed above.

These elements are not actually calculated, just crossed out—indicating zeroes (in fact, it is easier to follow the method with these crossed out numbers than it would be with zeroes).

Stage 1
Input
det.

$$\begin{vmatrix} 1.00000 & 0.00000 & -3.00000 & 1.00000 & 2.00000 \\ -2.00000 & 1.00000 & \boxed{5.00000} & -2.00000 & -2.00000 \\ -1.00000 & 1.00000 & 3.00000 & 1.00000 & -3.00000 \\ 0.00000 & 0.00000 & -1.00000 & -1.00000 & 3.00000 \\ 1.00000 & 1.00000 & -4.00000 & 3.00000 & 5.00000 \end{vmatrix}$$

Stage 2

$$\begin{vmatrix} 5.00000 & 1.00000 & -2.00000 & -2.00000 & -2.00000 \\ \cancel{-3.00000} & 0.60000 & -0.20000 & -0.20000 & 0.80000 \\ \cancel{3.00000} & 0.40000 & 0.20000 & 2.20000 & -1.80000 \\ \cancel{-1.00000} & 0.20000 & -0.40000 & -1.40000 & 2.60000 \\ \cancel{-4.00000} & 1.80000 & -0.60000 & 1.40000 & \boxed{3.40000} \end{vmatrix}$$

Stage 3

$$\begin{vmatrix} 5.00000 & -2.00000 & -2.00000 & -2.00000 & 1.00000 \\ \cancel{-4.00000} & 3.40000 & -0.60000 & 1.40000 & 1.80000 \\ \cancel{3.00000} & \cancel{-1.80000} & -0.11765 & \boxed{2.94118} & 1.35294 \\ \cancel{-1.00000} & \cancel{2.60000} & 0.05882 & -2.47059 & -1.17647 \\ \cancel{-3.00000} & \cancel{0.80000} & -0.05882 & -0.52941 & 0.17647 \end{vmatrix}$$

Stage 4

$$\begin{vmatrix} 5.00000 & -2.00000 & -2.00000 & -2.00000 & 1.00000 \\ \cancel{-4.00000} & 3.40000 & 1.40000 & -0.60000 & 1.80000 \\ \cancel{3.00000} & \cancel{-1.80000} & 2.94118 & -0.11765 & 1.35294 \\ \cancel{-1.00000} & \cancel{2.60000} & \cancel{-2.47059} & -0.04000 & -0.04000 \\ \cancel{-3.00000} & \cancel{0.80000} & \cancel{-0.52941} & -0.08000 & \boxed{0.42000} \end{vmatrix}$$

Stage 5

$$\begin{vmatrix} 5.00000 & -2.00000 & -2.00000 & 1.00000 & -2.00000 \\ \cancel{-4.00000} & 3.40000 & 1.40000 & 1.80000 & -0.60000 \\ \cancel{3.00000} & \cancel{-1.80000} & 2.94118 & 1.35294 & -0.11765 \\ \cancel{-3.00000} & \cancel{0.80000} & \cancel{-2.47059} & 0.42000 & -0.08000 \\ \cancel{-1.00000} & \cancel{2.60000} & \cancel{-0.52941} & -0.04000 & \boxed{-0.04762} \end{vmatrix}$$

At stage 3, the new pivot needs only a column exchange to arrive at the main diagonal. Then, the determinant value must be given a leading negative sign (at all other stages, both a row and a column exchange are required): $|\mathbf{A}| = -\{5.0 \times 3.4 \times 2.94118 \times 0.42 \times (-0.04762)\} = 1.0$.

Gaussian reduction is easy to program, and it is an efficient method. In the next chapter it will be seen again in developing the inverse matrix, and in the solution to linear equation sets. If the method is to be used in hand calculations, it is easier and less confusing to use "partial pivoting," where the pivots are chosen from successive columns, 1, then 2, and so on. In this way, column exchanges are not necessary. For small determinants, where roundoff will not be a problem, pivoting can be avoided altogether (but, zero pivots must be avoided).

Pivotal condensation is also efficient, and is especially easy to use in hand calculations. It lacks the "extension" to be used in the solution to equation sets. Since it is necessary to keep track of deleted rows and columns, the program is handy to use in calculating minors—selected rows and columns are marked as deleted at the outset.

2.9.3   RANK OF THE DETERMINANT LESS THAN n

When the rank of $|\mathbf{A}|$ is $n - 1$ (then $|\mathbf{A}| = 0$), the procedure (algorithm) described above calculates a 0 in the n, n position. For example, if the last (fifth) row of the preceding matrix is replaced with the sum of the first two rows, stage 5 finds a zero in the 5,5 position. The determinant is zero:

$$
\begin{array}{c}
\text{Rank = 4,} \\
\text{Stage 5}
\end{array}
\quad
\begin{vmatrix}
5.00000 & -2.00000 & -2.00000 & 1.00000 & -2.00000 \\
-1.00000 & 2.60000 & -1.40000 & 0.20000 & -0.40000 \\
-3.00000 & -1.80000 & 1.23077 & 0.53846 & -0.07692 \\
-3.00000 & 0.80000 & 0.23077 & 0.43750 & -0.06250 \\
2.00000 & 0.80000 & 0.23077 & 0.43750 & \boxed{0.00000}
\end{vmatrix}
$$

There is obviously a non-zero 4th order determinant.

If the rank of the original $n \times n$ matrix is $n - q$ then the algorithm will result in a q-by-q array of zero elements at the lower right. The non-zero determinant at upper left will be $n - q \times n - q$.

2.10   EXAMPLES

2.10.1   CRAMER'S RULE

At the beginning of this chapter, Cramer's rule was invoked in the discussion of the solution to three equations in three unknowns. In the light of the later discussion of the "adjoint" matrix in Section 2.8, we can revisit this rule. Given the n-dimensional $\mathbf{Ax} = \mathbf{c}$, premultiply both sides by $\mathbf{A}_{adj}$.

$$
\mathbf{A^a Ax} = |\mathbf{A}|\,\mathbf{x} = \mathbf{A^a c} =
\begin{bmatrix}
A_{11} & A_{21} & \cdots & A_{n1} \\
A_{12} & & \cdots & \\
\cdots & \cdots & \cdots & \cdots \\
A_{1n} & A_{2n} & \cdots & A_{nn}
\end{bmatrix}
\{c\} \ .
$$

(Note that $\mathbf{A}_{adj} = \mathbf{A^a}$). In the equation, the elements of $\mathbf{A}_{adj}$ are the transposed, signed, first minors of $|\mathbf{A}|$. The product of $\mathbf{A}$ times its adjoint (from Section 2.8) is $|\mathbf{A}|\mathbf{I}$. The result on the left, then, is $|\mathbf{A}|$ multiplying each $\mathbf{x}$ element.

Looking at x_1 for example, $|\mathbf{A}|x_1 = A_{11}c_1 + A_{21}c_2 + \cdots + A_{n1}c_n$. But, the expression on the right is just the LaPlace expansion of $|\mathbf{A}|$ with its first column replaced by the {c} vector.

Then, each x_i is obtained as the ratio of two determinants. The determinant in the numerator is $|\mathbf{A}|$ with its ith column replaced by the {c} vector, and the denominator is $|\mathbf{A}|$ itself. This is Cramer's rule.

2.10.2 AN EXAMPLE COMPLEX DETERMINANT

In Chapter 1 the sum of two matrices is given as the sum of the individual elements. Then

$$\mathbf{C} = [a_{ik} + jb_{ik}] = \mathbf{A} + j\mathbf{B}$$

and we can think of the matrix as a single one with complex elements or as two separate matrices. (Note that the notation $j = \sqrt{-1}$ "interferes with" the notation of referring to columns with the subscript "j").

The objective in this example is to determine $|\mathbf{C}| = |a_{ik} + jb_{ik}|$. If the routine available handles complex numbers, then $|\mathbf{C}|$ is evaluated without further complication. But, it is possible to evaluate $|\mathbf{C}|$ using only real arithmetic. This will be illustrated in the simplest case—a 2X2.

We will use "vector notation" $|\mathbf{C}| = |\mathbf{c}_1\mathbf{c}_2|$.

$$\mathbf{c}_k = \left\{ \begin{array}{c} a_{1k} + jb_{1k} \\ a_{2k} + jb_{2k} \end{array} \right\} = \left\{ \begin{array}{c} a_{1k} \\ a_{2k} \end{array} \right\} + j \left\{ \begin{array}{c} b_{1k} \\ b_{2k} \end{array} \right\} = \mathbf{a}_k + j\mathbf{b}_k \ .$$

Now, using determinant property 7:

$$|\mathbf{c}_1\mathbf{c}_2| = |\mathbf{a}_1\mathbf{c}_2| + j\,|\mathbf{b}_1\mathbf{c}_2|$$
$$|\mathbf{a}_1\mathbf{c}_2| = |\mathbf{a}_1\mathbf{a}_2| + j\,|\mathbf{a}_1\mathbf{b}_2|$$
$$j\,|\mathbf{b}_1\mathbf{c}_2| = j\,|\mathbf{b}_1\mathbf{a}_2| - |\mathbf{b}_1\mathbf{b}_2|$$

and therefore $|\mathbf{C}| = |\mathbf{a}_1\mathbf{a}_2| - |\mathbf{b}_1\mathbf{b}_2| + j\{|\mathbf{a}_1\mathbf{b}_2| + |\mathbf{b}_1\mathbf{a}_2|\}$.

The same method can be used in expanding any complex nXn determinant. The result will be 2^n determinants to expand, but, at least they will be real.

2.10.3 THE "CHARACTERISTIC DETERMINANT"

Associated with a matrix $\mathbf{A}(nXn)$ is a special determinant with a single variable, usually denoted λ. The matrix $\mathbf{A}(\lambda) = \mathbf{A} - \lambda\mathbf{I}$ is simply formed by subtracting λ from its main diagonal elements. The determinant of $\mathbf{A}(\lambda)$ is an nth order polynomial in the parameter. Again using a 2X2:

$$|\mathbf{A}(\lambda)| = \left| \begin{array}{cc} a_{11} - \lambda & a_{12} \\ a_{21} & a_{22} - \lambda \end{array} \right| .$$

In this particular case, the use of property 7 is "the hard way," but for higher order determinants it is easier, and can be programmed.

$$\begin{vmatrix} a_{11} - \lambda & a_{12} \\ a_{21} & a_{22} - \lambda \end{vmatrix} = \begin{vmatrix} a_{11} & a_{12} \\ a_{21} & a_{22} \end{vmatrix} - \begin{vmatrix} a_{11} & 0 \\ a_{21} & \lambda \end{vmatrix} - \begin{vmatrix} \lambda & a_{12} \\ 0 & a_{22} \end{vmatrix} + \begin{vmatrix} \lambda & 0 \\ 0 & \lambda \end{vmatrix}.$$

The "characteristic polynomial" is, then: $p(\lambda) = \lambda^2 - (a_{11} + a_{22})\lambda + |\mathbf{A}|$.

2.11 EXERCISES

2.1. Find the inversions in the digit sequences below:

$$5741326 \quad 35421 \quad 123465 \quad 654321.$$

2.2. Determine which of the terms, below, are terms in the expansion of a determinant. For those that are legal, determine the leading sign.

$$a_{34}a_{33}a_{14}a_{21} \quad a_{41}a_{32}a_{21}a_{14}a_{55} \quad b_{13}b_{24}b_{33}b_{42} \quad b_{44}b_{12}b_{31}b_{23} \quad c_{43}c_{22}c_{14}c_{51}c_{35}.$$

2.3. Expand the following determinants

$$\mathbf{A} = \begin{vmatrix} 3 & 1 & 2 \\ 0 & -4 & 1 \\ -1 & 2 & -2 \end{vmatrix} \quad \mathbf{B} = \begin{vmatrix} 1 & 2 & 3 & 4 \\ 0 & 1 & 2 & 3 \\ 0 & 2 & 0 & 1 \\ 0 & 3 & 0 & 5 \end{vmatrix}$$

2.4. Expand $|\mathbf{B}|$, above, using 2X2 complements from rows 3 and 4.

2.5. Expand $|\mathbf{B}|$ above by completing its transformation to triangular form.

2.6. Expand $|\mathbf{A}|$ above using pivotal condensation.

2.7. Given $\mathbf{A}$ and $\mathbf{B}$: $\mathbf{A} = \begin{bmatrix} 1 & 3 \\ 0 & -1 \\ 2 & 2 \\ 1 & 4 \end{bmatrix}$ and $\mathbf{B} = \begin{bmatrix} 2 & -3 & 0 & 1 \\ -1 & 1 & 2 & -2 \end{bmatrix}$.

Find $|\mathbf{A}\,\mathbf{B}|$ and $|\mathbf{B}\,\mathbf{A}|$.

2.8. Expand the determinant $\mathbf{C} = \begin{vmatrix} a_{11} + jb_{11} & a_{12} + jb_{12} & a_{13} + jb_{13} \\ a_{21} + jb_{21} & a_{22} + jb_{22} & a_{23} + jb_{23} \\ a_{31} + jb_{31} & a_{32} + jb_{32} & a_{33} + jb_{33} \end{vmatrix}$

Use the method given in Section 2.10.2.

2.9. Given $\mathbf{A}(12\text{X}12)$. How many terms are in the term-by term expansion of $|\mathbf{A}|$? How many factors are in each term? How long would it take your PC to calculate $|\mathbf{A}|$, term-by-term?

2.10. A (5X5) determinant is to be expanded by complements from its first 3 rows. One such complement is

$$\begin{vmatrix} a_{11} & a_{13} & a_{14} \\ a_{21} & a_{23} & a_{24} \\ a_{31} & a_{33} & a_{34} \end{vmatrix}$$. What leading sign should be placed on this term?

2.11. Determine the rank of the matrix $\mathbf{A}_1$

$$\mathbf{A}_1 = \begin{bmatrix} 1 & 2 & 3 & 4 \\ 2 & 4 & 6 & 8 \\ 3 & 6 & 9 & 12 \\ 4 & 8 & 12 & 16 \end{bmatrix} \qquad \mathbf{A}_2 = \begin{bmatrix} 5 & -3 & 11 & -5 \\ -2 & 0 & -4 & 1 \\ -1 & -3 & -1 & -2 \\ 3 & 3 & 5 & 1 \end{bmatrix}.$$

2.12. Using Gaussian Reduction methods, reduce $\mathbf{A}_2$ to triangle form, and determine its rank. Use "partial pivoting" (i.e., select pivots such that column interchanges are not required).

2.13. Given a 3X3 determinant made up of differentiable functions $y_{ij}(x)$, show that the derivative of the determinant is given by:

$$\frac{d}{dx} |\mathbf{y}_1\mathbf{y}_2\mathbf{y}_3| = |\mathbf{y}'_1\mathbf{y}_2\mathbf{y}_3| + |\mathbf{y}_1\mathbf{y}'_2\mathbf{y}_3| + |\mathbf{y}_1\mathbf{y}_2\mathbf{y}'_3|;$$

$$\mathbf{y}_j = \left\{ \begin{array}{c} y_{1j} \\ y_{2j} \\ y_{3j} \end{array} \right\}; \text{ and } \frac{d\mathbf{y}_j}{dx} = \mathbf{y}'_j = \left\{ \begin{array}{c} y'_{1j} \\ y'_{2j} \\ y'_{3j} \end{array} \right\}.$$

CHAPTER 3

Matrix Inversion

3.1   INTRODUCTION

This chapter will discuss matrix inversion, and the very closely related subject of the solution of simultaneous equation sets. The inversion matrix arrays will necessarily be square, (nXn), for which the inversion process is defined — and for which the determinant is defined.

Emphasis is placed on the mechanical methods used in the inversion process. The next chapter will consider simultaneous equation sets as "vector transformations," and is oriented toward a geometric interpretation, and considerations of compatibility.

In Chapter 2, Section 2.8, it was shown that a square matrix, whose determinant, $|\mathbf{A}|$, is other than zero, possesses an "inverse matrix," $\mathbf{A}^{-1}$, such that:

$$\mathbf{A}^{-1}\mathbf{A} = \mathbf{A}\mathbf{A}^{-1} = \mathbf{I} \tag{3.1}$$

where $\mathbf{I}$ is defined as the (nXn) unit matrix. The elements of the inverse matrix are the "cofactors" of $\mathbf{A}$ divided by $|\mathbf{A}|$; the cofactors being arranged into the "adjoint matrix."

$$\mathbf{A}^{-1} = [\mathbf{A}_{\text{adj}}]/|\mathbf{A}| \quad (|\mathbf{A}| \neq 0) . \tag{3.2}$$

The adjoint matrix is the transpose of the matrix of cofactors; its columns contain the row cofactors of $\mathbf{A}$. The cofactor of a_{ij} is the signed first minor of a_{ij}, the leading sign being determined negative if $i + j$ is odd, positive if it is even. ***Then, the inverse matrix is composed entirely of determinants***; the minor is the (n-1Xn-1) determinant formed by deleting the row and column of the a_{ij} term. Therefore, the inverse could be determined by this definition. But, these calculations are quite lengthy. Instead, pivotal reduction methods will be discussed — including the Gauss Reduction which was discussed in the previous chapter. This simple method will be shown to be an amazingly effective tool for inverting matrices and solving simultaneous linear equations.

As a preliminary step, the "elementary transformation matrices" (Chapter 1, Section 1.4) will be revisited, to provide further insight, and some justification for later methods.

3.2   ELEMENTARY OPERATIONS IN MATRIX FORM

Three elementary operations were used in the previous chapter, in diagonalizing a determinant. They are: (1) To any row (column) is added a multiple of another row (column). (2) A row (column) is divided by some factor. (3) Two rows (columns) are interchanged (this occurs when a pivot element is brought to the main diagonal). These operations can be put into matrix form.

Operation 1. $\mathbf{Q}_{ij}(k)$. Starting with the (nXn) unit matrix, replace the ijth element ($i \neq j$) with a factor k_{ij}. Now if a matrix, $\mathbf{A}$, is premultiplied by this "transform" matrix:

$$\mathbf{Q}_{ij}(k)\mathbf{A} = \mathbf{B}; \quad i \neq j . \tag{3.3}$$

The matrix $\mathbf{B}$ is the same as $\mathbf{A}$, except that to its ith row is added k times its jth row. Note the 3X3 example $\mathbf{Q}_{23}(k)$:

$$\begin{bmatrix} 1 & 0 & 0 \\ 0 & 1 & k \\ 0 & 0 & 1 \end{bmatrix} \begin{bmatrix} 3 & 1 & 2 \\ 0 & 2 & -1 \\ 1 & -1 & 2 \end{bmatrix} = \begin{bmatrix} 3 & 1 & 2 \\ 0+k & 2-k & -1+2k \\ 1 & -1 & 2 \end{bmatrix}. \tag{3.4}$$

The reader should try other examples – with the factor k in all the nondiagonal locations of $\mathbf{Q}_{ij}$.

Note that in every case (wherever the k factor is — as long as it is not on the main diagonal), the determinant, $|\mathbf{Q}_{ij}|$, is 1. Furthermore, from the previous chapter on determinants, the value of $|\mathbf{A}|$ is unchanged by this fundamental operation, i.e., $|\mathbf{B}| = |\mathbf{Q}_{ij}||\mathbf{A}| = |\mathbf{A}|$.

Now, in the case $\mathbf{B} = \mathbf{A}\mathbf{Q}_{ij}(k)$, (postmultiplication of $\mathbf{A}$ by the same type of transformation):

$$\begin{bmatrix} 3 & 1 & 2 \\ 0 & 2 & -1 \\ 1 & -1 & 2 \end{bmatrix} \begin{bmatrix} 1 & 0 & 0 \\ 0 & 1 & k \\ 0 & 0 & 1 \end{bmatrix} = \begin{bmatrix} 3 & 1 & 2+k \\ 0 & 2 & -1+2k \\ 1 & -1 & 2-k \end{bmatrix}. \tag{3.5}$$

In this case, to the jth column is added k times the ith column – where i and j are the row, column positions of k. Note the difference, compared to premultiplication.

Again note that $|\mathbf{Q}_{ij}| = 1$, and that $|\mathbf{A}| = |\mathbf{B}|$.

Operation 2. $\mathbf{Q}_j(k)$: Beginning with the unit matrix, replace the jth main diagonal element with a factor, k. It should be obvious that premultiplying $\mathbf{A}$ with this $\mathbf{Q}_j(k)$ will multiply elements of the jth row of $\mathbf{A}$ by k:

$$Q_2(k) \quad \begin{bmatrix} 1 & 0 & 0 \\ 0 & k & 0 \\ 0 & 0 & 1 \end{bmatrix} \begin{bmatrix} 3 & 1 & 2 \\ 0 & 2 & -1 \\ 1 & -1 & 2 \end{bmatrix} = \begin{bmatrix} 3 & 1 & 2 \\ 0 \times k & 2 \times k & -1 \times k \\ 1 & -1 & 2 \end{bmatrix} \tag{3.6}$$

and, in postmultiplication the jth column is multiplied:

$$\begin{bmatrix} 3 & 1 & 2 \\ 0 & 2 & -1 \\ 1 & -1 & 2 \end{bmatrix} \begin{bmatrix} 1 & 0 & 0 \\ 0 & 1 & 0 \\ 0 & 0 & k \end{bmatrix} = \begin{bmatrix} 3 & 1 & 2 \times k \\ 0 & 2 & -1 \times k \\ 1 & -1 & 2 \times k \end{bmatrix}. \tag{3.7}$$

In this case, $|\mathbf{Q}_j(k)| = k$, and $|\mathbf{B}| = |\mathbf{Q}_j(k)||\mathbf{A}| = k|\mathbf{A}|$.

Operation 3. $\mathbf{Q}_{i\sim j}$. Interchange row (or column) i with row (or column) j of the unit matrix. Now, premultiply $\mathbf{A}$ by this $\mathbf{Q}$ matrix:

$$\begin{bmatrix} 0 & 0 & 1 \\ 0 & 1 & 0 \\ 1 & 0 & 0 \end{bmatrix} \begin{bmatrix} 3 & 1 & 2 \\ 0 & 2 & -1 \\ 1 & -1 & 2 \end{bmatrix} = \begin{bmatrix} 1 & -1 & 2 \\ 0 & 2 & -1 \\ 3 & 1 & 2 \end{bmatrix}. \tag{3.8}$$

In this case, with $\mathbf{Q}_{i\sim j}$ formed by interchanging rows one and three of the unit matrix, the result of $\mathbf{B} = \mathbf{Q}_{i\sim j}\mathbf{A}$ is that the same rows of $\mathbf{A}$ are interchanged. In postmultiplication:

$$\begin{bmatrix} 3 & 1 & 2 \\ 0 & 2 & -1 \\ 1 & -1 & 2 \end{bmatrix} \begin{bmatrix} 1 & 0 & 0 \\ 0 & 0 & 1 \\ 0 & 1 & 0 \end{bmatrix} = \begin{bmatrix} 3 & 2 & 1 \\ 0 & -1 & 2 \\ 1 & 2 & -1 \end{bmatrix}. \tag{3.9}$$

Not surprisingly, in this case, with rows (columns) 2 and 3 of $\mathbf{I}$ interchanged, these same columns of $\mathbf{A}$ are interchanged.

The determinant $|\mathbf{Q}_{i\sim j}| = -1$, and $|\mathbf{B}| = -|\mathbf{A}|$. This is analogous to the property that interchanging two rows (columns) of a matrix changes the sign of the determinant.

3.2.1   DIAGONALIZATION USING ELEMENTARY MATRICES

The diagonalization or triangularization of a matrix, $\mathbf{A}$, can be accomplished by a series of these elementary operations. These, in turn, can be visualized as pre-, and/or postmultiplication of $\mathbf{A}$ by the elementary transform matrices. Note: In the equations below, the symbol $\mathbf{Q}$ is used without indication of its type. This is done so that the final transformation is more clearly shown as the product of the individual operations.

$$\mathbf{B} = \mathbf{QA} \tag{3.10}$$

where $\mathbf{Q} = \mathbf{Q}_m\mathbf{Q}_{m-1} .. \mathbf{Q}_2\mathbf{Q}_1$ (a series of m elementary operations, each of which is of a type discussed above) and $\mathbf{B}$ is, optionally, diagonal, or triangular. Then

$$\mathbf{A}^{-1}\mathbf{Q}^{-1} = \mathbf{B}^{-1}, \text{ and therefore} \tag{3.11}$$
$$\mathbf{A}^{-1} = \mathbf{B}^{-1}\mathbf{Q}. \tag{3.12}$$

The $\mathbf{B}$ matrix, whether diagonal or triangular, is easy to invert. The $\mathbf{Q}$ matrix is developed during the procedure — *and note that its inverse is not required*. Then, the method is a good learning tool, it provides the basis for the very practical inversion tools, and is not an unreasonable one to use for small matrices, by hand.

As an example of the method, the (3X3) used above, as the $\mathbf{A}$ matrix, will be transformed, by means of a premultiplier $\mathbf{Q}$ matrix, to diagonal form. The $\mathbf{Q}$ matrices are:

$\mathbf{Q}_{31}(-1/3)$; Unit matrix with element (3,1) replaced with $-1/3$; changes the 3rd row of $\mathbf{A}$ to

$$\begin{bmatrix} 0 & 4/3 & -4/3 \end{bmatrix}$$

$\mathbf{Q}_{32}(2/3)$; Unit matrix with element (3,2) replaced with $2/3$; changes the 3rd row of $\mathbf{A}$ to

$$\begin{bmatrix} 0 & 0 & 2/3 \end{bmatrix}$$

$\mathbf{Q}_{12}(-1/2)$; Unit matrix with element (1,2) replaced with $-1/2$; changes the 1st row of $\mathbf{A}$ to

$$\begin{bmatrix} 3 & 0 & 5/2 \end{bmatrix}$$

$Q_{13}(-15/4)$; Unit matrix with element (1,3) replaced with $-3\ 3/4$; changes the 1st row of **A** to

$$\begin{bmatrix} 3 & 0 & 0 \end{bmatrix}$$

$Q_{23}(3/2)$; Unit matrix with element (2,3) replaced with $1\ 1/2$; changes the 2nd row of **A** to

$$\begin{bmatrix} 0 & 2 & 0 \end{bmatrix}$$

Note that these changes "drive the off-diagonal elements of **A** to zero." Now, to find the accumulated **Q** matrix, the above must be multiplied *in the order*

$$\mathbf{Q} = \mathbf{Q}_{23}(3/2)\mathbf{Q}_{13}(-15/4)\mathbf{Q}_{12}(-1/2)\mathbf{Q}_{32}(2/3)\mathbf{Q}_{31}(-1/3)\ .$$

Note that each of the **Q** matrices are of type 1, $(\mathbf{Q}_{ij})$ whose determinant = 1. None of the unit matrix elements replaced is on the main diagonal.

$$\mathbf{Q} = \begin{bmatrix} 9/4 & -3 & -15/4 \\ -1/2 & 2 & 3/2 \\ -1/3 & 2/3 & 1 \end{bmatrix}. \tag{3.13}$$

The reader may want to verify that the determinant $|\mathbf{Q}| = 1$.

$$\mathbf{B} = \mathbf{QA} = \begin{bmatrix} 3 & 0 & 0 \\ 0 & 2 & 0 \\ 0 & 0 & 2/3 \end{bmatrix}. \tag{3.14}$$

Now, the inversion of **A** is simply given by $\mathbf{B}^{-1}\mathbf{Q}$, as shown in Equation (3.12).

Inversion of a Diagonal Matrix

Of course, the inversion of the diagonal **B** matrix is very simple. For example, if we premultiply **B** by a unit matrix with its (1,1) element replaced with 1/3 (i.e., operation 2, $\mathbf{Q}_1(1/3)$), then the first row of **B** is divided by 3. Just exactly what is needed.

Then, to invert a diagonal matrix **B** premultiply by a unit matrix whose diagonal elements are replaced by the reciprocals of the corresponding diagonal elements of **B**. Premultiplying both sides of (3.14) by such a matrix, **B** becomes the unit matrix, while a new **Q** matrix (say, **Q**') is formed on the right. From (3.15), it can be seen that this new **Q**' is the inverse of **A** (i.e., $\mathbf{I} = \mathbf{Q}'\mathbf{A}$).

$$\begin{bmatrix} \frac{1}{3} & 0 & 0 \\ 0 & \frac{1}{2} & 0 \\ 0 & 0 & \frac{3}{2} \end{bmatrix} \mathbf{B} = \begin{bmatrix} \frac{1}{3} & 0 & 0 \\ 0 & \frac{1}{2} & 0 \\ 0 & 0 & \frac{3}{2} \end{bmatrix} \mathbf{QA} \Rightarrow \mathbf{I} = \mathbf{Q}'\mathbf{A} = \begin{bmatrix} 3/4 & -1 & -5/4 \\ -1/4 & 1 & 3/4 \\ -1/2 & 1 & 3/2 \end{bmatrix} \mathbf{A}. \tag{3.15}$$

During the formation of the elementary operations, we could have decided to reduce the diagonal elements of **A** to unity as the operations progressed, rather than waiting to do it at the end. The results would obviously be the same.

3.3   GAUSS-JORDAN REDUCTION

Matrix inversion can be thought of as an algorithm — a series of elementary operations which result in the inverse of the input. The foregoing shows that the inverse is a product of those elementary operations in matrix form. Gauss-Jordan is the name of the method whose objective is specifically to operate on the input (using the elementary operations, but not in matrix form), until the unit matrix emerges. If these same operations are concurrently performed on a unit matrix, it will emerge as the inverse of the input.

To emphasize the concurrency of these operations, they are performed on an "augmented matrix" as shown in (3.16). In partitioning these two matrices side by side, no matrix operation is implied. The columns of $\mathbf{I}$ are simply added on to those of the input, forming an nX2n matrix.

$$\mathbf{A} \,|\, \mathbf{I} = \begin{bmatrix} 3 & 1 & 2 & 1 & 0 & 0 \\ 0 & 2 & -1 & 0 & 1 & 0 \\ 1 & -1 & 2 & 0 & 0 & 1 \end{bmatrix} \tag{3.16}$$

If this matrix were to be multiplied by $\mathbf{A}^{-1}$, the result would obviously be $\mathbf{I}|\mathbf{A}^{-1}$. However, the inverse is not yet known, so we must think in terms of an algorithm, a method by which $\mathbf{A}$ can be "reduced" to the unit matrix. If these operations succeed in this reduction, then — taken together — they must be $\mathbf{A}^{-1}$. If that is true, then their operation on the "augmented" columns will cause this inverse to appear on the right.

The method is basically the same as that used in all the methods of this chapter. "Pivots" are to be (re)located along the main diagonal. In general, row and column interchanges are required. However, these will be omitted in this discussion for reasons of clarity. (Note that if a row interchange is to be made, ***the interchange would include the augmented elements***.) Column interchanges are only between columns of $\mathbf{A}$. These must be taken into account, later.

The "pivot row" is then divided by this element, and this row is used to eliminate (reduce to zero) all other elements in the "pivot column."

$$\begin{bmatrix} 1.0 & a_{12}^1 & a_{13}^1 & b_{11}^1 & 0 & 0 \\ a_{21} & a_{22} & a_{23} & 0 & 1 & 0 \\ a_{31} & a_{32} & a_{33} & 0 & 0 & 1 \end{bmatrix}. \tag{3.17}$$

In (3.17), the augmented matrix is shown just after the first row is divided by a_{11}. Note that all elements in the row are changed (so they are shown with the superscript "1"). In particular, the 1,1 element of the unit matrix is no longer 1.0, since it has been divided as well.

Just as in the previous chapter, the elements below this first pivot will be reduced to zero by subtracting the proper multiple of row 1 from the other rows. The result is shown in (3.18). At this point, the first "elimination" step is complete. To begin the second step, the 2,2 element is taken as the pivot. Row 2 will be divided by this element and the new row will be used to eliminate all the elements in column two — both above and below the main diagonal.

$$\begin{bmatrix} 1.0 & a_{12}^1 & a_{13}^1 & b_{11}^1 & 0 & 0 \\ 0 & a_{22}^1 & a_{23}^1 & b_{21}^1 & 1 & 0 \\ 0 & a_{32}^1 & a_{33}^1 & b_{31}^1 & 0 & 1 \end{bmatrix}. \tag{3.18}$$

Using the augmented matrix from (3.16), the procedure is shown in 3 decimal places (rather than fractions). The 1st pivot element is a_{11} (i.e., 3.0). Dividing the 1st row (including augmenting columns) by this element:

$$\begin{Vmatrix} 1.000 & 0.333 & 0.667 & 0.333 & 0.000 & 0.000 \\ 0.000 & 2.000 & -1.000 & 0.000 & 1.000 & 0.000 \\ 1.000 & -1.000 & 2.000 & 0.000 & 0.000 & 1.000 \end{Vmatrix}.$$

Subtracting row 1 from row 3

$$\begin{Vmatrix} 1.000 & 0.333 & 0.667 & 0.333 & 0.000 & 0.000 \\ 0.000 & 2.000 & -1.000 & 0.000 & 1.000 & 0.000 \\ 0.000 & -1.333 & 1.333 & -0.333 & 0.000 & 1.000 \end{Vmatrix}.$$

The new pivot is a_{22} (2.000). After dividing row 2 by a_{22}, the other two elements in the second column are eliminated in the following two steps:

$$\begin{Vmatrix} 1.000 & 0.000 & 0.833 & 0.333 & -0.167 & 0.000 \\ 0.000 & 1.000 & -0.500 & 0.000 & 0.500 & 0.000 \\ 0.000 & -1.333 & 1.333 & -0.333 & 0.000 & 1.000 \end{Vmatrix}$$

$$\begin{Vmatrix} 1.000 & 0.000 & 0.833 & 0.333 & -0.167 & 0.000 \\ 0.000 & 1.000 & -0.500 & 0.000 & 0.500 & 0.000 \\ 0.000 & 0.000 & 0.667 & -0.333 & 0.667 & 1.000 \end{Vmatrix}.$$

The last pivot is a_{33} (0.667). The third row is divided by this amount, and then the other elements in column 3 are eliminated:

$$\begin{Vmatrix} 1.000 & 0.000 & 0.000 & 0.750 & -1.000 & -1.250 \\ 0.000 & 1.000 & -0.500 & 0.000 & 0.500 & 0.000 \\ 0.000 & 0.000 & 1.000 & -0.500 & 1.000 & 1.500 \end{Vmatrix}$$

$$\begin{Vmatrix} 1.000 & 0.000 & 0.000 & 0.750 & -1.000 & -1.250 \\ 0.000 & 1.000 & 0.000 & -0.250 & 1.000 & 0.750 \\ 0.000 & 0.000 & 1.000 & -0.500 & 1.000 & 1.500 \end{Vmatrix}.$$

The last 3 columns of the above augmented matrix are the inverse of the given matrix, $\mathbf{A}$.

$$\mathbf{A} = \begin{bmatrix} 3 & 1 & 2 \\ 0 & 2 & -1 \\ 1 & -1 & 2 \end{bmatrix}; \qquad \mathbf{A}^{-1} = \begin{bmatrix} 3/4 & -1 & -5/4 \\ -1/4 & 1 & 3/4 \\ 1/2 & 1 & 3/2 \end{bmatrix}. \tag{3.19}$$

In the event that the given problem requires the solution to $\mathbf{Ax} = \mathbf{c}$, the inverse is not needed. In this case, the augmented matrix would contain the single column, $\mathbf{c}$, or perhaps multiple columns, if several solutions are to be found. The method would be exactly the same — the input $\mathbf{A}$ would be reduced to $\mathbf{I}$ while the given column(s) develop into the required solution vectors.

3.3.1   SINGULAR MATRICES

If the $\mathbf{A}$ matrix is singular, zero (or near zero) elements will appear on, and to the right of, the main diagonal. Results from Gauss-Jordan reduction of a (6X6) are shown here, to illustrate the condition:

x_1	x_2	x_3	x_4	x_5	x_6
1.00000	0.00000	0.00000	0.00000	x.x	x.x
0.00000	1.00000	0.00000	0.00000	x.x	x.x
0.00000	0.00000	1.00000	0.00000	x.x	x.x
0.00000	0.00000	0.00000	1.00000	x.x	x.x
0.00000	0.00000	0.00000	0.00000	0.00000	−0.00000
0.00000	0.00000	0.00000	0.00000	−0.00000	0.00000

In the above case the upper left 4X4 diagonalizes normally — pivot elements within the expected range of the problem. Then, suddenly, the 5,5 pivot value drops to (near) zero (note the underlined values). Care must be taken in the programming for this condition — roundoff errors prevent the pivot from being exactly zero. Note the terms −0.00000. These indicate a negative value which is zero to five decimal places, but apparently not exactly zero. The point is that a sudden drop in absolute value must be sensed (i.e., well below the range of expected values).
The elements above these pivots, indicated by 'x.x," will not be zero.

In general, if the rank of $\mathbf{A}$(nXn) is n, the procedure completes normally ($\mathbf{A}$ is non-singular). If the rank of $\mathbf{A}$ is r < n, then an rXr unit matrix is calculated normally, in the upper left of the augmented matrix, but a qXq (q = n − r) array of (near) zeros will appear at lower right.

In the case where the inverse of $\mathbf{A}$ is required, obviously, the procedure and the problem are at an end – since no inverse exists. In the case $\mathbf{Ax} = \mathbf{c}$, no *unique* solution exists. However, a "general solution" may be found if the equation set is "compatible." This possibility will be discussed in more detail in the following chapter.

The Gauss-Jordan method as a matrix inverter will not be pursued further because it is inefficient compared to other methods. However, it is a marvelous tool for determining many characteristics of vector sets and matrices — the subject of Chapter 4.

3.4   THE GAUSS REDUCTION METHOD

The objective of this method is a triangular matrix form (rather than the unit matrix) emerging from the input. In other respects it is the same as Gauss-Jordan. In particular, the pivot elements are always on the main diagonal, and in general, row/column interchanges are necessary to put them there.

$$
\begin{bmatrix}
\begin{array}{c|ccc}
1.0 & a_{11}^1 & a_{12}^1 & a_{13}^1 \\
\hline
a_{21} & a_{22} & a_{23} & a_{24} \\
a_{31} & a_{32} & a_{33} & a_{34} \\
a_{41} & a_{42} & a_{43} & a_{44}
\end{array}
\end{bmatrix}
\begin{bmatrix}
c_1^1 \\
c_2 \\
c_3 \\
c_4
\end{bmatrix} .
$$

The diagram above shows a 4X4 with one single augmenting column. This column is the right-hand side of $\mathbf{Ax} = \mathbf{c}$. The first row (including c_1) has already been divided by the pivot, a_{11}. To indicate the changes of value, the elements in row 1 are given a superscript. The elements under a_{11} are now to be reduced to zero. This can be accomplished in row 2 by subtracting from it a_{21} times row 1. And, the leading elements of the other rows are eliminated in this same fashion. *Note that row 1 includes the c_1 element, and when the a_{i1} multiples of row 1 are subtracted from the lower rows, the c_i elements will be changed.* Also, there may be several, even many, augmenting columns (the n columns of a unit matrix, perhaps). These additional columns would take part in the operations in the same way that the $\mathbf{c}$ column, above, does. In this discussion, the $\mathbf{c}$ columns occupy a separate matrix, $\mathbf{C}(nXm)$, rather than be the "augmenting columns" of $\mathbf{A}$. Today, it is unlikely that these operations are to be performed by hand; so the visualization of the "side-by-side" columns is unnecessary. In the computer program, moving these columns into the A matrix would be a wasted effort.

Of course, it is not necessary to actually calculate any of the elements in column 1. The top (pivot) value will be 1.0, and the elements below it will be 0.0.

Define k to be the index to the pivot. Then, k sequences from 1 to $n-1$, where n is the order of the matrix (when $k = n$, there are no elements to "eliminate." However, the nth row of the augmented matrix must be divided by this n, n pivot value.). At any stage $k < n$, the method described above can be written into the "Pascal-like" code shown below.

The steps shown are within an outer loop which steps k from 1 to $n - 1$. *Note that in every stage, the elements operated upon are those to the right of, and below, the pivot.* The elements above the pivot are not affected. Two (identical) loops are shown in the code — one for $\mathbf{A}$ (j = k+1 to n) and the other for the augmented, $\mathbf{c}$, columns (j = 1 to m). The code shown emphasizes that the same operations are carried out on the augmenting rows (the variables c_{ij}).

If the data is truly in a single augmented $\mathbf{A}(nXn+m)$ matrix, the code could be written with just one loop indexed from $k+1$ to $n + m$. However, the "augmented matrix" concept need not to be taken literally as far as data storage in the computer is concerned.

The triangular objective is reached when the a_{nn} element is chosen as the pivot. No further reduction is necessary at this point; however, the nth row must be divided by a_{nn}.

```
                    Gauss Reduction Method "code"
for j = k+1 to n do
begin
    akj  = akj /akk
    for i = k+1 to n do
    begin
        aij = aij - aik * akj
    end;
end;
for j = 1 to m do
begin    {Note: m = number of augmenting columns}
    ckj = ckj /akk
    for i = k+1 to n do
    begin
        cij = cij - aik * ckj
    end;
end;
```

As a simple example $\mathbf{Ax = c}$:

$$\left[\mathbf{A}|\mathbf{c}\right] = \begin{bmatrix} 3 & 1 & 2 & 4 \\ 0 & 2 & -1 & 0 \\ 1 & -1 & 2 & 2 \end{bmatrix}, \text{ where } \mathbf{c} = \{4, 0, 2\} \tag{3.20}$$

the method quickly produces the triangular form:

$$\begin{bmatrix} 1 & 1/3 & 2/3 & 4/3 \\ 0 & 1 & -1/2 & 0 \\ 0 & 0 & 1 & 1 \end{bmatrix}. \tag{3.21}$$

Now, the solution for x_3 is apparent, and from there, each unknown can be obtained in "reverse order":

$$x_3 = 1.$$

$$x_2 - \tfrac{1}{2}x_3 = 0; \quad x_2 = \tfrac{1}{2} \tag{3.22}$$

$$x_1 + \tfrac{1}{3}x_2 + \tfrac{2}{3}x_3 = \tfrac{4}{3}; \quad x_1 = \tfrac{1}{2}.$$

This reverse order solution method is often called "back substitution."

3.4.1   GAUSS REDUCTION IN DETAIL

The method, including full pivoting, is described here. Row and column exchanges will be accomplished by exchanging indexes in row and column lists, rather than exchanging data rows/columns.

Three lists are used: a row list, "rlist," a column list, "clist," and a second column list, "blist." The blist remembers the column exchanges, ***and the order in which they occur***.

One method change is made here: The pivot rows will not be divided by the pivot element, as is done in the earlier description. However, the rows below the pivot are operated on by the same values as before (the division step is included in these row subtractions. See the variable x in step 3 below). The example problem given below will be followed more easily in doing this, and also this change in method converts more directly into LU decomposition.

In the steps, below, the term "condensed determinant" refers to the square array $|a_{kk}, a_{nn}|$, from the pivot (k,k) to the (n,n) term in the given **A** matrix.

1. ***Initialization***. If the data rows and columns are not actually going to be exchanged, the lists through which the data is accessed must be initialized. rlist[j] = j and clist[j] = j are set, and the blist is set to all zeros for j = 1 to N (the order of the matrix).

2. ***Maximum Element***. At each stage, k (a total of $n - 1$ stages for an nXn matrix), the largest element in the condensed determinant is chosen. It is found in the pth row, qth column. In general, p is not in the pivot (kth) row, and q is not in the kth column.

   Then rlist[k]$\Leftrightarrow$rlist[p] and clist[k]$\Leftrightarrow$clist[q] (The symbol $\Leftrightarrow$ indicates "exchange"). Also, if a column (clist) exchange did occur, blist[k] is set to q.

3. ***Central Operation Loop***. At each stage, k, the objective is to zero the elements under the pivot element a_{kk}. The following "Pascal-like" code is the best way to describe this. In particular, ***the pivot rows are not divided by the pivot elements***. Instead, the variable x is employed to contain the ratio of beginning element value to pivot, as shown here:

```
for i = k+1 to N do { N is the order of input matrix }
begin
    x = a_ik / a_kk
    for j = k+1 to N do a_ij = a_ij - x · a_kj
end;
```

The index k is the row/column of the pivot and the indexing deserves special attention. The element a_{ij}, for example, would ordinarily be accessed by A[i,j]. However, because of row and column interchanges it becomes A[rlist[i],clist[j]]. Then the temporary variable x, above, is

$$x = A[rlist[i],clist[k]]/A[rlist[k],clist[k]] .$$

This is the price that is paid for being able to exchange the list indexes rather than the data rows/columns. Notice also that both i and j run from k+1 to N.

The operations 2 and 3 are repeated for the stages k = 1 to k = N–1 (when the pivot is the (n,n) element, the matrix is already triangular).

4. **Back Substitution.** An upper triangular set of equations, $\mathbf{Ax} = \mathbf{c}$ is solved from x_n back up to x_1 according to the following (easily verified) relations:

$$x_i = \frac{1}{a_{ii}} \left\{ c_i - \sum_{k=i+1}^{n} a_{ik} x_k \right\}; \quad i = n, n-1, \cdots, 1; \quad \text{Note: } x_n = \frac{c_n}{a_{nn}}. \qquad (3.23)$$

If there are multiple {c} columns (for example, the augmented matrix includes a unit matrix), then (3.23) is executed for each row of each column. See the back substitution code in Section 3.5.1.

In the computer implementation, the x-vector overwrites {c}. Then, in (3.23) just replace x_i with c_i. Note that $c_n = c_n / a_{nn}$. Since the c values are found (overwritten) in reverse order, each c_i depends only upon c_k values where $k > i$, which have just been overwritten.

5. **Unscramble rlist.** Because of full pivoting, column interchanges occur. When they do, the solution variables, though calculated correctly, come out in a scrambled order. To rectify this, the blist was kept, which remembers the column exchange (if any) and in which stage it occurred.

The initialized blist contains all zeros. If a column exchange occurs at stage k, then blist[k] is set to the column, q, in which the new pivot was found. After the Gauss reduction, unscrambling of the rlist must be done in the reverse order:

for i = N−1 downto 1 do if blist[i] $\neq$ 0 then rlist[i] $\Leftrightarrow$ rlist[blist[i]]

again, the symbol $\Leftrightarrow$ indicates interchange.

6. **Unscramble data.** At this point, the rlist order is correct but this is not 1, 2, 3 order. It is then necessary to physically arrange the {c} data columns into 1, 2, 3 order (the user of the routine cannot be expected to view the output solution vectors "through" the rlist).

Data Storage

In the computer implementation of the above, the input $\mathbf{A}$ matrix is operated upon directly. The input is thus destroyed in favor of the triangular form. Similarly, the {c} vectors are destroyed, becoming the output solution vectors. If the input {c} vectors are the unit matrix, then of course this matrix is replaced by $\mathbf{A}^{-1}$.

3.4.2   EXAMPLE GAUSS REDUCTION

The Gaussian reduction of a 5X5 set of equations is presented as an example. Its data is given with little discussion — intended as check values for the reader's own programmed solution.

$$
\begin{bmatrix}
1 & 0 & -3 & 1 & 2 \\
-2 & 1 & 5 & -2 & -2 \\
-1 & 1 & 3 & 1 & -3 \\
0 & 0 & -1 & -1 & 3 \\
1 & 1 & -4 & 3 & 5
\end{bmatrix}
\begin{Bmatrix}
x_1 \\ x_2 \\ x_3 \\ x_4 \\ x_5
\end{Bmatrix}
=
\begin{Bmatrix}
-2 \\ 1 \\ 4 \\ -3 \\ 5
\end{Bmatrix}.
\tag{3.24}
$$

The determinant of the $\mathbf{A}$(5X5) matrix is 1.0, chosen for clarity (so that the solution vector $\{x\}$ would have integer values). The following table lists the pivots chosen during the procedure, their p,q locations, and their values. Also in the table are the resultant rlist and blist values (i.e., the rlist and blist are shown with their *final values*, at the end of the Gaussian reduction):

p	q	value	clist	rlist	blist	Un Scrambled rlist
2	3	5.00000	3	2	3	4
5	5	3.40000	5	5	5	1
3	4	2.94118	4	3	4	2
5	5	0.42000	2	1	5	3
5	5	-0.04762	1	4	0	5

The output augmented matrix, at the termination of the triangularization process, is shown in the following table. Note: This is a printout of A[rlist[i], clist[j]].

Matrix A after Gauss Triangularization					c-column
5.0000	−2.0000	−2.0000	1.00000	−2.0000	1.00000
−4.0000	3.4000	1.4000	1.80000	−0.6000	5.80000
3.00000	1.8000	2.94118	1.35294	−0.1176	6.47059
−3.00000	0.8000	0.5294	0.42000	−0.0800	−1.60000
−1.00000	2.6000	−2.4706	−0.0400	-0.0476	−1.95238

The back substitution starts at the bottom of this augmented matrix. For example:

$$x[\text{rlist}[5]] = -1.95238/-0.04762 = 41$$
$$x[\text{rlist}[4]] = (0.08*41-1.6)/0.42 = 4$$

The next table shows the completed results of back substitution, and the unscrambling of the rlist. At the left of the table is a copy of the c-column printout.

The column is accessed via rlist, so for example, the first c value is that which is "pointed to" by the first index in rlist (i.e., rlist[1]). Since rlist[1] = 2, that first value must occupy location 2 in the c-column. By looking back "through rlist" in this way, the data can be placed in its actual locations.

c-column c[rlist[i]]	rlist	Actual Data Location	Data After Back Substitution	Un-Scrambled rlist	Final Data
1.00000 → 2		-1.60000	4.00	4 →	41.00
5.80000 → 5		1.00000	19.00	1 →	4.00
6.47059 → 3		6.47059	2.00	2 →	19.00
-1.60000 → 1		-1.95238	41.00	3 →	2.00
-1.95238 → 4		5.80000	6.00	5 →	6.00

Next to this data are the $\{x\}$ values after back substitution. Then, using the corrected (unscrambled) rlist, the correct order of the data can be obtained.

The Pascal-like code for unscrambling the rlist, then the data, is given below. Note that the rlist becomes scrambled in the reduction process because column interchanges occur (due to full pivoting). If only partial pivoting is used, the rlist would not need to be "unscrambled."

```
{      UNSCRAMBLE rlist - - - }
for i:=N-1 downto 1 do if blist[i] <> 0 then
begin j:=rlist[i]; rlist[i]:=rlist[blist[i]];
rlist[blist[i]]:=j; end;
```

Next, the output vector(s) must be unscrambled, to cease dependence upon the rlist. Note that the output vectors are in the same storage space as the input c-vectors — thus the code still refers to them as **c**-vectors.

```
{ UNSCRAMBLE the ROWS of the c-vectors - - - }
for p:=1 to N do if rlist[p] <> p then
begin
  for i:=p to N do if rlist[i] = p then k:=i;
  for j:=1 to m do { NOTE: m is the number of c-vectors  }
 begin { Exchange c[rlist[k]] with c[rlist[p]]   }
    x:=c[rlist[k],j];
    c[rlist[k],j]:=c[rlist[p],j];
    c[rlist[p],j]:=x;
  end;
  rlist[k]:=rlist[p]; rlist[p]:=p;
 end; { c MATRIX NOW CONTAINS THE SOL'N VECTORS IN ORDER  }
```

Partial Pivoting

Step 2 of the Gauss reduction procedure outlined above describes a "maximum element" routine which chooses the largest (absolute value) element in the reduced matrix. If it is desired that no column exchanges occur in the transfer of the pivot element to the pivot position, the maximum element search could be confined to the pivot column only. The largest element in this column is found and a row exchange then occurs. This method is called "partial pivoting."

The method outlined here accommodates partial pivoting by simply changing the maximum element routine. Of course, the rlist will not have to be unscrambled, the blist is now superfluous, its content remaining at all zeroes (see the rlist unscramble routine, above).

3.5   LU DECOMPOSITION

With a couple of very minor changes, the foregoing Gaussian method can become "LU decomposition." These two methods are fundamentally the same, and achieve exactly the same numerical results. Nevertheless, there is reason for our interest in LU.

This method finds a very clever use for the lower element positions (below the main diagonal) as the input matrix is being reduced. Remember that in Gaussian reduction (and LU decomposition as well), all elements below the main diagonal are reduced to zero — this is the objective of the method. In LU (decomposition) these element positions are stored with data that can be used *later* to "reduce" the input c-vectors. ***Then, the initial input to LU is just the Amatrix, without any "augmenting columns."*** The initial output is the "decomposition" of A into L (lower) and U (upper) triangular matrices, as shown here (a 4X4 example):

$$
\begin{bmatrix}
a_{11} & a_{12} & a_{13} & a_{14} \\
a_{21} & a_{22} & a_{23} & a_{24} \\
a_{31} & a_{32} & a_{33} & a_{34} \\
a_{41} & a_{42} & a_{43} & a_{44}
\end{bmatrix}
\Rightarrow
\begin{bmatrix}
u_{11} & u_{12} & u_{13} & u_{14} \\
l_{21} & u_{22} & u_{23} & u_{24} \\
l_{31} & l_{32} & u_{33} & u_{34} \\
l_{41} & l_{42} & l_{43} & u_{44}
\end{bmatrix}.
\tag{3.25}
$$

The lower triangular matrix, L, *has all unity (1.0) main diagonal elements*. And (3.25) can be taken literally in equation form: $A = LU$. The u_{ij} elements are exactly those that are calculated in Gaussian reduction — *given that the pivot rows are __not__ divided by the pivots*.

The advantage in all this is that once A is decomposed, any number of c-vector columns can be input and solved without reducing A again. In effect, the l_{ij} elements remember the operations that are to be made on the augmented vectors. Since pivoting must be used, the row and column interchanges must be remembered as well, of course. In the case of full pivoting with index switching rather than actual row/column exchanges — as in the previous section — the rlist, clist, and blist must all be saved. The solution to $Ax = c$ proceeds:

$$Lc' = c \quad \text{(Forward Substitution)} \tag{3.26}$$
$$Ux = c' \quad \text{(Back Substitution)} . \tag{3.27}$$

For every input **c** vector, (3.26) must be solved to obtain **c'**. Then, (3.27) is solved to find the solution **x**, of the given equation set, given that **c** vector. The **c'** vector *is the same as that which would have emerged from Gaussian reduction, prior to back substitution*. Since both **L** and **U** are triangular, these equations solve easily. The solution of (3.27) (back substitution) has already been discussed, and is a basic algorithm in Gaussian reduction. The solution to the upper triangular set (3.26) is very similar, called "forward substitution" whose algorithm is almost identical to back substitution.

3.5.1   LU DECOMPOSITION IN DETAIL

The detailed description of LU follows that for Gaussian reduction, almost exactly. Full pivoting will be used again in this method (partial pivoting is a viable alternative). Then, the first 2 steps are the same as previous method, and the important step 3 is only trivially different:

```
for i = k+1 to N do { N is the order of input matrix }
begin
    l_ik = a_ik/a_kk   {Note this difference from Gauss}
    for j = k+1 to N do a_ij = a_ij − l_ik · a_kj
end;
```

Remember that the above code is within an outer loop whose index is k, running from 1 to N-1. Thus, the l_{ij} elements are nothing more than the ratios of the pivots divided into the leading elements of each row — immediately below the pivot. Rather than form the ratio $\frac{a_{ik}}{a_{kk}}$ in a temporary variable, x, (as in Gauss reduction) these ratios are simply stored into the "unused" below-diagonal element positions Again, the indexing is not simple, as implied above. As before, the element a_{ij} is indexed:

$$a_{ij} = A[\text{rlist}[i], \text{clist}[j]] \ .$$

All of the subscripted variables in the code must be accessed through rlist and clist.

Forward Substitution

The solution to (3.26) is known as "forward substitution." It is the solution to a lower triangular set of equations; (3.28) gives a 4X4 example:

$$\begin{bmatrix} 1 & 0 & 0 & 0 \\ l_{21} & 1 & 0 & 0 \\ l_{31} & l_{32} & 1 & 0 \\ l_{41} & l_{42} & l_{43} & 1 \end{bmatrix} \begin{bmatrix} c'_1 \\ c'_2 \\ c'_3 \\ c'_4 \end{bmatrix} = \begin{bmatrix} c_1 \\ c_2 \\ c_3 \\ c_4 \end{bmatrix} . \tag{3.28}$$

In this case, (as contrasted with back substitution), the solution proceeds in 1, 2, 3 order, i.e., from c'_1. Obviously, $c'_1 = c_1$; and $c'_2 = c_2 - l_{21}c_1$. In general, (since main diagonal elements are 1.0):

$$c'_i = c_i - \sum_{j=1}^{i-1} l_{ij}c'_j \quad \text{for } i = 1, 2, 3, \ldots n \ . \tag{3.29}$$

The l_{ij} elements over write the below-diagonal a_{ij} elements. Then, in a computer program, these would still be accessed as A[rlist[i],clist[j]]. Also, the c' values overwrite the c values. In (3.29), l_{ij} could be written a_{ij}, and there is no real need for the "primes" on c.

Note how similar this is to back substitution. This forward substitution is to be done on every input c-vector (or each column of the input unit matrix, if the routine is to calculate an inverse).

From this point, the LU method is again the same as Gauss reduction. The data must be unscrambled. The unscramble method depends on the pivoting that was used. Assuming full pivoting with the index lists, as described before, the unscrambling is the same as before.

3.5.2 EXAMPLE LU DECOMPOSITION

When the LU changes are made to the Gauss reduction, the resulting LU matrix is not triangular as shown in (3.30). **This matrix is not LU**, it is "LU[rlist[i], clist[j]]" just like the one given in the Gauss example, Section 3.4.2 (in fact, note the similarity).

$$
\begin{vmatrix}
\boxed{5.0000} & -2.0000 & -2.0000 & 1.00000 & -2.0000 \\
-0.80000 & \boxed{3.40000} & 1.40000 & 1.80000 & -0.60000 \\
0.60000 & -0.52941 & \boxed{2.94118} & 1.35294 & -0.11765 \\
0.60000 & 0.23529 & -0.17999 & \boxed{0.42000} & 0.08000 \\
-0.20000 & 0.76471 & -0.84000 & -0.95238 & \boxed{-0.04762}
\end{vmatrix}
\tag{3.30}
$$

If the **L*U** matrix product is taken (remembering that the data must be accessed via rlist and clist) the result is the original **A** matrix. Its rows/columns will not be scrambled.

```
// ---------------------------------------- FORWARD SUBSTITUTION
    for i:=2 to N do {A is LU matrix }
    begin {N is matrix order}
       sum:=0; p:=rlist[i]; {c is the right side vector}
       for j:=1 to i-1 do
       begin
          q:=clist[j]; sum:=sum+A[p,q]*c[rlist[j]];
       end;
       c[p]:=c[p]-sum;
    end;
// -------------------------- BACK SUBSTITUTION
    p:=rlist[N]; q:=clist[N];
    c[p]:=c[p] / A[rlist[N],clist[N]];
    for i:=N-1 downto 1 do
    begin
       sum:=0; p:=rlist[i]; for j:=i+1 to N do
       begin
          q:=clist[j]; sum:=sum+A[p,q]*c[rlist[j]];
```

```
    end;
      c[p]:=(c[p]-sum)/A[p,clist[i]];
    end;
```

Note: At the end of forward substitution, the problem is exactly like the Gauss example. The $\mathbf{c}'$ column is the same as that given in the augmented matrix of the Gauss reduction. The back substitution is the same as was done in that example. The unscambling routines are also the same.

3.6   MATRIX INVERSION BY PARTITIONING

When the order of the inversion matrix is large, roundoff error is an especially important consideration due to the huge number of operations involved. If a large matrix inversion could be attacked in a series of smaller inversions, with iterative improvement at each step, the possibility is that the roundoff error might be held at an acceptable level.

Partitioning the large matrix affords the ability of such an attack.

$$\mathbf{M}(nXn) = \left[\begin{array}{c|c} \mathbf{A}\,(n_1 X n_1) & \mathbf{D}\,(n_1 X n_2) \\ \hline \mathbf{G}\,(n_2 X n_1) & \mathbf{B}\,(n_2 X n_2) \end{array} \right]$$

$$n = n_1 - n_2$$

Inversion by partitioning can be regarded as a generalization of reduction (elimination) methods. In Gaussian reduction, for example, each stage "reduces" the given matrix one unknown by solving it in terms of the remaining ones. Now we consider eliminating whole sets of unknowns. The diagram above shows a matrix $\mathbf{M}$(nXn). It is partitioned into 4 submatrices—not usually of the same size. In this case, $n = n_1 + n_2$. The diagram implies $n_2 > n_1$, but that need not be true. It is required that $\mathbf{A}$ be square, however, since the first step is to obtain its inverse.

Consider the equation set which incorporates these partitions:

$$\begin{cases} \mathbf{Ax} + \mathbf{Dy} = \mathbf{c}_1 \\ \mathbf{Gx} + \mathbf{By} = \mathbf{c}_2 \end{cases} \quad \text{Note that } \mathbf{x} \text{ and } \mathbf{c}_1 \text{ are } (n_1 X1); \mathbf{y} \text{ and } \mathbf{c}_2 \text{ are } (n_2 X1) . \tag{3.31}$$

Solving these matrix equations just like any 2-by-2 set will result in

$$\left[\mathbf{M}^{-1}\right][\mathbf{c}] = \left[\begin{array}{c} \mathbf{x} \\ \mathbf{y} \end{array} \right] = \begin{cases} \mathbf{A}_1\mathbf{c}_1 + \mathbf{D}_1\mathbf{c}_2 = \mathbf{x} \\ \mathbf{G}_1\mathbf{c}_1 + \mathbf{B}_1\mathbf{c}_2 = \mathbf{y} \end{cases} \tag{3.32}$$

whose partitions $\mathbf{A}_1$, $\mathbf{D}_1$, $\mathbf{G}_1$, and $\mathbf{B}_1$ are those of the inverse matrix, $\mathbf{M}^{-1}$. The results are

$$\begin{cases} \mathbf{A}_1\,(n_1 X n_1) &= \mathbf{A}^{-1} + \mathbf{A}^{-1}\mathbf{D}\mathbf{H}^{-1}\mathbf{G}\mathbf{A}^{-1} \\ \mathbf{D}_1\,(n_2 X n_1) &= -\mathbf{A}^{-1}\mathbf{D}\mathbf{H}^{-1} \\ \mathbf{G}_1\,(n_1 X n_2) &= -\mathbf{H}^{-1}\mathbf{G}\mathbf{A}^{-1} \\ \mathbf{B}_1\,(n_2 X n_2) &= \mathbf{H}^{-1} \end{cases} \tag{3.33}$$

The price that is paid for being able to invert the nXn matrix by inverting the two smaller matrices is the large amount of matrix multiplication, and the roundoff errors that are bound to accrue. Nevertheless, the method should be considered for the inversion of large matrices.

A numerical example of the method is given in Section 3.8.1.

3.7   ADDITIONAL TOPICS

Both Gauss reduction or the LU method are excellent tools for determining inverses, or for solving linear equation sets. Since LU offers the advantage that additional solutions can be obtained from additional c-vectors, its "forward substitution" is preferred.

In both of these methods, when an inverse is required the "c-vector" input consists of the n columns of a unit matrix. In most programs, the input **A** matrix is overwritten during the inversion process and the unit matrix input is overwritten with the inverse.

Because the LU method can be entered with just the **A** matrix without any augmenting columns, it is efficient in the calculation of determinants as well. The one precaution is in the row/column interchanges. In the general case both row and column are exchanged to place the largest element at the pivot position. In this case the determinant is unchanged since there are two sign changes. However, if just a row or a column exchange occurs, the determinant value must be multiplied by -1. Note that in the 5X5 example, above, the product of the diagonal elements is -1, but the correct determinant value is $+1$. In the 3rd stage, only a column exchange occurred, thereby multiplying the product of diagonal elements by -1.

The "essential computer effort" in matrix inversion is the number of lengthy floating point operations required.

Usually only multiplications and divisions are counted, although additions (subtractions) are sometimes included. The number of these operations required for an LU decomposition can be determined with reference to the inset diagram, below.

```
for k:=1 to N-1
    1 mult for determinant
    for i:=k+1 to N
        1 division
        for j:=k+1 to N
            1 mult and 1 add
        end;
    end;
end;
```

Figure 3.1: Floating point operations in a LU decomposition.

At every cycle, k, a new element is moved to the main diagonal. This element is multiplied by the accumulated value of the determinant at that point. Underneath the new pivot there are $N-k$

elements which must be divided by the pivot (to determine the l_{ij} elements). Adjacent to each of these are $N-k$ elements, a_{ij}, from whom are subtracted a product. Then, for the k_{th} execution of the outer loop there are $N-k$ divisions, and $(N-k)^2$ multiplications and subtractions. The multiplication for the determinant value is neglected since it is not an essential part of the method. Then, the number of divisions:

$$\text{div} = \sum_{k=1}^{N-1} (N-k) = \tfrac{1}{2}N(N-1) \,.$$

The number of multiplications and subtractions are

$$\text{mult} = \sum_{k=1}^{N-1} (N-k)^2 = \tfrac{1}{6}N(N-1)(2N-1) \,.$$

The sum of divisions plus multiplications

$$\tfrac{1}{2}N(N-1) + \tfrac{1}{6}N(N-1)(2N-1) = \tfrac{1}{3}N(N^2-1) \,.$$

In a matrix inversion there are N "c-vectors" most of whose elements are zero. However, it is very unusual for the program to take advantage of this fact. Therefore, we will consider the general case in which N c-vectors are input. In this case, entirely similar reasoning leads to

$$\text{Forward Substitution ops} = \tfrac{1}{2}N^2(N-1)$$
$$\text{Backward Substitution ops} = \tfrac{1}{2}N^2(N+1) \,.$$

The total inversion process, then, requires N^3 operations. With the speed and precision of modern computers this numbers is a problem only when the matrix is very large. Although the inversion of these very large systems is outside the scope of this work, several of the following paragraphs speak to the problem by discussing column normalization, improving the inverse, and inversion by orthogonalization.

3.7.1 COLUMN NORMALIZATION

If the determinant of the matrix is "ill-conditioned," the inversion process may accumulate error or even fail. In Section 2.7 of Chapter 2 it was shown that a determinant value, $|\mathbf{A}|$, is geometrically related to the n-dimensional volume enclosed within the column vectors of $\mathbf{A}$. If one or more of these vectors is disproportionately small, the determinant value will be small. The condition is easily spotted, and easily fixed. Simply write:

$$\mathbf{Ax} = \mathbf{c} \Rightarrow \mathbf{a}_1 x_1 + \mathbf{a}_2 x_2 + \mathbf{a}_3 x_3 + \cdots = \mathbf{c}$$

Now, change the variables, $x_j = \alpha_j y_j$ and set the α value such that its vector, $\mathbf{a}$, is normalized to unit length.

The other source of problem is determinant "skew." In the worst case one or more of the column vectors is a linear combination of the others—the matrix is singular, no inverse exists. In less severe cases the input matrix may be resolved into the product of an orthogonal matrix and a triangular one; see Section 3.7.4.

3.7.2 IMPROVING THE INVERSE

Matrix inversion is characterized by a large number of simple arithmetic operations (in fact, on the order of N^3 of them). It is not unusual for the inverse process to lose precision due to the accumulation of roundoff error. The accumulation is greater the larger the matrix, of course, and is particularly troublesome when the matrix is nearly singular.

In general, the input matrix is not known exactly, with element values the result of measurement. Then, an exact inverse is rarely required. Instead, we invoke a clever iterative process which can usually restore all the precision that is meaningful to the problem.

The matrix equation $\mathbf{AX} = \mathbf{I}$ defines $\mathbf{X}$ as $\mathbf{A}^{-1}$. Each column of $\mathbf{I}$ is the product of $\mathbf{Ax}$, where $\mathbf{x}$ is the corresponding column within $\mathbf{X}$. For simplicity, then, consider $\mathbf{Ax} = \mathbf{b}$, where the vector $\mathbf{b}$ is any one of the vectors in $\mathbf{I}$. This equation set is to be solved for $\mathbf{x}$, using LU decomposition followed by forward and back substitution.

Of course the set has an exact solution, $\mathbf{x}$, but the accumulation of roundoff error produces a somewhat different vector, $\mathbf{x_0} = \mathbf{x} + \Delta\mathbf{x}$. The (hopefully small) $\Delta\mathbf{x}$ is the departure from the exact solution, and it produces a "residual" vector, $\Delta\mathbf{b}$. That is:

$$\mathbf{A}(\mathbf{x} + \Delta\mathbf{x}) = \mathbf{b} + \Delta\mathbf{b}, \quad \text{and since } \mathbf{Ax} = \mathbf{b} \tag{3.34}$$
$$\mathbf{A}\Delta\mathbf{x} = \Delta\mathbf{b} . \tag{3.35}$$

Now, $\Delta\mathbf{b}$ is simply $\mathbf{Ax_0} - \mathbf{b}$, and $\mathbf{b}$ is known — *it's one of the columns of* $\mathbf{I}$. Then the iterative process is:

1. Save the input $\mathbf{A}$ matrix. Use LU and forward, back substitution $\rightarrow \mathbf{x_0}$.

2. Multiply $\mathbf{Ax_0}$ and subtract $\mathbf{b} \rightarrow \Delta\mathbf{b}$. If the elements of $\Delta\mathbf{b}$ are small enough, then stop, else:

3. Use forward back substitution with $\Delta\mathbf{b}$ as input $\rightarrow \Delta\mathbf{x}$.

4. Subtract $\Delta\mathbf{x}$ from $\mathbf{x_0} \rightarrow$ Defines a new $\mathbf{x_0}$.

5. Go back to step 2.

When a stop occurs at step 2, the $\Delta\mathbf{b}$ is within the required precision and the $\mathbf{x}$ vector is the improved solution.

Especially note that step 3 does not involve LU decomposition. The LU matrix already exists, having been produced in step 1. Remember that this is the primary advantage of LU, compared to Gauss reduction — the ability to input any number of vectors, after the input matrix has been decomposed.

In step 1, the input **A** matrix is saved because the LU decomposition overwrites the **A** input. It is the saved version that is used in the multiplication in step 2.

In step 2, it is desirable, and may be necessary, to use greater precision in the calculation for $\Delta\mathbf{b}$. This could be very difficult, since it is likely that the original $\mathbf{x}_0$ was done with the longest floating point data length. This is only necessary when trying to attain the precision of the computers data length. In the usual case, iteration improves the solution. It cannot hurt the solution as long as the $\Delta\mathbf{b}$ vectors are decreasing.

It is meaningless to require greater precision in the inverse than that in the input **A** matrix. If a matrix **B** is found such that:

$$\mathbf{AB} = \mathbf{I} + \mathbf{R} \tag{3.36}$$

and **R** (residual matrix) is beyond the practical precision of **A**, then **B** is the inverse of **A**. In general, the precision of **B** will be less than that of **A**. In most cases, 3 or 4 iterations will be enough. Of course, the entire procedure must be repeated for all the vectors in the inverse, changing the location of the unit value in the input **b** column.

3.7.3 INVERSE OF A TRIANGULAR MATRIX

The algorithm for the inversion of a triangular matrix is much more direct than that for the general matrix. Consider an upper triangular matrix, **P**. Its elements below the main diagonal are all zero; those on the main diagonal are all nonzero; and those above it, are not (all) zero. The determinant, $|\mathbf{P}|$, is given by the product of its main diagonal elements (hence, none of these may be zero). The inverse of **P** is, say, **Q**. *It will also be an upper triangular matrix.* Its main diagonal elements are the reciprocals of those of **P**.

Now, we consider the product $\mathbf{QP} = \mathbf{I}$. As in any matrix product, the ijth element of **I** is given by $\mathbf{q}_i \bullet \mathbf{p}_j$, the dot product of the ith row of **Q** by the jth column of **P**. Using a 4X4 example, we have:

$$\begin{bmatrix} q_{11} & q_{12} & q_{13} & q_{14} \\ 0 & q_{22} & q_{23} & q_{24} \\ 0 & 0 & q_{33} & q_{34} \\ 0 & 0 & 0 & q_{44} \end{bmatrix} \begin{bmatrix} p_{11} & p_{12} & p_{13} & p_{14} \\ 0 & p_{22} & p_{23} & p_{24} \\ 0 & 0 & p_{33} & p_{34} \\ 0 & 0 & 0 & p_{44} \end{bmatrix} = \mathbf{I} \tag{3.37}$$

$[\mathbf{q}_1]\{\mathbf{p}_1\} = q_{11}p_{11} = 1$; then, $q_{11} = 1/p_{11}$;
$[\mathbf{q}_1]\{\mathbf{p}_2\} = q_{11}p_{12} + q_{12}p_{22} = 0$; Solve for q_{12} ;
$[\mathbf{q}_1]\{\mathbf{p}_3\} = q_{11}p_{13} + q_{12}p_{23} + q_{13}p_{33} = 0$; Solve for q_{13} ;
$[\mathbf{q}_2]\{\mathbf{p}_3\} = q_{22}p_{23} + q_{23}p_{33} = 0$; Solve for q_{23} .

The above may be generalized to:

$$q_{ij} = -\frac{1}{p_{jj}} \sum_{k=i}^{j-1} q_{ik}p_{kj} \tag{3.38}$$

where nXn is the order of the matrix , *and in the given order:*

$$i = 1, 2, 3 \ldots n;$$
$$j = i, i + 1, i + 2, \ldots n; \ (j > i);$$
$$k = i, i + 1, \ldots j - 1 .$$

(3.39)

A "Pascal-like" description is:

```
for i:=1 to n do
for j:=i to n do
begin
  if j = i then qjj:=1/pjj else
  begin
    qij:=0;
    for k:=i to j-1 do qij:=qij + qik * Pkj;
    qij:= - qij/Pjj;
  end;
end;
```

An algorithm for the inversion of a lower triangular matrix, **P**, is given below. In this case, the elements of **P** above the main diagonal are all zero. The inverse matrix, **Q**, will also be lower triangular. Then, $q_{ij} = 0$, if i < j. Further, $q_{ii} = 1/p_{ii}$, and also, the determinant of both **P** and **Q** is given by the product of their diagonal elements.

For the lower triangular elements (i.e., i > j):

$$q_{ij} = -q_{jj} \sum_k q_{ik} p_{kj}$$

(3.40)

where, in the given order (and n is the order of the matrix):

$$i = n, n - 1, n - 2, \ldots 1.$$
$$j = i - 1, i - 2 \ldots 1.$$
$$k = j + 1, \ldots i .$$

(3.41)

The elements of **Q** are calculated from the lower right corner toward the upper left corner. That is, the nth row is calculated from the (n,n-1) element to the (n,1) element. Then, the n-1st row (not including the main diagonal, since it is already defined as the reciprocal of the **P** main diagonal), and so on. As an example of the method, consider the following **P** matrix:

$$\mathbf{P} = \begin{bmatrix} 1 & 0 & 0 & 0 \\ 2 & 2 & 0 & 0 \\ 3 & 5 & 3 & 0 \\ 4 & 7 & 8 & 4 \end{bmatrix} .$$

(3.42)

Its inverse, $\mathbf{Q}$, is:

$$\mathbf{Q} = \begin{bmatrix} 1 & 0 & 0 & 0 \\ -1 & 1/2 & 0 & 0 \\ 2/3 & -5/6 & 1/3 & 0 \\ 7/12 & 19/24 & -2/3 & 1/4 \end{bmatrix}. \tag{3.43}$$

A few sample calculations are:

$$q_{43} = -q_{33}(q_{44}p_{43}) = -1/3(1/4)(8) = -2/3$$
$$q_{41} = -q_{11}(q_{42}p_{21} + q_{43}p_{31} + q_{44}p_{41})$$
$$q_{41} = -1[1/4\,(4) + (-2/3)(3) + (19/24)(2)] = 7/12$$
$$q_{32} = -q_{22}(q_{33})(p_{32}) = -1/2(1/3)(5) = -5/6.$$

3.7.4   INVERSION BY ORTHOGONALIZATION

It is a remarkable fact that a general square, nonsingular matrix can be resolved into the product of an orthogonal matrix, say $\mathbf{V}$, times a triangular matrix, $\mathbf{P}$. Both $\mathbf{V}$ and $\mathbf{P}$ are easy to invert!

The news isn't all rosy, however. The method is susceptible to roundoff error, so it is not recommended as a matrix inverter. But, it does work, given enough precision, and besides, the method is a very interesting one to develop.

Given $\mathbf{A}(n\mathrm{X}n)$, we set about deriving the orthogonal matrix in the following way. Consider $\mathbf{A}$ as an assemblage of column vectors $\mathbf{a}_1$, $\mathbf{a}_2, \cdots \mathbf{a}_k, \cdots, \mathbf{a}_n$, where $\mathbf{a}_k$ is the kth column of $\mathbf{A}$. Select the first column and normalize it to unit length. This new unit vector will be $\mathbf{v}_1$:

$$\mathbf{v}_1 = \frac{\mathbf{a}_1}{l_1}; \quad l_1 = \sqrt{a_1^2 + a_2^2 + \cdots + a_n^2}. \tag{3.44}$$

The second vector, $\mathbf{v}_2$, is chosen to be in the same plane as $\mathbf{v}_1$ and $\mathbf{a}_2$, a linear combination of these two vectors: $\mathbf{v}'_2 = c_1\mathbf{v}_1 + c_2\mathbf{a}_2$. The prime merely indicates an unnormalized vector. Since $\mathbf{v}_1$ and $\mathbf{v}_2$ must be orthogonal we dot $\mathbf{v}_1$ with $\mathbf{v}'_2$ and solve for c_1 (c_2 can be set to 1).

$$\mathbf{v}_1 \bullet \mathbf{v}'_2 = c_1\mathbf{v}_1 \bullet \mathbf{v}_1 + c_2\mathbf{a}_2 \bullet \mathbf{v}_1 = 0$$
$$c_2 = 1, \ c_1 = -\mathbf{a}_2 \bullet \mathbf{v}_1 \tag{3.45}$$
$$\mathbf{v}'_2 = \mathbf{a}_2 - (\mathbf{a}_2 \bullet \mathbf{v}_1)\mathbf{v}_1; \quad \mathbf{v}_2 = \frac{\mathbf{v}'_2}{l_2}.$$

Note that $\mathbf{v}_1$ and $\mathbf{v}_2$ are orthogonal.

In the same manner $\mathbf{v}'_3 = \mathbf{a}_3 - (\mathbf{v}_1 \bullet \mathbf{a}_3)\mathbf{v}_1 - (\mathbf{v}_2 \bullet \mathbf{a}_3)\mathbf{v}_2$ and in general

$$\mathbf{v}'_j = \mathbf{a}_j - \sum_{i=1}^{j-1} p_{ij}\mathbf{v}_i \quad \text{where} \tag{3.46}$$

$$p_{ij} = \mathbf{v}_i \bullet \mathbf{a}_j. \tag{3.47}$$

The p_{ij} factors can be arranged into an upper triangular matrix, with the main diagonal elements being the normalization lengths of the vectors, l_j. Note that the jth column of $\mathbf{P}$ provides the p_{ij} factors in (3.46).

Further, solving (3.46) for $\mathbf{a}_j$

$$\mathbf{a}_j = p_{jj}\mathbf{v}_j + p_{(j-1)j}\mathbf{v}_{j-1} + \cdots + p_{1j}\mathbf{v}_1$$
$$p_{jj} = l_j = \sqrt{v_{1j}^2 + v_{2j}^2 + \cdots + v_{nj}^2} \ .$$

That is:

$$\mathbf{A} = \mathbf{VP} \ . \tag{3.48}$$

The inversion of $\mathbf{A}$ is now a relatively simple matter. The triangular $\mathbf{Q} = \mathbf{P}^{-1}$ has been discussed earlier, and the inverse of $\mathbf{V}$ is obtained by transposition ($\mathbf{V}^{-1} = \mathbf{V}^t$). Then

$$\mathbf{A}^{-1} = \mathbf{QV}^t \ . \tag{3.49}$$

3.7.5   INVERSION OF A COMPLEX MATRIX

The Gauss reduction method, and any other method that will successfully invert a real matrix, will work equally well on a matrix whose elements are complex — given that the routines used support complex arithmetic. Some minor adjustments must be made. For example, the routine which chooses the largest element now must be made to determine the absolute value of a complex number.

Complex arithmetic can be difficult to do if the compiler itself does not recognize the complex type. Also, the need for inversion of a complex matrix may not arise often enough. So, whatever the reason, it may be required to invert the complex matrix using only real arithmetic, and real numbers: Find a complex matrix $\mathbf{B}$ such that $\mathbf{AB} = \mathbf{I}$, where $\mathbf{A}$ is complex. Then:

$$(\mathbf{A}_r + j\mathbf{A}_i)(\mathbf{B}_r + j\mathbf{B}_i) = \mathbf{I}; \quad j^2 = -1 \ . \tag{3.50}$$

Equating real (subscript, r) and imaginary (subscript, i) parts:

$$\mathbf{A}_r\mathbf{B}_r - \mathbf{A}_i\mathbf{B}_i = \mathbf{I} \ \text{and} \ \mathbf{A}_r\mathbf{B}_i = -\mathbf{A}_i\mathbf{B}_r \ .$$

Then, assuming that $\mathbf{A}_r$ has an inverse, $\mathbf{B}_i = -\mathbf{A}_r^{-1}\mathbf{A}_i\mathbf{B}_r$ and $(\mathbf{A}_r + \mathbf{A}_i\mathbf{A}_r^{-1}\mathbf{A}_i)\mathbf{B}_r = \mathbf{I}$. And the elements of the complex matrix $\mathbf{B}_r + j\mathbf{B}_i$ are

$$\left\{ \begin{array}{l} \mathbf{B}_r = (\mathbf{A}_r + \mathbf{A}_i\mathbf{A}_r^{-1}\mathbf{A}_i)^{-1} \\ \quad\quad \mathbf{B}_i = -\mathbf{A}_r^{-1}\mathbf{A}_i\mathbf{B}_r \ . \end{array} \right. \tag{3.51}$$

Notice that the increased difficulty of complex numbers cannot be avoided. Although just two matrices must be inverted, both just nXn, there is a lot of matrix multiplication involved.

3.8   EXAMPLES

3.8.1   INVERSION USING PARTITIONS

This example intends to simulate the inversion of a large matrix. For reasons of clarity and lack of space, this "large" matrix, **M**, is only 8X8. *Its inversion will be affected by inverting no larger than a 3X3 array*. The process is straightforward, but the "bookkeeping" becomes cumbersome. To begin, **M** is partitioned as shown, with a 3X3 in the upper left.

$$
\left\|
\begin{array}{ccc|ccccc}
1.00 & -2.00 & 3.00 & 4.00 & 0.00 & 1.00 & 0.00 & 0.00 \\
3.00 & -1.00 & 2.00 & 5.00 & -2.00 & -1.00 & 1.00 & 0.00 \\
2.00 & 4.00 & -5.00 & 1.00 & -1.00 & 2.00 & 3.00 & 1.00 \\
\hline
4.00 & 2.00 & -1.00 & 3.00 & 0.00 & 3.00 & -2.00 & 0.00 \\
-2.00 & 0.00 & 2.00 & -2.00 & 5.00 & 1.00 & -1.00 & 1.00 \\
3.00 & 1.00 & 3.00 & 4.00 & 2.00 & 1.00 & 0.00 & -4.00 \\
1.00 & 3.00 & 0.00 & -1.00 & -2.00 & 0.00 & 2.00 & 0.00 \\
0.00 & -1.00 & -1.00 & 2.00 & 4.00 & -2.00 & 1.00 & 2.00
\end{array}
\right\|
$$

The four partitions of **M** are named according to the set of Equations in (3.31):

$$
\begin{bmatrix} \mathbf{A}(3x3) & \mathbf{D}(3x5) \\ \mathbf{G}(5x3) & \mathbf{B}(5x5) \end{bmatrix}
\begin{bmatrix} \mathbf{x} \\ \mathbf{y} \end{bmatrix}
=
\begin{bmatrix} \mathbf{c}_1 \\ \mathbf{c}_2 \end{bmatrix}
=
\begin{array}{l} \mathbf{Ax} + \mathbf{Dy} = \mathbf{c}_1 \\ \mathbf{Gx} + \mathbf{By} = \mathbf{c}_2 \end{array}.
$$

If **A**(3X3) is inverted the vector, **x**, can be solved for in terms of the remaining unknowns:

$$
\mathbf{x} = \mathbf{A}^{-1}(\mathbf{c}_1 - \mathbf{Dy})
$$

$$
\mathbf{A} = \left\|
\begin{array}{ccc}
1.00 & -2.00 & 3.00 \\
3.00 & -1.00 & 2.00 \\
2.00 & 4.00 & -5.00
\end{array}
\right\|
\qquad
\mathbf{A}^{-1} = \left\|
\begin{array}{ccc}
-3.00 & 2.00 & -1.00 \\
19.00 & -11.00 & 7.00 \\
14.00 & -8.00 & 5.00
\end{array}
\right\|
$$

Plugging the **x** value back into the second equation, then solving for **y**, yields:

$$
(\mathbf{B} - \mathbf{GA}^{-1}\mathbf{D})\mathbf{y} = \mathbf{c}_2 - \mathbf{GA}^{-1}\mathbf{c}_1
$$

$$
\mathbf{Hy} = \mathbf{c}_2 - \mathbf{GA}^{-1}\mathbf{c}_1; \quad \text{where } \mathbf{H} = \mathbf{B} - \mathbf{GA}^{-1}\mathbf{D}.
$$

Note that, with **A**$^{-1}$ known, **H** is known — and it has the dimensions of **B**(5X5).

$$
\mathbf{H} = \left\|
\begin{array}{ccc|cc}
-20.00 & -7.00 & -25.00 & -11.00 & -5.00 \\
-50.00 & -23.00 & -77.00 & -17.00 & -11.00 \\
-78.00 & -37.00 & -118.00 & -28.00 & -23.00 \\
\hline
-82.00 & -44.00 & -125.00 & -27.00 & -20.00 \\
51.00 & 30.00 & 74.00 & 18.00 & 14.00
\end{array}
\right\|
$$

It was a coincidence that A^{-1} has an integer inverse. Because of this coincidence, H is also an integer matrix. H^{-1} will surely not be "so lucky," and in order to proceed, we must have H^{-1}. Since the largest array that can be inverted is 3X3, H must be partitioned — again with a 3X3 in the upper left position.

The partitions of H will be named A_2, D_2, G_2, and B_2, occupying the same positions as those in the original matrix, M. Proceeding as before:

$$A_2(3x3)w + D_2(3x2)z = d_1; \quad w = A_2^{-1}(d_1 - D_2z)$$
$$G_2(2x3)w + B_2(2x2)z = d_2$$

$$A_2 = \begin{Vmatrix} -20.00 & -7.00 & -25.00 \\ -50.00 & -23.00 & -77.00 \\ -78.00 & -37.00 & -118.00 \end{Vmatrix} \qquad A_2^{-1} = \begin{Vmatrix} -0.2419 & 0.1774 & -0.0645 \\ 0.1900 & 0.7348 & -0.5197 \\ 0.1004 & -0.3477 & 0.1971 \end{Vmatrix}$$

As before, $H_2 = (B_2 - G_2A_2^{-1}D_2)$, which can now be written because A_2^{-1} is known. This time, H_2 will be 2X2, the same dimensions as B_2, and will be easy to invert.

$$\mathbf{H_2} \qquad\qquad\qquad \mathbf{H_2^{-1}}$$

$$\begin{Vmatrix} 1.6129 & 17.9355 \\ -2.3907 & -21.8244 \end{Vmatrix} \qquad \begin{Vmatrix} -2.8427 & -2.3361 \\ 0.3114 & 0.2101 \end{Vmatrix}$$

The cumbersome part is that H^{-1} must be found, which requires a complete solution to the above equation set

$$w = A_2^{-1}(d_1 - D_2z)$$
$$z = H_2^{-1}(d_2 - G_2A_2^{-1}d_1)$$

The value for z must now be plugged back into the expression for w. With some algebra, and rearrangement, the results are like those given in Equations (3.33):

$$\begin{bmatrix} w \\ z \end{bmatrix} = \begin{bmatrix} H^{-1} \end{bmatrix}\begin{bmatrix} d \end{bmatrix} = \begin{bmatrix} w \\ z \end{bmatrix} = \begin{bmatrix} A_2^{-1} + A_2^{-1}D_2H_2G_2A_2^{-1} & -A_2^{-1}D_2H_2^{-1} \\ -H_2^{-1}G_2A_2^{-1} & H_2^{-1} \end{bmatrix}\begin{bmatrix} d_1 \\ d_2 \end{bmatrix}.$$

This equation defines H^{-1}. Since A_2^{-1} and H_2^{-1} are known, each of the 4 partitions of H^{-1} can be calculated — for example, its upper left 3X3 is $A_2^{-1} + A_2^{-1}D_2H_2^{-1}G_2A_2^{-1}$. See M^{-1}, below.

With H^{-1} known, using the format of Equation (3.33) M^{-1} becomes:

$$M^{-1} = \begin{Vmatrix} -6.0672 & 8.2063 & 2.8721 & -1.7526 & 5.2418 & -0.1438 & -5.3707 & -4.3445 \\ 4.7815 & -6.6349 & -2.4435 & 1.6097 & -4.241 & 0.1438 & 4.6564 & 3.6303 \\ 0.1513 & -0.0476 & -0.1289 & 0.0266 & 0.0392 & -0.0098 & 0.2507 & 0.0252 \\ 4.4118 & -5.8889 & -2.0915 & 1.3595 & -3.8562 & 0.1307 & 3.8954 & 3.2353 \\ -1.0840 & 1.3968 & 0.5345 & -0.3296 & 0.9967 & 0.0425 & -0.9911 & -0.6807 \\ -1.4706 & 2.2222 & 0.9935 & -0.5458 & 1.5817 & -0.0621 & -1.6503 & -1.4118 \\ -3.0168 & 4.3016 & 1.7180 & -1.1881 & 2.8105 & -0.0359 & -2.8427 & -2.3361 \\ 0.2605 & -0.1746 & -0.1293 & 0.1662 & -0.0621 & -0.1928 & 0.3114 & 0.2101 \end{Vmatrix}$$

For example, the lower right 5X5 of $\mathbf{M}^{-1}$ is $\mathbf{B}_1 = \mathbf{H}^{-1}$. In turn, its lower right 2X2 is $\mathbf{H}_2^{-1}$ as written above.

If the inverse of $\mathbf{M}$ is not required, just the solution to the equation set (3.31), many of the tedious matrix operations can be avoided.

$$\mathbf{y} = \mathbf{H}^{-1}(\mathbf{c}_2 - \mathbf{GA}^{-1}\mathbf{c}_1)$$
$$\mathbf{x} = \mathbf{A}^{-1}(\mathbf{c}_1 - \mathbf{Dy})$$

The solution, above, does require that both $\mathbf{A}$ and $\mathbf{H}$ must be inverted. However, when this is done, fewer operations remain, and these equations can be solved using "matrix-times-vector" operations rather than "matrix-times-matrix". *The savings are considerable.*

As this example indicates, when the matrix $\mathbf{B}$ (and consequently, $\mathbf{H}_1$) is still too large to be inverted directly, additional "partitioning" is required until that lower right matrix is within the range to be inverted—possibly a lengthy process. But, the final result will retain greater precision than a direct approach.

3.9   EXERCISES

3.1. Given the matrix, $\mathbf{A}$, below, determine $\mathbf{Q}$ such that the product $\mathbf{QA}$ produces zero element values in the first column of $\mathbf{A}$, except, a_{11}.

What is the determinant value, $|\mathbf{A}|$?

$$\mathbf{A} = \begin{bmatrix} 1 & 3 & -1 \\ 3 & 11 & 1 \\ 2 & 6 & -1 \end{bmatrix}.$$

3.2. With the $\mathbf{A}$ given in problem 1, determine the solution to $\mathbf{Ax} = \mathbf{c} = \{4, 12, 7\}$. Note that $\{c\}$ is a column vector.

3.3. Find the inverse of the complex matrix $\mathbf{A} = \begin{bmatrix} 1 + j0 & 1 + j2 & 1 + j8 \\ 0 + j1 & -1 + j1 & -6 + j3 \\ 1 + j1 & 0 + j5 & -8 + j15 \end{bmatrix}.$

3.4. Find the inverse of the complex matrix $\mathbf{A} = \mathbf{A}_r + j\mathbf{A}_i$

$$\mathbf{A}_r = \begin{bmatrix} 1 & -1 & 0 \\ 0 & 0 & 0 \\ 2 & 0 & 1 \end{bmatrix} \text{ and } \mathbf{A}_i = \begin{bmatrix} 1 & -2 & 3 \\ 3 & -1 & 2 \\ 2 & 4 & -5 \end{bmatrix}.$$

3.5. Given the equations $\mathbf{Ax} = \mathbf{c}$, and $\mathbf{A}^{-1} = \mathbf{B}$. If two columns in $\mathbf{A}$ are exchanged, how is the solution, $\mathbf{x}$, affected? How is $\mathbf{B}$ affected?

3.6.   Perform an LU decomposition on the 5X5 matrix in Section 3.4.2. Do not use pivoting. Show that L*U does not equal the input S matrix.

3.7.   Using the result from exercise 3.6, solve the example problem using the same right-side c-vector from Equation (3.24).

CHAPTER 4

Linear Simultaneous Equation Sets

4.1   INTRODUCTION

This chapter turns to an interpretation of the solution to linear equation sets, using a geometric approach and insight. We will look at an equation set in several different (and perhaps new) ways, and consider the *solvability* and *compatibility* of an equation set. Most of the mechanics of solution have already been discussed. This chapter intends to be largely conceptual.

Many applications in mechanics, dynamics, and electric circuits depend on the insights gained, and presented here.

We begin by defining the equation set $\mathbf{Ax} = \mathbf{b}$ as "nonhomogeneous" because the $\mathbf{b}$ vector is assumed to be nonzero. Associated with this set is the "homogeneous" set, $\mathbf{Ax} = \mathbf{0}$; the same set, but with the $\mathbf{b}$ vector replaced by the zero vector. In the event that matrix $\mathbf{A}$ is nonsingular, and has an inverse, the homogeneous set plays no part. But, when $\mathbf{A}$ is singular, we will find interest in both $\mathbf{Ax} = \mathbf{0}$, and in $\mathbf{A'x} = \mathbf{0}$ (the transposed homogeneous set).

4.2   VECTORS AND VECTOR SETS

In order to gain greater insight into its solution, the equation set will be interpreted as a "vector transformation." The equation $\mathbf{Ax} = \mathbf{y}$ "transforms" the columns of $\mathbf{A}(nXm)$ into the *vector* $\mathbf{y}$. Alternatively, $\mathbf{y}$ is "synthesized" as a linear vector sum of the column vectors of A.

$$\mathbf{Ax} = \mathbf{y} = \left\{\begin{matrix} a_{11} \\ a_{21} \\ \vdots \\ a_{n1} \end{matrix}\right\} x_1 + \left\{\begin{matrix} a_{12} \\ a_{22} \\ \vdots \\ a_{n2} \end{matrix}\right\} x_2 + \cdots + \left\{\begin{matrix} a_{1m} \\ a_{2m} \\ \vdots \\ a_{nm} \end{matrix}\right\} x_m = \left\{\begin{matrix} y_1 \\ y_2 \\ \vdots \\ y_n \end{matrix}\right\}, \text{ or:} \quad (4.1)$$

$$= \mathbf{a}_1 x_1 + \mathbf{a}_2 x_2 + \cdots + \mathbf{a}_m x_m = \mathbf{y}. \quad (4.2)$$

In this quite general example, $\mathbf{A}$ is (nXm); there are m vectors (the columns of $\mathbf{A}$), each with n coordinates (dimensions — the rows of $\mathbf{A}$). It is often instructive to draw the same vector picture of the transposed matrix, i.e., $\mathbf{A}'$, whose n column vectors are the rows of $\mathbf{A}$.

To begin, the discussion of vectors in Chapter 1 is reviewed and enlarged upon in the paragraphs, below.

In two dimensions, a vector, $\mathbf{v}$, is described as $\{v_x, v_y\}$, where the subscripts "x" and "y" refer to unit vectors in a rectangular coordinate set. These unit vectors could be written as $\{1,0\}$ and $\{0,1\}$, showing both their orthogonality and their unit length. v_x and v_y are the components of $\mathbf{v}$ along the coordinate axes. Extension to three dimensions is simply: $\mathbf{v} = \{v_x, v_y, v_z\}$, and in either 2 or 3 dimensions, it is convenient to use subscripted letters (e.g., "x," "y," and "z") to refer to the unit vectors (the coordinate axes).

The two and three-dimensional cases are familiar, and easily visualized. But, in generalizing to greater than 3 dimensions, visualization is lost. For this reason, the plan will be to view the various concepts in the 2 and 3 dimensional cases; then simply extend the reasoning into n-dimensions. For example, the definition of a vector:

$$\mathbf{v} = \{v_x, v_y, v_z\} \text{ in three dimensions} \tag{4.3}$$
$$\mathbf{v} = \{v_1, v_2, \ldots v_n\} \text{ in "}n\text{" dimensions} \tag{4.4}$$

is extended to n-dimensions with only a relatively minor change in notation: The coordinate axes are now given numbers, rather than "x, y, z, . ." letters. *But these n coordinate axes are still perceived as rectangular axes, in an "n dimensional space," and the values, v_j are the components of $\mathbf{v}$ along these axes*. In fact, the component v_j is defined as the product of the (n-dimensional) length of $\mathbf{v}$ multiplied by the cosine of the angle between $\mathbf{v}$ and the jth coordinate axis (i.e., the concept of the "direction cosine," in n dimensions).

If $\mathbf{v}$ is composed of real components, its length is defined as

$$|\mathbf{v}| = \text{sqrt}(v_1^2 + v_2^2 + \cdots + v_n^2) = \sqrt{v_1^2 + v_2^2 + \cdots + v_n^2} \tag{4.5}$$
$$|\mathbf{v}| = \text{sqrt}(\mathbf{v} \bullet \mathbf{v}) = \text{sqrt}(\mathbf{v}'\mathbf{v}) . \tag{4.6}$$

In (4.6), the notation $(\mathbf{v} \bullet \mathbf{v})$ and $(\mathbf{v}'\mathbf{v})$ or $(\mathbf{v}^t\mathbf{v})$ denote the dot product of $\mathbf{v}$ into itself, in n dimensions ("n-space"). In general, the dot product of two vectors, $\mathbf{u} \bullet \mathbf{v}$, is simply the sum of products of the respective components of the vectors. Equivalently, this (scalar) dot product can be expressed as the product of their magnitudes, multiplied by the cosine of the angle between them. *Also, two (nonzero) vectors are said to be "orthogonal" in n dimensions if their dot product is zero*:

$$\mathbf{v}'\mathbf{u} = \mathbf{v}^t\mathbf{u} = (v_1 u_1 + v_2 u_2 + \cdots + v_n u_n) = 0 . \tag{4.7}$$

A vector whose length, $|\mathbf{v}|$, is unity is called a "unit vector."

If the vectors consist of complex numbers, the definitions must be modified. For this purpose, a new notation is introduced: If $c = a + jb$ is a complex number, its "complex conjugate" (i.e., the number $a - jb$) is denoted $\tilde{c}$.

Then the length of a complex vector, v, is $|\mathbf{v}| = \text{sqrt}(\tilde{\mathbf{v}} \bullet \mathbf{v}) = \text{sqrt}(\tilde{\mathbf{v}}'\mathbf{v})$. Similarly, the "Hermitian" scalar product between two complex vectors, u and v, is $\tilde{\mathbf{u}} \bullet \mathbf{v}$ (which is generally complex, and not equal to $\tilde{\mathbf{v}} \bullet \mathbf{u}$).

4.2.1   LINEAR INDEPENDENCE OF A VECTOR SET

A set of vectors, $\mathbf{a}_1, \mathbf{a}_2, \mathbf{a}_3, \ldots \mathbf{a}_m$, is said to be "linearly independent" if **no** (scalar) constants, c_k, can be found which relates them in the following way:

$$\mathbf{a}_1 c_1 \mathbf{a}_2 c_2 \mathbf{a}_3 c_3 + \cdots + \mathbf{a}_m c_m = \mathbf{0} \equiv A(n \times m)\mathbf{c}(m \times 1) = \mathbf{0}(n \times 1) . \tag{4.8}$$

Note that there are m vectors, each with n coordinates (dimensions).

In 2-space, and with two vectors, Equation (4.8) becomes: $c_1\mathbf{a}_1 + c_2\mathbf{a}_2 = 0$. In this simple case, if nonzero values for c_1 and c_2 can be found, it means that the two vectors are scalar multiples of one another. The (dependent) vectors are collinear. Such vectors "use" only 1 of the two dimensions available (although these vectors may not be parallel to either of the coordinate axes). Conversely, in 2-space, any two vectors that are not collinear, are linearly independent, and are said to "fill" the space—two constants cannot be found which relate them in the sense of (4.8). Furthermore, the determinant of the square **A(2X2)** matrix formed of the vector components will be non-zero (**A** will not be singular).

Note that in 2-space, three vectors are necessarily dependent, whether or not they fill the space. In general, in an m-space, more than m vectors form a dependent set.

In 3-space, three vectors which do not lie in a plane are linearly independent, the case in Fig. 4.1, i.e., $\mathbf{a}_2$ and $\mathbf{a}_3$ lie within plane-p, $\mathbf{a}_1$ does not. It is clearly not possible to derive any one of

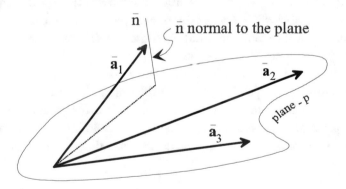

Figure 4.1:

the **a** vectors as a linear sum of the other two. The equation:

$$A\{\mathbf{c}\} = c_1\mathbf{a}_1 + c_2\mathbf{a}_2 + c_3\mathbf{a}_3 = \mathbf{0} \tag{4.9}$$

has no solution (except $\{\mathbf{c}\} = \{0\}$).

Now, slide the tip of the $\mathbf{a}_1$ down the normal until the vector lies in the plane-p. Clearly, any one of the vectors can now be obtained as a linear sum of the other two by a simple vector addition and (4.9) has a non-trivial solution. The 3 vectors do not fill the 3-space (the term "3-space" is used to describe a 3 dimensioanl space. Then, the term "n-space" will refer to a space of n dimensions). With all three vectors in plane-p it is possible to find a fourth vector orthogonal to all three; for example, the vector **n**. In general, this circumstance is determined by the existence of a non-trivial solution to the transposed set,

$$\mathbf{A}'\{\mathbf{z}\} = \mathbf{0}. \qquad \mathbf{A}'(\mathbf{mXn})\mathbf{z}(\mathbf{nX1} = \mathbf{0}(\mathbf{mX1}) \ . \qquad (4.10)$$

The original m vectors are row vectors in $\mathbf{A}'$. If non-trivial **z** vectors can be found, they are orthogonal to the original set.

Summarizing: the linear (in)dependence of the m vectors in n-space is determined by investigating the possible (non-trival) solutions of equations (4.8) and (4.10). Gauss-Jordan reduction (Section 3.3) is often used in this investigation.

4.2.2   RANK OF A VECTOR SET

The **rank** of a *vector set*, $\mathbf{A}$(mXn), is equal to the order of the ***largest nonvanishing determinant*** that can be formed from the *matrix* $\mathbf{A}$(nXm); and the largest non-vanishing determinant cannot be greater then the smaller of n and m.

In the event that $m < n$ (more dimensions than vectors), and the rank is $r < m$, the set is dependent and there will be $m - r$ solutions to the equation set (4.8). If $r = m$ then the vector set is independent, and (4.8) has only the trivial solution. This is also true for the "square" case, $m = n$.

If $m > n$, the rank of $\mathbf{A}$ cannot be greater than n. Necessarily, the m vectors are dependent, and non-trivial solutions will be found for (4.8). Again, the rank could be less than n, in which case the (many) m vectors still do not fill the n-space.

An obvious example is the $\mathbf{A}$(4X3), shown below, with three 4-dimensional unit vectors. Clearly, the three vectors are independent, although there are only 3 vectors, and the 4-space is not filled. Because the 4-space is not filled, there must be a vector orthogonal to all the 3 (unit) vectors shown — one independent solution to $\mathbf{A}'\mathbf{x} = \mathbf{0}$. Clearly, that solution is the fourth unit vector. The "rank" of $\mathbf{A}$ is 3 — the size of the largest non-zero determinant that can be formed from the elements of the vectors.

$$\begin{bmatrix} 1 & 0 & 0 \\ 0 & 1 & 0 \\ 0 & 0 & 1 \\ 0 & 0 & 0 \end{bmatrix} .$$

Also, note that given a **y** vector: $\mathbf{Ax} = \{y_1, y_2, y_3, 0\}$ (which, obviously, lies in the same subspace), the solution to $\mathbf{Ax} = \mathbf{y}$ is $\mathbf{x} = \{y_1, y_2, y_3\}$. But, if **y** has $y_4 \neq 0$, the set has no solution.

Not quite so obvious is the next example, again 3 vectors, in 4-space.

$$\begin{bmatrix} 2 & -5 & 5 \\ 3 & -3 & 6 \\ 1 & 2 & 1 \\ -1 & -8 & 1 \end{bmatrix} .$$

As in the previous case, there are only 3 vectors, and $\mathbf{A'z = 0}$ *must* have at least one nontrivial solution. If the vectors are independent, then $\mathbf{Ax = 0}$ has only the solution $\mathbf{x = 0}$; however, if they are dependent, then $\mathbf{Ax = 0}$ has a solution, and the transposed set, $\mathbf{A'z = 0}$, has more than one solution. Note that $\mathbf{z}$ is a 4 dimensional vector, while $\mathbf{x}$ is 3 dimensional.

The Gauss-Jordan method, introduced in Chapter 3, provides an important tool for determining the (in)dependence of these vectors, and the solutions to both the $\mathbf{Ax = 0}$ set, and the $\mathbf{A'z = 0}$, if any exist. Gauss-Jordan operates on the input matrix with only elementary operations, thus not altering the rank of the given set. For this example:

$$
\begin{Vmatrix} 2 & -5 & 5 \\ 3 & -3 & 6 \\ 1 & 2 & 1 \\ -1 & -8 & 1 \end{Vmatrix} \quad \text{Gauss-Jordan} \rightarrow \quad \begin{Vmatrix} 1 & 0 & 1.667 \\ 0 & 1 & -0.333 \\ 0 & 0 & 0 \\ 0 & 0 & 0 \end{Vmatrix}
$$

The 2X2 unit matrix formed at the upper left of the reduced set indicates that the rank is 2 (the largest non-zero determinant). Also, the reduction gives the solution to (4.9). The value of x_3 can be set arbitrarily (say, $x_3 = k$), and:

$$
\begin{Bmatrix} x_1 \\ x_2 \\ x_3 \end{Bmatrix} = \begin{Bmatrix} -5/3 \\ 1/3 \\ 1 \end{Bmatrix} k \quad \text{(a single infinity of solutions)}.
$$

It is instructive to continue this example by solving the transposed set. Since the rank is two we expect a two-fold infinity of solutions. The Gauss-Jordan of the transposed set is

$$
\begin{Vmatrix} 2 & 3 & 1 & -1 \\ -5 & -3 & 2 & -8 \\ 5 & 6 & 1 & 1 \end{Vmatrix} \quad \text{Gauss-Jordan} \rightarrow \quad \begin{Vmatrix} 1 & 0 & -1 & 3 \\ 0 & 1 & 1 & -2.333 \\ 0 & 0 & 0 & 0 \end{Vmatrix}
$$

The rank is two, so $\mathbf{z}_3$ and $\mathbf{z}_4$ can be set arbitrarily (say k_1 and k_2).

Then
$$
\begin{Bmatrix} z_1 \\ z_2 \end{Bmatrix} = \begin{Bmatrix} 1 \\ -1 \end{Bmatrix} k_1 + \begin{Bmatrix} -3 \\ 7/3 \end{Bmatrix} k_2 .
$$

And so
$$
\begin{Bmatrix} z_1 \\ z_2 \\ z_3 \\ z_4 \end{Bmatrix} = \begin{Bmatrix} 1 \\ -1 \\ 1 \\ 0 \end{Bmatrix} k_1 + \begin{Bmatrix} -3 \\ 7/3 \\ 0 \\ 1 \end{Bmatrix} k_2 .
$$

When the set is square, $\mathbf{A}(nXn)$, probably the most important case, if the determinant, $|\mathbf{A}|$, is zero then the vectors are dependent. There will be an independent, non-unique solution for each level of "degeneracy" (i.e., $n - r = 1, 2, \ldots$) where r is the rank.

4.3 SIMULTANEOUS EQUATION SETS

This section considers equation sets, $\mathbf{Ax} = \mathbf{c}$ in which the right-hand side, $\mathbf{c}$, is non-zero. The equation set can be viewed as a vector transformation in which $\{\mathbf{c}\}$ is to be synthesized by a linear weighted sum of the left-hand column vectors (if possible). The problem is to find the weight factors (the elements of the $\mathbf{x}$ column).

4.3.1 SQUARE EQUATION SETS

Writing Equation (4.1) as a vector equation, with $m = n$ ("Square"):

$$\mathbf{A}(n \times n)\mathbf{x} = \mathbf{y} \Rightarrow \begin{Bmatrix} a_{11} \\ a_{21} \\ \vdots \\ a_{n1} \end{Bmatrix} x_1 + \begin{Bmatrix} a_{12} \\ a_{22} \\ \vdots \\ a_{n2} \end{Bmatrix} x_2 + \cdots + \begin{Bmatrix} a_{1n} \\ a_{2n} \\ \vdots \\ a_{nn} \end{Bmatrix} x_n = \mathbf{y} = \begin{Bmatrix} y_1 \\ y_2 \\ \vdots \\ y_n \end{Bmatrix}. \qquad (4.11)$$

The columns of $\mathbf{A}$ are the vectors to be added, using weighting factors, x_j, resulting in an output vector $\mathbf{y}$. These equations are definitely "coupled" (into a single vector equation). But, if a vector, say, $\mathbf{v}_1$, could be found, that is simultaneously orthogonal to (i.e., perpendicular to) all the $\mathbf{a}$ vectors in (4.11) save the first — that is:

$$\mathbf{v}_1 \bullet \mathbf{a}_j = 0; \quad \text{for} \quad j=2, 3, \cdots n.$$

Then, we could dot $\mathbf{v}_1$ through (4.11):

$$(\mathbf{v}_1 \bullet \mathbf{a}_1)x_1 + (\mathbf{v}_1 \bullet \mathbf{a}_2)x_2 + \cdots + (\mathbf{v}_1 \bullet \mathbf{a}_n)x_n = (\mathbf{v}_1 \bullet \mathbf{y}). \qquad (4.12)$$

All the products $(\mathbf{v}_1 \bullet \mathbf{a}_j)$ are zero, except the first (i.e., $j = 1$). Then: $(\mathbf{v}_1 \bullet \mathbf{a}_1)x_1 = (\mathbf{v}_1 \bullet \mathbf{y})$, and:

$$x_1 = \frac{(\mathbf{v}_1 \bullet \mathbf{y})}{(\mathbf{v}_1 \bullet \mathbf{a}_1)}. \qquad (4.13)$$

Next, if a vector $\mathbf{v}_2$ could be found that is orthogonal to all except $\mathbf{a}_2$ then the same procedure could be used to uncouple x_2 from the rest. And, so on. Of course, it may not be easy to find successive vectors, $\mathbf{v}_j$, such that each is orthogonal to all but the jth $\mathbf{a}$-vector. But, in 2 and three dimensions it is easy. A "2-space" example is:

$$\text{given } \mathbf{Ax} = \mathbf{y} \Rightarrow \begin{bmatrix} -2 & 1 \\ 0 & 2 \end{bmatrix} \begin{Bmatrix} x_1 \\ x_2 \end{Bmatrix} = \begin{Bmatrix} -2 \\ 0 \end{Bmatrix} x_1 + \begin{Bmatrix} 1 \\ 2 \end{Bmatrix} x_2 = \mathbf{y} = \begin{Bmatrix} 1 \\ -2 \end{Bmatrix}. \qquad (4.14)$$

Choose $\mathbf{v}_1 = \{2, -1\}$ and $\mathbf{v}_2 = \{0, 1\}$. Then:

$$\begin{bmatrix} 2 & -1 \end{bmatrix} \begin{Bmatrix} -2 \\ 0 \end{Bmatrix} x_1 = \begin{bmatrix} 2 & -1 \end{bmatrix} \begin{Bmatrix} 1 \\ -2 \end{Bmatrix} \Rightarrow x_1 = -1$$

$$\text{and } \begin{bmatrix} 0 & 1 \end{bmatrix} \begin{Bmatrix} 1 \\ 2 \end{Bmatrix} x_2 = \begin{bmatrix} 0 & 1 \end{bmatrix} \begin{Bmatrix} 1 \\ -2 \end{Bmatrix} \Rightarrow x_2 = -1.$$

A 3-space example is much more interesting. Equation (4.15) shows a general (3X3) equation set, using the vector form. Figure 4.1 is reproduced below, for reference. If the set is independent, then $|A| \neq 0$, and $\mathbf{a}_1$ does not lie within (on) the plane, but has a component along $\bar{\mathbf{n}}$.

$$\left\{ \begin{array}{c} a_{11} \\ a_{21} \\ a_{31} \end{array} \right\} x_1 + \left\{ \begin{array}{c} a_{12} \\ a_{22} \\ a_{32} \end{array} \right\} x_2 + \left\{ \begin{array}{c} a_{13} \\ a_{23} \\ a_{33} \end{array} \right\} x_3 = \mathbf{y} = \left\{ \begin{array}{c} y_1 \\ y_2 \\ y_3 \end{array} \right\}. \tag{4.15}$$

As before, a vector, $\mathbf{v}_1$, is required, and it must be orthogonal to both $\mathbf{a}_2$ and $\mathbf{a}_3$. But, the figure already shows this; i.e., $\bar{\mathbf{n}}$ is clearly normal to $\mathbf{a}_2$ and $\mathbf{a}_3$. Then, dot-through (4.15) by a vector parallel to $\bar{\mathbf{n}}$, and only a coefficient on x_1 will remain on the left side of the equation. And, it is easy to define a vector along $\bar{\mathbf{n}}$. The vector cross product of $\mathbf{a}_2 \times \mathbf{a}_3$ will do nicely (the cross product $\mathbf{a}_3 \times \mathbf{a}_2$ would do just as well). Then, just as in the (2X2) case (since $\mathbf{v}_1 = \mathbf{a}_2 \times \mathbf{a}_3$):

$$x_1 = \frac{(\mathbf{a}_2 \times \mathbf{a}_3 \bullet \mathbf{y})}{(\mathbf{a}_2 \times \mathbf{a}_3 \bullet \mathbf{a}_1)}. \tag{4.16}$$

Note that Figure 4.2 is perfectly general. That is, any 2 of the 3 $\mathbf{a}$ vectors can be chosen to define a plane, then the remaining vector is viewed in terms of its projection onto the plane, and its component normal to it. From that point, the solution for each of the x values is the same as above, and will have the form of (4.16).

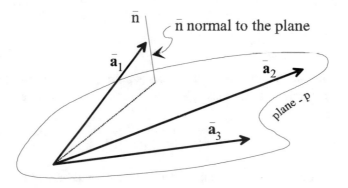

Figure 4.2: Redrawn of Figure 4.1.

When the dimensions are > 3, the ability to draw pictures, and visualize results is lost. *But, the approach is valid.* In fact, the above examples have really been an interpretation of the solution by premultiplication of the inverse matrix. Given that $\mathbf{B}$ is the inverse of $\mathbf{A}$, and $\mathbf{Ax} = \mathbf{y}$,

$$\mathbf{BA} = \mathbf{I}; \text{ and, therefore: } \mathbf{BAx} = \mathbf{x} = \mathbf{By} = \mathbf{A}^{-1}\mathbf{y}. \tag{4.17}$$

Clearly, the rows, $\mathbf{b}_i$, of $\mathbf{B}$ are orthogonal to the columns, $\mathbf{a}_j$, of $\mathbf{A}$ — except when $i = j$ (in which case the dot product is unity). And, since the product is commutative, the rows of $\mathbf{A}$ are in the same orthogonal relationship with the columns of **B**. Then, in any number of dimensions, premultiplication by the inverse matrix "uncouples" the given equation set.

In summary: ***Given a non singular matrix A, the equation set Ax = y has a unique solution, for any vector, y. That solution is obtained by premultiplying the equation by the inverse matrix. The solution vector, x, can be viewed as the set of coefficients in the synthesis of y by the column vectors within A, as "base vectors."***

　　　　Return to the 3 dimensional example discussed above. But, now slide the tip of $\mathbf{a}_1$ down the normal, until $\mathbf{a}_1$ lies in the plane-p (all 3 vectors now lie in the plane). See Figure 4.3. In this

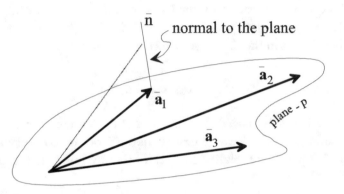

Figure 4.3:

case, when the cross product of any 2 of the **a** vectors is found, it will be orthogonal to **all three** of them. The method of solution clearly fails. The reason is that the **a** vectors are no longer linearly independent. The equation

$$c_1\mathbf{a}_1 + c_2\mathbf{a}_2 + c_3\mathbf{a}_3 = \mathbf{0}$$

now has a non-trivial, non-unique, solution; the transposed set will have at least one solution.

　　　　As an example, $\mathbf{A} = \begin{bmatrix} 1 & 1 & 1 \\ 0 & -2 & -1 \\ 2 & -4 & -1 \end{bmatrix}$. These column vectors lie in a plane whose normal

lies along a line $\{2, 3, -1\}$, which is the (only) solution to $\mathbf{A'z = 0}$. The solution to $\mathbf{Ax = 0}$ is $k\{-1, -1, 2\}$.

　　　　If the given non homogeneous set is $\mathbf{Ax = y} = \{0, 1, 3\}$, a solution may not be possible, unless the **y** vector also lies within the subspace occupied by the **A** column vectors. The test for this is that **y** must be orthogonal to all independent solutions to the transposed set. In this example, the test product $\mathbf{z \cdot y} = 0$, and the set is compatible.

Then, a total (complete) solution is

$$\mathbf{x} = k \left\{ \begin{array}{c} -1 \\ -1 \\ 2 \end{array} \right\} + \left\{ \begin{array}{c} 2 \\ 1 \\ -3 \end{array} \right\}.$$

Which is the sum of all solutions to the homogeneous set, plus any solution to the non homogeneous set.

The rank of the original equation set may be less than $n-1$:

$$
\begin{aligned}
x_1 + 2x_2 + 3x_3 &= y_1 \\
x_1 + 2x_2 + 3x_3 &= y_2 \\
x_1 + 2x_2 + 3x_3 &= y_3 .
\end{aligned}
\tag{4.18}
$$

Now the columns of $\mathbf{A}$ are collinear; the rank of $\mathbf{A}$ being $n-2$ ($n=3$). It can therefore be anticipated that there will be a double infinity of solutions to the homogeneous set (i.e., two arbitrary constants).

The two solutions to $\mathbf{Ax} = \mathbf{0}$ are $k_1\{-1, -1, 1\}$, and $k_2\{-5, 4, -1\}$. These solutions are not only independent, they are orthogonal (their dot product is zero). While this orthogonality is not necessary (just linear independence will do), it is not surprising that two orthogonal vectors could be found: because, two dimensions are not included in the columns of $\mathbf{A}$ — that is, a plane. Within this plane, there are an infinity of sets of orthogonal vectors.

The solution $k_1\{-1, -1, 1\}$ was found by inspection. The second solution can always be found that is orthogonal to both the first row of $\mathbf{A}$, and $\{-1, -1, 1\}$ by solving:

$$\left[\begin{array}{ccc} 1 & 2 & 3 \\ -1 & -1 & 1 \end{array} \right] \left\{ \begin{array}{c} z_1 \\ z_2 \\ z_3 \end{array} \right\} = \left\{ \begin{array}{c} 0 \\ 0 \end{array} \right\}$$

whose solution is $k_2\{-5, 4, -1\}$.

Given a $\mathbf{y}$ vector, in (4.18), which results in a compatible set, the solution will be:

$$\mathbf{x} = k_1 \left\{ \begin{array}{c} -1 \\ -1 \\ 1 \end{array} \right\} + k_2 \left\{ \begin{array}{c} -5 \\ 4 \\ -1 \end{array} \right\} + \left\{ \begin{array}{c} \text{Any solution to the} \\ \text{non-homogeneous set} \end{array} \right\}. \tag{4.19}$$

The $\mathbf{y}$ vector in (4.18) must be collinear with the direction of all the $\{\mathbf{a}\}$ vectors, $\{1, 1, 1\}$. Any $\mathbf{y}$ vector which is orthogonal to the plane whose normal is $\{1, 1, 1\}$ is necessarily in the direction $\{1, 1, 1\}$, and will hence, be compatible. Vectors that lie in this plane are solutions to:

$$\mathbf{A}'\mathbf{z} = \left[\begin{array}{ccc} 1 & 1 & 1 \\ 2 & 2 & 2 \\ 3 & 3 & 3 \end{array} \right] \{z\} = \{0\}. \tag{4.20}$$

There are two solutions, of course. They are $k_1\{-1, 1, 0\}$, and $k_2\{1, 1, -2\}$. Again, these solutions are orthogonal (not necessary, but this ensures linear independence). In (4.18), if a $\mathbf{y}$ vector is given that is orthogonal to *both* of these solutions to (4.20), then compatibility is assured; else, the given set of equations is incompatible, and has no solution.

In this simple (3X3) example, it is easy to see the compatibility requirement. In the general case it will not be possible to visualize geometrically. But, in the general (nXn) case: $\mathbf{Ax} = \mathbf{y}$, *when the rank of $\mathbf{A}$ is $r < n$, and n is the order of $\mathbf{A}$, there will be n-r solutions to the homogeneous equations $\mathbf{A}'\mathbf{z}=0$. If the given $\mathbf{y}$ vector is orthogonal to all of these solutions, then the given set is compatible.*

As was shown in the example, there will also be n-r solutions to the homogeneous set $\mathbf{Ax} = \mathbf{0}$. The complete solution to the original set is the sum of these latter solutions, and any solution of the nonhomogeneous set.

4.3.2   UNDERDETERMINED EQUATION SETS

Given $\mathbf{Ax} = \mathbf{y}$ in which $\mathbf{A}$ is nXm, and $n < m$, the set is "underdetermined" – i.e., there are an insufficient number of equations to determine the $\mathbf{x}$ vector uniquely. If the set is compatible, non-unique solutions will be possible.

When the set is viewed as a vector equation, two cases are apparent. First, if the rank of $\mathbf{A}$ is n, then the solution is much like the square, nonsingular set. Assuming that the first n columns of $\mathbf{A}$ have rank n (or renumbering the columns and $\mathbf{x}$ vector components so that this is so), these n vectors can be partitioned:

$$\mathbf{Bu} + \mathbf{Dv} = \mathbf{y} \tag{4.21}$$

where, now the $\mathbf{B}$ matrix comprises just the (nXn) first (nonsingular) columns of $\mathbf{A}$. The vector $\mathbf{u}$ is $\mathbf{u} = \{x_1, x_2, \ldots x_n\}$, the first n components of $\mathbf{x}$, and $\mathbf{v}$ is $\mathbf{v} = \{x_{n+1}, \ldots x_m\}$, the remaining components of $\mathbf{x}$. Matrix $\mathbf{D}$ holds the remaining columns of the original $\mathbf{A}$ matrix. Since $\mathbf{B}$ is nonsingular, then a solution for $\mathbf{u}$ can be found, in terms of $\mathbf{y}$, and $\mathbf{v}$ whose components can be assigned arbitrarily:

$$\mathbf{u} = \mathbf{B}^{-1}\mathbf{y} - \mathbf{B}^{-1}\mathbf{Dv} \quad (\mathbf{v} \text{ arbitrary}) . \tag{4.22}$$

That is, there are $m-n$ arbitrary constants in the solution (there is an $m-n$ fold infinity of solutions).

If the rank of $\mathbf{A}$ is less than n, there may be no solutions at all, unless the $\mathbf{y}$ vector lies within the same subspace as the $\mathbf{A}$ vector set. Consider the following (4X5) example:

$$\begin{bmatrix} 1 & -1 & 0 & -1 & 0 \\ 3 & 1 & 6 & -5 & -2 \\ -1 & 2 & 3 & -1 & 1 \\ 1 & 0 & 1 & -1 & -1 \end{bmatrix} x = \left\{ \begin{array}{c} -1 \\ 3 \\ 4 \\ 0 \end{array} \right\} . \tag{4.23}$$

The Gauss-Jordan reduction method terminates at:

$$
\begin{array}{ccccc|c}
x_1 & x_2 & x_3 & x_4 & x_5 & c \\
\hline
1 & 0 & 0 & 0 & -2 & -1 \\
0 & 1 & 0 & 1 & -2 & 0 \\
0 & 0 & 1 & -1 & 1 & 1 \\
0 & 0 & 0 & 0 & 0 & 0
\end{array}
\tag{4.24}
$$

where the column set apart at the right is the "augmenting" column, originally, the $\mathbf{y}$ vector. Since the final row is all zero (including the augmenting column) the set is compatible, and has the rank 3. Then x_4 and x_5 can be set arbitrarily (say, $x_4 = k_1$, and $x_5 = k_2$), and the complete solution is

$$
\{x\} =
\left\{
\begin{array}{c}
0 \\ -1 \\ 1 \\ 1 \\ 0
\end{array}
\right\} k_1 +
\left\{
\begin{array}{c}
2 \\ 2 \\ -1 \\ 0 \\ 1
\end{array}
\right\} k_2 +
\left\{
\begin{array}{c}
-1 \\ 0 \\ 1 \\ 0 \\ 0
\end{array}
\right\}.
\tag{4.25}
$$

The Gauss Jordan reduction shows the compatibility, and if compatible, shows the complete solution.

Although the Gauss-Jordan reduction solves the problem, it is instructive to derive it in the manner of the previous section, and show that the set is compatible. The homogeneous transposed set is:

$$
\begin{bmatrix}
1 & 3 & -1 & 1 \\
-1 & 1 & 2 & 0 \\
0 & 6 & 3 & 1 \\
-1 & -5 & -1 & -1 \\
0 & -2 & 1 & -1
\end{bmatrix}
\{z\} = \{0\}.
\tag{4.26}
$$

This set has the solution $\{z\} = \{1, -1, 1, 3\}$. The dot product of this solution vector, with the original $\mathbf{y}$ vector, $\{-1, 3, 4, 0\}$ must be zero for the set to be compatible. This is clearly so.

Incidentally, in this example the $\mathbf{z}$ vector can be found, by deleting the last equation of the transposed set, and calculating the adjoint matrix. Since it is known that the rank of both $\mathbf{A}$ and $\mathbf{A}'$ is 3, the adjoint matrix will be of rank 1. Then, at least one of its columns will be nonzero, and the solution to (4.26). If the rank of $\mathbf{A}$ were less than 3, the adjoint would be null, and this method could not have been used.

4.3.3 OVERDETERMINED EQUATION SETS

When the number of equations, n, is greater than the number of columns, m, the set is said to be "overdetermined." Stated the other way, interpreting the set as a vector equation, the set is "overdetermined" when the dimensionality of the vectors, n, is larger than their number, m. However, it is possible for a set to appear to be overdetermined, simply by having more equations than unknowns, when, in fact, it is underdetermined because the equations are not independent. That is, if the rank of $\mathbf{A}(nXm)$ is less than m, the set is really underdetermined.

Since $\mathbf{A}'$ is (mXn), whose rank cannot be greater than m, there will always be nontrivial solutions to $\mathbf{A}'\mathbf{z}=\mathbf{0}$. Therefore, there will always be compatibility conditions to be met. Thus, the determination of compatibility may become the larger problem. After the set is found to be compatible, the extra equations can be discarded (resulting in an mXm), and the set solved.

But, there is another way. From the "geometry" of the set, itself, it may appear worthwhile to premultiply the given set by $\mathbf{A}'$:

$$\mathbf{A}'\mathbf{A}\mathbf{x} = \mathbf{A}'\mathbf{y} . \tag{4.27}$$

The matrix $\mathbf{A}'\mathbf{A}$ is (mXm), the smaller of the two dimensions, and its rank should be the same as that of $\mathbf{A}$ itself. Surprisingly enough, this is one time that appearance does suggest an appropriate

approach. If $|\mathbf{A}'\mathbf{A}|$ exists, the equation set (4.27) is compatible whether or not the given set is compatible. If the given set is compatible, the solution to (4.27) yields the correct $\mathbf{x}$ vector. If the given set is incompatible, the solution to the above is "the best available" in the so-called "least squares sense." The following article will derive a solution to $\mathbf{A}\mathbf{x} = \mathbf{b}$ which minimizes the sum of squared error. It will be the same as the solution to (4.27).

Least Squares Solutions

Given $\mathbf{A}\mathbf{x} = \mathbf{b}$, where $\mathbf{A}$ is (nXm), and $n > m$, *any* given $\{x\}$, will yield an $\mathbf{A}\mathbf{x}$ vector with some amount of error, $\mathbf{e}$:

$$\mathbf{e} = \mathbf{A}\mathbf{x} - \mathbf{b} \quad \mathbf{A}(nXm); \quad \mathbf{e}, \mathbf{b} \text{ are } (nX1), \mathbf{x}(mX1), \text{ and } n > m . \tag{4.28}$$

If the original set is compatible, and $n - m$ of the equations are functions of the first m, then it is possible to derive an exact solution (with $\mathbf{e} = \mathbf{0}$). The least squares situation arises when the set is incompatible and *any* $\mathbf{x}$ vector results in errors. ***The least square criterion defines the "best" x solution as the one in which the sum of the squared error is minimized.*** The sum of squared error is given by $\mathbf{e}'\mathbf{e}$ (the *scalar* dot product of $\mathbf{e} \bullet \mathbf{e}$):

$$\mathbf{e}'\mathbf{e} = (\mathbf{A}\mathbf{x} - \mathbf{b})'(\mathbf{A}\mathbf{x} - \mathbf{b}), \text{ or}$$
$$\mathbf{e}'\mathbf{e} = \mathbf{x}'\mathbf{A}'\mathbf{A}\mathbf{x} - \mathbf{x}'\mathbf{A}'\mathbf{b} - \mathbf{b}'\mathbf{A}\mathbf{x} + \mathbf{b}'\mathbf{b} .$$

Both $\mathbf{x}'\mathbf{A}'\mathbf{b}$ and $\mathbf{b}'\mathbf{A}\mathbf{x}$ express the same dot product $(\mathbf{b} \bullet \mathbf{A}\mathbf{x})$. Then $\mathbf{b}'\mathbf{A}\mathbf{x} = \mathbf{x}'\mathbf{A}'\mathbf{b}$:

$$\mathbf{e}'\mathbf{e} = \mathbf{x}'\mathbf{A}'\mathbf{A}\mathbf{x} - 2\mathbf{x}'\mathbf{A}'\mathbf{b} + \mathbf{b}'\mathbf{b} . \tag{4.29}$$

The (scalar) term $\mathbf{x}'\mathbf{A}'\mathbf{A}\mathbf{x}$ is called a "quadratic" form, because in its expansion, the variables appear as a second degree product, $x_i x_j$, in every term. Also required in the definition is that the (necessarily square) matrix be symmetric. Note that $\mathbf{A}'\mathbf{A}$ is symmetric. The term $\mathbf{x}'\mathbf{A}'\mathbf{b}$ could be called a "bilinear form," if one considers the $\mathbf{b}$ vector as a variable. In that case, $x_i b_j$ appear as products (hence "bilinear"). It is not required that the matrix ($\mathbf{A}'$, in this case), be symmetric; and, indeed, $\mathbf{A}'$ is not (it's not even square).

The method is to take the partial derivatives of $\mathbf{e}'\mathbf{e}$ with respect to each of the $\mathbf{m}$ variables, $x_{i,}$ in turn, and equate them simultaneously to zero. The resultant $\mathbf{x}$ vector minimizes $\mathbf{e}'\mathbf{e}$.

$$\left\{ \begin{array}{c} \dfrac{\partial \mathbf{e}'\mathbf{e}}{\partial x_1} \\[2mm] \dfrac{\partial \mathbf{e}'\mathbf{e}}{\partial x_2} \\ \vdots \\ \dfrac{\partial \mathbf{e}'\mathbf{e}}{\partial x_m} \end{array} \right\} = \{0\}. \tag{4.30}$$

The solution to the equation set that results from (4.30) is the x vector which minimizes $\mathbf{e}'\mathbf{e}$. Appendix A discusses the partial differentiation of bilinear and quadratic forms. It begins by defining the *vector* differential operator, ∇.

$$\nabla = \left\{ \begin{array}{c} \dfrac{\partial}{\partial x_1} \\[2mm] \dfrac{\partial}{\partial x_2} \\ \vdots \\ \dfrac{\partial}{\partial x_m} \end{array} \right\}. \tag{4.31}$$

Using this definition (4.30) becomes $\nabla \mathbf{e}'\mathbf{e} = \mathbf{0}$ and from (4.29)

$$\nabla \mathbf{e}'\mathbf{e} = \nabla(\mathbf{x}'\mathbf{A}'\mathbf{A}\mathbf{x}) - \nabla(2\mathbf{x}'\mathbf{A}'\mathbf{b}) + \nabla(\mathbf{b}'\mathbf{b}) = \mathbf{0}. \tag{4.32}$$

The $\mathbf{b}$ vector is not a function of $\mathbf{x}$, so the last term is $\mathbf{0}$. Appendix A finds $\nabla(\mathbf{x}'\mathbf{A}'\mathbf{A}\mathbf{x}) = 2\mathbf{A}'\mathbf{A}\mathbf{x}$, and $\nabla(2\mathbf{x}'\mathbf{A}'\mathbf{b}) = 2\mathbf{A}'\mathbf{b}$. Then:

$$\nabla \mathbf{e}'\mathbf{e} = 2\mathbf{A}'\mathbf{A}\mathbf{x} - 2\mathbf{A}'\mathbf{b} = \{0\} \tag{4.33}$$

$$\mathbf{A}'\mathbf{A}\mathbf{x} = \mathbf{A}'\mathbf{b} \quad \text{(see Equation (4.27))}. \tag{4.34}$$

This remarkable result indicates that the minimum squared error will be obtained when the $\mathbf{x}$ vector is defined by solution to the square (mXm) set of (4.34). The original $\mathbf{A}$ is nXm, so $\mathbf{A}'\mathbf{A}$ is mXm). By hypothesis $\mathbf{A}$ has the rank m, so $\mathbf{A}'\mathbf{A}$ is nonsingular giving

$$\mathbf{x} = (\mathbf{A}'\mathbf{A})^{-1}\mathbf{A}'\mathbf{b}. \tag{4.35}$$

What's more, if the original set is compatible, (4.35) yields the unique solution!

4.4   LINEAR REGRESSION

The engineering sciences are based upon physical entities and the relationships between them. However, the relationships are most often expressed in exact equation form, implying a knowledge of the exact values of the variables they contain. Usually, this is not the case. Many physical variables are the result of empirical measurement. For example, in dynamics, a velocity or acceleration is known as a result of observations. It may be known accurately, but not exactly.

Over a limited range the relationship between variables, though not known, may be assumed to be linear. Then, "linear regression" is used to determine a "best" straight line relationship. Most often, a least squares fit to the data is chosen to define the "best fit." There are some good statistical reasons for this choice; and (perhaps the most compelling reason) the least squares analysis is easy to perform.

It has already been decided that the relationship between a dependent variable, y, known only by a set of observed data points, y_i, and an independent variable, x, is a linear curve, part of which is shown in Figure 4.4.

$$y = c_1 x + c_2 . \tag{4.36}$$

If exact (x, y) data could be obtained, it would only take two pairs to determine c_1 and c_2. But,

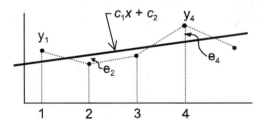

Figure 4.4:

the relationship between x and y is a complicated one and the data contains observation errors. The problem is, then, to determine a "best fit" curve so that other y-data can be predicted from given x-data. "Best" is determined to be a least squares fit to the data. In general, quite a few (x, y) measurements are taken over the range of interest, in an attempt to "average-out" as much observation error as possible. Thus, if an equation $y_i = c_1 x_i + c_2$ is written for every one of the observations, a very overdetermined equation set results.

The x_i data need not be equi-spaced (as implied by Fig. 4.4), and some (but not all) of the y_i points may be redundant measurements at the same value of x_i. The objective is, of course, to allow the error to "average out," yielding a regression line that is accurate to within the requirements of the physical problem.

Then, given the set of N observed (x, y) data points, write:

$$\mathbf{y} = \mathbf{Xc} = [X]\{c\} \tag{4.37}$$

where $y = \{y_i\}$, the y observed data, and $\mathbf{X}$ containing the x-data:

$$\mathbf{X} = \begin{bmatrix} x_1 & 1 \\ x_2 & 1 \\ \cdots & \cdots \\ x_N & 1 \end{bmatrix} \qquad \mathbf{c} = \begin{Bmatrix} c_1 \\ c_2 \end{Bmatrix} \qquad \mathbf{y} = \begin{Bmatrix} y_1 \\ y_2 \\ \vdots \\ y_N \end{Bmatrix}.$$

Equation (4.37) is an overdetermined (NX2) set of linear equations in the unknown variables c_1 and c_2. The previous article, and Equation (4.35), provide the solution:

$$\mathbf{c} = (\mathbf{X}'\mathbf{X})^{-1}[\mathbf{X}'\mathbf{y}]. \tag{4.38}$$

In (4.38), the $\mathbf{X}'\mathbf{X}$ matrix is 2X2, clearly symmetric and nonsingular, unless the data is all at the same x_i. The columns $\{\mathbf{c}\}$ and $[\mathbf{X}'\mathbf{y}]$ are 2X1 ($\mathbf{X}'$ is 2XN, times $\mathbf{y}$(NX1)).

$$\mathbf{X}'\mathbf{X} = \begin{bmatrix} \sum x_i^2 & \sum x_i \\ \sum x_i & N \end{bmatrix} \qquad \mathbf{X}'\mathbf{y} = \begin{bmatrix} \sum x_i y_i \\ \sum y_i \end{bmatrix}. \tag{4.39}$$

In these equations, the summations are to be taken over the index, i, from 1 to N. To avoid messy matrix terms, the inverse of $\mathbf{X}'\mathbf{X}$ will be expressed in terms of its adjoint and its determinant in the following:

$$[\mathbf{X}'\mathbf{X}]^{adj} = \begin{bmatrix} N & -\sum x_i \\ -\sum x_i & \sum x_i^2 \end{bmatrix} \qquad |\mathbf{X}'\mathbf{X}| = N\sum x_i^2 - \left(\sum x_i\right)^2. \tag{4.40}$$

Carrying out the product terms indicated in (4.38), the solutions for c_1 and c_2 are:

$$c_1 = \frac{N\sum x_i y_i - \sum x_i \sum y_i}{N\sum x_i^2 - (\sum x_i)^2} \tag{4.41}$$

$$c_2 = \frac{\sum x_i^2 \sum y_i - \sum x_i \sum x_i y_i}{N\sum x_i^2 - (\sum x_i)^2}. \tag{4.42}$$

Some additional algebraic work can be done on these two equations, which will result in an appearance that is much more appealing. First, define the average values of y_i and x_i as $\bar{x}$ and $\bar{y}$, where:

$$\bar{x} = \frac{\sum x_i}{N} \qquad \text{and} \qquad \bar{y} = \frac{\sum y_i}{N}.$$

To reduce c_2, subtract $\bar{y}$ from both sides of (4.42):

$$c_2 - \bar{y} = \frac{\sum x_i^2 \sum y_i - \sum x_i \sum x_i y_i}{N\sum x_i^2 - (\sum x_i)^2} - \frac{\sum y_i}{N}. \tag{4.43}$$

Now, on the right-hand side of (4.43), gather both terms over a common denominator, and note that a term, $N\sum x_i^2 \sum y_i$, cancels. The result is:

$$c_2 - \bar{y} = -\frac{\sum x_i (\sum x_i y_i - \sum x_i \sum y_i)}{N(N \sum x_i^2 - (\sum x_i)^2)}. \tag{4.44}$$

Compare the right side of (4.44) to (4.41), and write:

$$c_2 - \bar{y} = -\bar{x}c_1; \quad \text{or} \quad c_2 = \bar{y} - \bar{x}c_1. \tag{4.45}$$

To reduce c_1 (Equation (4.41)), first work on the denominator. Note that:

$$\sum (x_i - \bar{x})^2 = \sum x_i^2 - 2\bar{x} \sum x_i + N\bar{x}^2$$
$$= \sum x_i^2 - N\bar{x}^2.$$

Then the denominator is simply $N\sum(x_i - \bar{x})^2$. And in similar fashion, it is found that the numerator is $\sum(x_i - \bar{x})(y_i - \bar{y})$. This yields the final regression line equation:

$$y = \bar{y} + c_1(x - \bar{x}); \quad \text{where} \quad c_1 = \frac{\sum (x_i - \bar{x})(y_i - \bar{y})}{\sum (x_i - \bar{x})^2}. \tag{4.46}$$

Which is the final result.

4.4.1   EXAMPLE REGRESSION PROBLEM

As an example of the method, the following analysis determines the dependence of the diameter of a cylindrical part on the temperature of a heat treating process. Over the range of temperatures involved, this dependence is assumed to be linear:

$$d = c_1 t + c_2 = \bar{d} + c_1(t - \bar{t}) \tag{4.47}$$

where d is the diameter and t is temperature. The data obtained in the laboratory is tabulated and shown graphically in Figure 4.5. The temperature, t, is given in thousands of degrees; diameter, d, measured in inches. $\bar{d}$ is average diameter, and $\bar{t}$ is average process temperature.

There are 12 sets of (t,d) data points available—12 equations $d = c_1 t + c_2$—an overdetermined and incompatible 12X2 set in c_1 and c_2. Linear regression determines these unknowns using the least squares best fit of the data to a straight line, called "regression line."

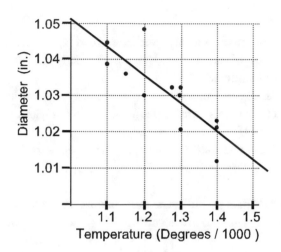

Test Data

T	d
1.10	1.039
1.10	1.045
1.15	1.037
1.20	1.030
1.20	1.049
1.28	1.033
1.30	1.033
1.30	1.030
1.30	1.020
1.40	1.023
1.40	1.021
1.40	1.012

Figure 4.5: Linear Regression Diagram.

From the given data, the following results are calculated:

$$\text{Average temperature, } \bar{t} = 1.2608$$
$$\text{Average diameter, } \bar{d} = 1.031$$
$$c_1 = -0.07986 \text{ inches per 1000 deg}.$$

The equation of the regression line drawn in Figure 4.5 is:

$$\mathbf{d} = \bar{\mathbf{d}} + \mathbf{c_1}(\mathbf{t} - \bar{\mathbf{t}}) . \tag{4.48}$$

4.4.2   QUADRATIC CURVE FIT

The regression method is not limited to a linear curve fit. The data may be fit to a quadratic equation. The starting point would be (compare this with 4.36):

$$y = a_1 x^2 + a_2 x + a_3 = \{x^2\}a_1 + \{x\}a_2 + \{1\}a_3 = \mathbf{Xa} \tag{4.49}$$

There are three columns in $\mathbf{X}$(nx3) and three variables, a_j in $\{\mathbf{a}\}$. Just as before, the least squares solution is obtained by premultiplying by $\mathbf{X'}$, this time resulting in a (3X3) matrix, $\mathbf{X'X}$. The subsequent inversion yields:

$$\mathbf{a} = \{a_1, a_2, a_3\} = \left[\mathbf{X'X}\right]^1 \left[\mathbf{X'}\right]\mathbf{y} . \tag{4.50}$$

4.5   LAGRANGE INTERPOLATION POLYNOMIALS

4.5.1   INTERPOLATION

The curve fitting problem of the previous section involves a very overdetermined equation set. The resulting best-fit curve is not expected to pass through any of the given points exactly. The very idea is to achieve "smoothing" of data obtained by measurement.

The objectives of the interpolation problem are quite different. A set of (x_k, y_k) values are given, and these represent the true values of a continuous, integrable function $y = f(x)$, and at each of the given points, $y_k = f(x_k)$. The function itself may or may not be known.

A relatively simple representation of $f(x)$ is desired, ***that will pass through the given points exactly*** and can be used to interpolate values of $f(x)$ at intermediate points, x, within the given range.

One approach is to simply "curve fit" the n data points in the same manner as in the previous section, but using an (nXn) matrix — not overdetermined. The result will of course be a polynomial of degree $n - 1$:

$$p(x) = c_1 + c_2 x + c_3 x^2 + \cdots + c_n x^{n-1} \tag{4.51a}$$

whose coefficients, c, are to be determined by:

$$\mathbf{Xc} = \mathbf{y} = \begin{bmatrix} 1 & x_1 & x_1^2 & \cdots & x_1^{n-1} \\ 1 & x_2 & x_2^2 & \cdots & x_2^{n-1} \\ 1 & \cdots & \cdots & \cdots & x_k^{n-1} \\ 1 & \cdots & \cdots & \cdots & \cdots \\ 1 & x_n & x_n^2 & \cdots & x_n^{n-1} \end{bmatrix} \begin{bmatrix} c_1 \\ c_2 \\ c_k \\ \cdots \\ c_n \end{bmatrix} = \begin{bmatrix} y_1 \\ y_2 \\ y_k \\ \cdots \\ y_n \end{bmatrix}. \tag{4.51b}$$

This is similar to the least square fit problem, but the set is obviously not overdetermined. The indicated approach to determine the c coefficients, is to "simply" invert the $\mathbf{X}$ matrix. The resulting function $p(x)$ will pass through the given (x_j, y_j) points.

The matrix, $\mathbf{X}$, has some interesting characteristics. Note that if x_1 were to take on any of the values, $x_2, \ldots x_n$, the determinant, $|\mathbf{X}|$, vanishes because $|\mathbf{X}|$ then would have two identical rows. For the same reason, the determinant vanishes if x_2 assumes any of the values $x_3, \ldots x_n$. And so on. Apparently, $|\mathbf{X}|$ is some function of the x_k values which vanishes if any two of the values are the same. This is such a powerful characteristic that we might deduce a product of all the possible differences of the x_k values (Equation (4.52)). An additional factor, f, is added, since the product of differences can only be deduced as proportional to $|\mathbf{X}|$.

$$|\mathbf{X}| = f(x_n - x_{n-1}) \cdots (x_n - x_1)(x_{n-1} - x_{n-2}) \cdots (x_{n-1} - x_1) \cdots \cdots (x_2 - x_1). \tag{4.52}$$

In the general case, there will be $\frac{n(n-1)}{2}$ terms in (4.52). As an example, if $n = 4$, its determinant must have the factors:

$$f(x_4 - x_3)(x_4 - x_2)(x_4 - x_1)(x_3 - x_2)(x_3 - x_1)(x_2 - x_1). \tag{4.53}$$

Note that the x with the lower valued index is subtracted from that with the higher index regardless of the respective numeric values of the two.

To determine the value of f, note that the main diagonal term in the determinant expansion is $(1 \times x_2 \times x_3^2 \cdots x_n^{n-1})$. But, in (4.52), the very first term will be just that, when the products are multiplied out. Therefore, the factor is $f = 1$, and the determinant is simply the product of the difference terms.

Unfortunately, the elements of the adjoint matrix are not so easily found — although these, too, contain factors of the type $(x_j - x_i)$. Further, the $\mathbf{X}$ matrix is usually ill-conditioned. Note that there could be huge differences in the $[\ x_{ij}\ ]$ terms and may be difficult to accurately invert in the "normal" way. For such reasons, Equation (4.51b) is rarely attacked directly.

4.5.2   THE LAGRANGE POLYNOMIALS

The Lagrange interpolation polynomial is defined as

$$p(x) = \frac{(x - x_2)(x - x_3) \cdots (x - x_n)}{(x_1 - x_2)(x_1 - x_3) \cdots (x_1 - x_n)} y_1 + \frac{(x - x_1)(x - x_3) \cdots (x - x_n)}{(x_2 - x_3)(x_2 - x_4) \cdots (x_2 - x_n)} y_2 +$$
$$+ \cdots + \frac{(x - x_1)(x - x_2) \cdots (x - x_{n-1})}{(x_n - x_2)(x_n - x_3) \cdots (x_n - x_{n-1})} y_n. \tag{4.54}$$

It's a bit messy looking, but it does the job. $p(x)$ is a continuous function and $p(x_k) = y_k$. Each of the terms in (4.54) is, itself, an $n - 1$ degree polynomial and can be written compactly as:

$$q_i(x) y_i = \prod_{\substack{j=1 \\ j \neq i}}^{n} \frac{(x - x_j)}{(x_i - x_j)} y_i \tag{4.55}$$

and $p(x)$ is the sum of the (4.55) terms.

When attacked this way, there is no matrix or matrix inversion. The Equations (4.54) and (4.55) can be used directly (there are ways to do the numerical calculations efficiently). But both approaches arrive at the same result, so there must be a very close relationship between them. In order to show this, write the polynomial $q_i(x)$ as

$$q_i(x) = a_{1i} + a_{2i} x + \cdots + a_{ni} x^{n-1} \text{ and } q_i(x_k) = a_{1i} + a_{2i} x_k + \cdots + a_{ni} x_k^{n-1} = \delta_{ik}. \tag{4.56}$$

Note that in (4.56), the Kronecker delta is used because $q_i(x_k) = 0$ unless $k = i$, where $q_i(x_i) = 1$. The equation for $q_i(x_k)$ can be written as a vector dot product

$$\mathbf{x}_k \bullet \mathbf{a}_i = \delta_{ik}. \tag{4.57}$$

In (4.57) the vector $\mathbf{a}_i$ is formed from the n coefficients, a_{ik}; the vector $\mathbf{x}_k = \{1 \ \ x_k \ \ \cdots \ \ x_k^{n-1}\}$ is the ith row vector of $\mathbf{X}$. The two are orthogonal unless $i = k$, as shown in (4.57).

For clarity, consider the 4th order problem, and the following matrix product:

$$\mathbf{XA} = \begin{bmatrix} 1 & x_1 & x_1^2 & x_1^3 \\ 1 & x_2 & x_2^2 & x_2^3 \\ 1 & x_3 & x_3^2 & x_3^3 \\ 1 & x_4 & x_4^2 & x_4^3 \end{bmatrix} \begin{bmatrix} a_{11} & a_{12} & a_{13} & a_{14} \\ a_{21} & a_{22} & a_{23} & a_{24} \\ a_{31} & a_{32} & a_{33} & a_{34} \\ a_{41} & a_{42} & a_{43} & a_{44} \end{bmatrix}. \tag{4.58}$$

The *columns* of $\mathbf{A}$ are the coefficients of the $q_i(x)$ polynomial. For example:

$$q_1(x) = \frac{(x - x_2)(x - x_3)(x - x_4)}{(x_1 - x_2)(x_1 - x_3)(x_1 - x_4)} = a_{11} + a_{21}x + a_{31}x^2 + a_{41}x^3 \tag{4.59}$$

$$\begin{cases} a_{11} = \dfrac{-x_2 x_3 x_4}{(x_1 - x_2)(x_1 - x_3)(x_1 - x_4)} \\[3mm] a_{21} = \dfrac{x_4 x_2 + x_4 x_3 + x_3 x_2}{(x_1 - x_2)(x_1 - x_3)(x_1 - x_4)} \\[3mm] a_{31} = \dfrac{-(x_2 + x_3 + x_4)}{(x_1 - x_2)(x_1 - x_3)(x_1 - x_4)} \\[3mm] a_{41} = \dfrac{1}{(x_1 - x_2)(x_1 - x_3)(x_1 - x_4)} \end{cases} \quad \text{See footnote}^{[1]}. \tag{4.60}$$

This column vector $\mathbf{a_1} = \{a_{11}\ a_{21}\ a_{31}\ a_{41}\}$ is orthogonal to $\{1\ x_k\ x_k^2\ x_k^3\}$ unless $k = 1$, in which case the dot product is 1. Check it out. Since the other columns of $\mathbf{A}$ are similarly constructed, it must be true that $\mathbf{A}$ is the inverse of $\mathbf{X}$. Then, returning to Equations (4.51a) and (4.51b), the final interpolation polynomial is

$$\text{(rewrite (4.51a))} \qquad p(x) = c_1 + c_2 x + c_3 x^2 + \cdots + c_n x^{n-1}$$

where $\mathbf{c} = \mathbf{Ay}$, with the elements of $\mathbf{A}$ determined as in Equations (4.58) through (4.60).

4.6   EXERCISES

4.1.   Given the 3 vectors: $\mathbf{a}_1 = \{-1, 2, 5\}$, $\mathbf{a}_2 = \{2, -1, 0\}$, and $\mathbf{a}_3 = \{-5, 2, 3\}$, expressed by their coordinates along rectangular axes, find the length of each and the direction cosines of each with respect to the coordinate system base vectors.

Are these vectors linearly independent?

4.2.   Find the solution to $\mathbf{Ax} = \mathbf{c}$ with $\mathbf{A}(3\text{X}3)$ and $\mathbf{c}$ given below, by purely vector operations.

$$\mathbf{Ax} = \mathbf{c} = \begin{bmatrix} -1 & 2 & -5 \\ 2 & -1 & 2 \\ 5 & 0 & 3 \end{bmatrix} \mathbf{x} = \begin{Bmatrix} 1 \\ 0 \\ 1 \end{Bmatrix}.$$

---

[1] The numerators of these Equations (4.60) can be written directly. See Appendix B, "Polynomials," Equations (B.3) and (B.4) describing the relationships between the roots of a polynomial and its coefficients.

4.3. Given the vectors from problem 1, form three vectors:

$$\mathbf{b}_1 = \mathbf{a}_2 - \mathbf{a}_1, \mathbf{b}_2 = \mathbf{a}_3 - \mathbf{a}_2, \text{ and } \mathbf{b}_3 = \mathbf{a}_1 - \mathbf{a}_3 .$$

Are the **b** vectors linearly independent? Is there a non trivial solution to **Bx** = **c**, where **B** is formed using the new **b**-vectors, and **c** is defined in problem 2? Explain your answers.

4.4. Find the rank of the 3X5 matrix, **M**:

$$\mathbf{M} = \begin{bmatrix} 3 & 0 & -1 & 2 & 5 \\ 1 & -1 & 2 & 0 & 1 \\ 1 & 2 & -5 & 2 & 3 \end{bmatrix}.$$

4.5. With the **M**(3X5) matrix above

(a) Determine whether or not the columns of **M** are independent.

(b) Determine whether or not the rows of **A** are independent.

(c) Find the solutions (if any) to **Mx** = **0.**

(d) What are the conditions necessary for **Mx** = **y** to have a solution?

4.6. Given the matrix, **A**(5X4), below

$$\begin{bmatrix} 3 & -1 & 1 & 0 \\ -6 & -7 & 2 & -3 \\ -4 & -5 & 1 & -2 \\ 3 & 2 & 0 & 1 \\ -1 & 1 & -4 & -2 \end{bmatrix}.$$

(a) Determine whether or not the columns of **A** are independent.

(b) Are the rows of **A** independent?

(c) Find a column **z**(5X1) that is orthogonal to all the columns of **A**. If such a column cannot be found, explain why.

(d) Given a column vector **y**={−8, 18, 11, −8, −2}, determine whether or not **Ax** = **y** is compatible. If so, solve for **x.**

(e) Given a column vector **y** ={−6, 0, 13, 6, −3}, determine whether or not **Ax** = **y** is compatible. If so, solve for **x.**

4.7. Determine which is the better fit: (a) The linear fit, or (b) The quadratic fit in the diameter vs temperature problem.

4.8. Given **Ax** = **c** (**A** non-singular, **B** = **A**$^{-1}$), discuss the following in a "vector sense:"

(a) $\mathbf{A}$ columns i and j are interchanged, how is $\mathbf{B}$ affected? How is the solution, $\mathbf{x}$, affected?

(b) $\mathbf{A}$ rows i and j are interchanged, how is $\mathbf{B}$ affected? How is the solution, $\mathbf{x}$, affected?

(c) If rows i and j of the vector, $\mathbf{c}$, are interchanged, how are $\mathbf{B}$ and $\mathbf{x}$ affected?

4.9. Determine whether or not there are values of the λ parameter for which a solution exists in the equation set below.

$$\begin{bmatrix} \lambda & -1 & 3 \\ -4 & 2 & 0 \\ 8 & -\lambda & 2\lambda \end{bmatrix} \begin{Bmatrix} x_1 \\ x_2 \\ x_3 \end{Bmatrix} = \begin{Bmatrix} 0 \\ 0 \\ 0 \end{Bmatrix}.$$

(a) How many such values exist?

(b) For each one, find the general solution to the set.

4.10. Find the least squares best solution for c_1 and c_2 in the equation set below.

$$1.00c_1 + c_2 = 1.83$$
$$1.50c_1 + c_2 = 1.98$$
$$1.80c_1 + c_2 = 2.09$$
$$2.00c_1 + c_2 = 2.17$$
$$3.10c_1 + c_2 = 2.52$$
$$3.20c_1 + c_2 = 2.56$$
$$3.30c_1 + c_2 = 2.59 .$$

4.11. Using Equations (4.60) show that $\mathbf{q_1} \bullet \mathbf{x}_j = \delta_{1j}$ where $\mathbf{q_1}$ is the vector formed from the coefficients of the $q_1(x)$ polynomial and $\mathbf{x}_j = \begin{Bmatrix} 1 & x_j & x_j^2 & x_j^3 \end{Bmatrix}$, the jth row of $\mathbf{X}$.

4.12. Show that

$$|X| = \begin{vmatrix} 1 & x_1^2 & x_1^3 \\ 1 & x_3^2 & x_3^3 \\ 1 & x_4^2 & x_4^3 \end{vmatrix} = (x_1x_3 + x_1x_4 + x_3x_4)(x_4 - x_1)(x_4 - x_3)(x_3 - x_1).$$

Hint: Note the subscript numbering in $|X|$. Start with the X(4X4), and delete row and column 2.

4.13. In the polynomial $\prod_{j=1}^{6} (x - x_j) = x^6 + c_1x^5 + \cdots + c_n$ find c_2 and c_3. Describe the formation of each of the coefficients.

CHAPTER 5

Orthogonal Transforms

5.1   INTRODUCTION

This chapter will explore other uses and characteristics of the ***transform*** equation $\mathbf{Ax} = \mathbf{y}$. In this new case, however, the transform matrix will be an orthogonal one (see definition in Chapter 1); and so, it will not be denoted by the letter "$\mathbf{A}$," but, by some (hopefully more descriptive) letter — usually "$\mathbf{T}$," "$\mathbf{P}$," or "$\mathbf{Q}$."

This chapter will be largely "conceptual," with emphasis on three dimensional thinking. We will be concerned with physical displacements and motions in the real world; three linear displacement coordinates, plus angular displacement, and motion, about the three coordinates. There will not be much extension of concept into n-space, although orthogonal transforms are certainly not limited to 3-space. The next chapter will include some very interesting examples in n-space.

These are relatively simple concepts. But, they are of great value to the engineer, who is often required to conceptualize in three dimensions. The transform matrix will be seen to provide an invaluable framework for his thinking, and approach to problem solving.

5.2   ORTHOGONAL MATRICES AND TRANSFORMS

The definition of an orthogonal matrix is one whose transpose is equal to its inverse. Then, given the orthogonal matrix, $\mathbf{T}$:

$$\mathbf{T}' = \mathbf{T}^{-1}$$
$$\mathbf{TT}' = \mathbf{T}'\mathbf{T} = \mathbf{I} \,. \tag{5.1}$$

The dot product of any two columns (rows) of an orthogonal matrix is zero. The dot product of the column (row) by itself is 1. Then, the orthogonal matrix is also "orthonormal."

The usual function of such a matrix is to describe rotation in a 2 or 3-dimensional system. The transform equation $\mathbf{x} = \mathbf{Ty}$ relates the coordinates of a vector as measured in two rectangular coordinate systems. In two dimensions, consider two coordinate sets, x and y, that are collinear (superimposed). In this case, any vector, say $\mathbf{r}$, has identical components when represented in either the x-set, or the y-set. The transform relating coordinates in the two sets is $\mathbf{y} = \mathbf{Ix}$, where x and y are 2X1 vectors representing coordinates in the x and y sets, and the transform matrix is the 2X2 unit matrix (note that the unit matrix, I, is orthogonal). However, this case is trivial.

Next, the y-set is rotated in the + direction (counterclockwise) by an amount θ. Now, the coordinates of $\mathbf{r}$ are different in the y-set, and there is a nontrivial, orthogonal, "transformation" between the two sets. We will define this transform.

In the x-set, the vector is described as (Figure 5.1):

$$r_{x1} = r_m \cos(\theta + \varphi)$$
$$r_{x2} = r_m \sin(\theta + \varphi)$$

where r_m is the absolute magnitude (length) of $\mathbf{r}$. Then:

$$r_{x1} = r_m(\cos\theta \cos\varphi - \sin\theta \sin\varphi)$$
$$r_{x2} = r_m(\cos\theta \sin\varphi + \sin\theta \cos\varphi).$$

Since $r_m \cos\varphi$ and $r_m \sin\varphi$ are the coordinates of $\mathbf{r}$ in the y-set:

$$r_{x1} = r_{y1} \cos\theta - r_{y2} \sin\theta$$
$$r_{x2} = r_{y1} \sin\theta + r_{y2} \cos\theta \quad \text{or}$$

$$\mathbf{x} = \begin{bmatrix} \cos\theta & -\sin\theta \\ \sin\theta & \cos\theta \end{bmatrix} \mathbf{y} \quad \text{or} \quad \mathbf{x} = \mathbf{Ty}. \tag{5.2}$$

In (5.2), since the transform represents *any* vector, the reference to $\mathbf{r}$ is omitted. This equation set

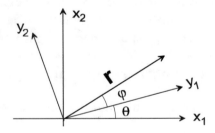

Figure 5.1:

defines the x-set coordinates of a vector in terms of its y-set coordinates. ***Note that the transform matrix, T, is orthogonal. Its columns (rows) are mutually perpendicular, with the dot product of zero. Furthermore, the columns and rows are normalized to unity***:

Matrices of the type $\mathbf{T}$ are the subject of this entire chapter. Such transforms preserve both linear and angular measurement. For example, in (5.2), the squared length of a vector in the x-set can be denoted $\mathbf{x'x}$. Since $\mathbf{x} = \mathbf{Ty}$, then

$$\mathbf{x'x} = \mathbf{y'T'Ty} = \mathbf{y'y} \text{ (since } \mathbf{T'T} = \mathbf{I}) \ .$$

That is, the length is the same in either set. The very same reasoning shows that, given two **unit** vectors $\mathbf{u}$ and $\mathbf{v}$, known in the x-set as $\mathbf{u}_x$ and $\mathbf{v}_x$, the cos of the angle between them is $\mathbf{u}_x \bullet \mathbf{v}_x$. This same value results when the dot product of the two is taken in the y-set.

In order to define the inverse transform of (5.2), we need only transpose the $\mathbf{T}$ matrix:

$$\mathbf{y} = \left[\begin{array}{cc} \cos\theta & \sin\theta \\ -\sin\theta & \cos\theta \end{array} \right] \mathbf{x} \quad \text{or} \quad \mathbf{y} = \mathbf{T}'\mathbf{x}. \tag{5.3}$$

Note that which matrix is called $\mathbf{T}$, and which $\mathbf{T}'$, is largely a matter of choice.

So, of what use is this transform matrix? To see the answer, just consider the $\mathbf{r}$ vector in motion. Conceptually, *we attach this vector to the y-set*. Its rotations are those of its coordinate set. And, we can simply describe any rectilinear motion in this set. Then, to see the total motion, we just transform the vector back into the "inertial" (fixed) x-set.

The 3-dimensional case is a trivial extension of (5.2) and (5.3). From the Figure 5.1, above, just include the $+x_3$ and $+y_3$ axes coming directly upward – out of the plane of the page. Especially note that the rotation θ occurs around these axes; they therefore remain collinear (and the coordinate of any vector in this direction is measured the same in both x and y-sets). Then:

$$\mathbf{y} = \left[\begin{array}{ccc} \cos\theta & \sin\theta & 0 \\ -\sin\theta & \cos\theta & 0 \\ 0 & 0 & 1 \end{array} \right] \mathbf{x}. \tag{5.4}$$

Equation (5.4) is the transform matrix between the inertial x-set, and a y-set, which has been rotated by a $+$ angle θ about the x_3 axis. The inverse transform is simply the transpose of the matrix in (5.4).

5.2.1   RIGHTHANDED COORDINATES, AND POSITIVE ANGLE

One must be careful to describe a 3-dimensional coordinate set by the so called "right hand rule," and to define positive angle in the same way. In Figure 5.1, the positive x_1-axis is directed toward the right, the positive x_2 axis upward (from the bottom of the page toward the top). Then, the positive x_3 axis necessarily must come out of the page, toward you (the negative x_3 axis is, then, directed away from you, into the pages of the book). *All of the coordinate sets constructed in this chapter will follow this rule.*

Another way to see this is: Curl the fingers of your *right* hand from the +1-axis to the +2-axis. Then, your thumb will point in the direction of the positive 3-axis. Now, do the same with the 2-axis, toward the 3-axis. The thumb will point to the positive 1-axis. Finally, assure yourself by curling the right fingers from the +3-axis toward the +1-axis. The thumb will now point toward positive 2-axis. See the next section, where the vector (cross) product is discussed.

Positive angle will be measured in the same sense: Rotation about any positive axis will itself be plus in the direction of the curled fingers of the right hand – counterclockwise, when the positive axis is in the same direction as the thumb.

These rules are very important. An incorrect sign can easily occur, and be very difficult to trace to a coordinate set improperly constructed.

Now, consider any orthogonal transform in which we regard the x-set as "stationary," with the y-set having undergone some series of rotations. In 3 dimensions, define unit vectors in both sets, as follows: $\mathbf{1}_x, \mathbf{2}_x, \mathbf{3}_x$ are the defined unit vectors in the x-set, in the directions along the x_1, x_2, x_3 axes respectively. In the same way, define the unit vectors $\mathbf{1}_y, \mathbf{2}_y,$ and $\mathbf{3}_y$ in the y-set. Then

$$\mathbf{y} = \mathbf{Tx} = \begin{bmatrix} \mathbf{1}_y \bullet \mathbf{1}_x & \mathbf{1}_y \bullet \mathbf{2}_x & \mathbf{1}_y \bullet \mathbf{3}_x \\ \mathbf{2}_y \bullet \mathbf{1}_x & \mathbf{2}_y \bullet \mathbf{2}_x & \mathbf{2}_y \bullet \mathbf{3}_x \\ \mathbf{3}_y \bullet \mathbf{1}_x & \mathbf{3}_y \bullet \mathbf{2}_x & \mathbf{3}_y \bullet \mathbf{3}_x \end{bmatrix} \mathbf{x} . \tag{5.5}$$

That is, the elements of $\mathbf{T}$ are the dot products of the respective unit vectors, as shown. In the specific case of the transform (5.4), comparison of (5.5) with (5.4) shows that (see Figure 5.1):

$$t_{11} = \mathbf{1}_y \bullet \mathbf{1}_x = \cos\theta$$
$$t_{12} = \mathbf{1}_y \bullet \mathbf{2}_x = \cos(90 - \theta) = \sin\theta$$
$$t_{13} = \mathbf{1}_y \bullet \mathbf{3}_x = \cos(90) = 0$$

> The components of $\mathbf{1}_y$ in the x-set

$$t_{21} = \mathbf{2}_y \bullet \mathbf{1}_x = \cos(90 + \theta) = -\sin\theta$$
$$t_{22} = \mathbf{2}_y \bullet \mathbf{2}_x = \cos\theta$$
$$t_{23} = \mathbf{2}_y \bullet \mathbf{3}_x = \cos(90) = 0$$

> The components of $\mathbf{2}_y$ in the x-set

$$t_{31} = \mathbf{3}_y \bullet \mathbf{1}_x = \cos(90) = 0$$
$$t_{32} = \mathbf{3}_y \bullet \mathbf{2}_x = \cos(90) = 0$$
$$t_{33} = \mathbf{3}_y \bullet \mathbf{3}_x = \cos(0) = 1$$

> The components of $\mathbf{3}_y$ in the x-set

In the above, the reference to "90" implies angular measurement in degrees – i.e., 90 degrees. In (5.5), the first row dots the $\mathbf{1}_y$ unit vector into the x set unit vectors – each in turn. The second row dots the $\mathbf{2}_y$ vector; the third row, the $\mathbf{3}_y$ vector, into the x-set unit vectors, in turn. If the inverse transform is required, then just transpose (5.5). If one cares to memorize these dot products, the transform matrix can be written directly, rather than going through the development that precedes (5.2). These transform matrices will be found all through this chapter, so it is well to see clearly the manner of their construction. It is very simple, but it can be "tricky," and sign errors can result.

5.3　EXAMPLE COORDINATE TRANSFORMS

In a practical case, the "complete" transform is usually the result of a series of simple transforms – each one being a rotation about one of the coordinate axes, with a transform equation similar to (5.4). For example, we may start by a rotation of a y-set relative to the fixed x-set:

$$\mathbf{y} = \mathbf{T}_1\mathbf{x}$$

where $\mathbf{T}_1$ is an orthogonal matrix of the type in (5.4). Next, we may have a rotation of another coordinate set, say a z-set, relative to the y-set:

$$\mathbf{z} = \mathbf{T}_2\mathbf{y} .$$

Then, the (final) combined transform, between the z- and x-sets is:

$$z = T_2 T_1 x = Tx; \quad T = T_2 T_1 \;.$$

Both T_1 and T_2 are orthogonal. It is easy to show that the product, T, is also orthogonal, by multiplying $T_2 T_1$ by its transpose

$$T'T = [T_1'T_2'T_2 T_1] = I \;.$$

5.3.1 EARTH-CENTERED COORDINATES

A very practical, yet simple, example is the construction of earth-centered coordinates. To define the motion of a rocket or orbiting body, the observations of position and velocity taken at a station located at the surface of the earth must be transformed to a coordinate set located at earth center. The example given here will be to develop the transform of a station located at longitude θ and latitude φ back to an earth-centered set.

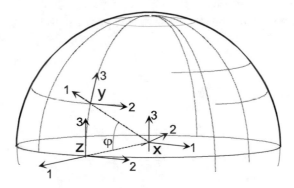

Figure 5.2: Earth-centerd Coordinates.

It will be assume that the earth is a perfect sphere of radius, r, although this is actually not the case – the earth radius is some 10 miles less at the poles than at the equator. The x-set will be at earth center, with x_1 pointing at the zeroth longitude. The $x_1 x_2$ plane lies in the equatorial plane; the x_3 axis points from earth center toward the north pole.

An intermediate z-set is constructed at longitude θ, but with zero latitude; i.e., located along the equator. We will first relate the z-set to the x-set, then relate the y-set to the z-set, and, finally, combine the two.

Looking down upon the $x_1 x_2$ (equatorial) plane, the z-set has its z_1 axis pointing directly skyward, z_2 points east, z_3 northward. The radius of the earth is r. Since these are the same conditions

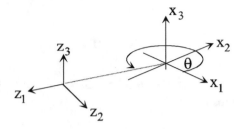

as those of Equation (5.4), we can write directly:

$$\mathbf{x} = \mathbf{T}_1 \mathbf{z} \;\Rightarrow\; \mathbf{x} = \begin{bmatrix} \cos\theta & -\sin\theta & 0 \\ \sin\theta & \cos\theta & 0 \\ 0 & 0 & 1 \end{bmatrix} \mathbf{z}\,. \tag{5.6}$$

The z-set and x-set are not collocated. ***Nevertheless, Equation*** (5.6) ***accurately represents the angular displacement between the two sets***. Now, superimpose a y-set onto the z coordinates, and then slip the new y-set directly north, remaining at longitude, θ, and keeping the $y_2 y_3$ plane tangent to the sphere. When the y-set has been slipped through an angle φ, Figure 5.3 can be used to develop a transform between the two coordinate sets. Note that y_1 points skyward, y_2 east, and y_3 north. Also,

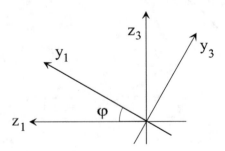

Figure 5.3: z-y transform.

the z_2 and y_2 axes continue to be parallel.

$$\mathbf{z} = \mathbf{T}_2 \mathbf{y} = \begin{bmatrix} \cos\varphi & 0 & -\sin\varphi \\ 0 & 1 & 0 \\ \sin\varphi & 0 & \cos\varphi \end{bmatrix} \mathbf{y}\,. \tag{5.7}$$

Then, the overall transform is given by eliminating z between (5.6) and (5.7). That is

$$\mathbf{x} = \mathbf{T}_1 \mathbf{T}_2 \mathbf{y} = \mathbf{T}\mathbf{y} = \begin{bmatrix} \cos\theta\cos\varphi & -\sin\theta & -\sin\varphi\cos\theta \\ \cos\varphi\sin\theta & \cos\theta & -\sin\theta\sin\varphi \\ \sin\varphi & 0 & \cos\varphi \end{bmatrix} \mathbf{y}\,. \tag{5.8}$$

Note that θ is measured eastward from zero degrees longitude to 360 degrees, not the usually given East Longitude and West Longitude (wherein θ is an angle between 0 and 180 degrees). In this measure, then, points in the United States will have θ values greater than 230 degrees. The latitude is measured in the usual way, from zero degrees at the equator, to 90 degrees at the north pole.

The radius, $\mathbf{r}$ from earth center to the station, is given by $\{r, 0, 0\}$, measured in the y-set. We transform $\mathbf{r}$ into the x-set via $\mathbf{T}$ in (5.8). The result is:

$$\mathbf{r}_x = \left\{ \begin{array}{c} r\cos\theta\cos\varphi \\ r\sin\theta\cos\varphi \\ r\sin\varphi \end{array} \right\}.$$

These are the well known polar coordinates of the vector. Note that although the y- and x-sets do not have the same origin, vectors known in either set can be transformed to the other. More importantly, the above vector $\mathbf{r}_x$ must be added to position vector observations taken at the station, (y-set) and then transformed to the x-set. For example, radar data, taken from several stations is transformed first to a single station. This data defines, say, the instantaneous position of an orbiting body in its local coordinates. Its position relative to the inertial coordinates is $\mathbf{r}_x$ plus the transformed position into the x-set. That is (with $\mathbf{T}$ taken from (5.8)):

$$\mathbf{p}_x = [\mathbf{T}]\mathbf{p}_y + \mathbf{r}_x . \tag{5.9}$$

The time derivative of (5.9) defines velocity. In cases wherein the rotation of the earth must be taken into account, θ becomes a time dependent variable. Thus, the matrix $\mathbf{T}$ must be differentiated. We will consider the differentiation of a matrix in a later section.

As a check of the transform, $\mathbf{T}$, plug all the y-set unit vectors, in turn, into (5.8). The results in each case, of course, would be the columns of $\mathbf{T}$ – and the direction cosines of each of the y-set unit vectors, expressed in the x-set. For example, note that column 2 of $\mathbf{T}$ depends only upon θ. That does check: the unit vector $\{0, 1, 0\}_y$ is parallel to the x_1x_2 plane, and it projects onto that plane as $\{\cos(90 + \theta), \sin(90 + \theta), 0\}$. The point is that if this same reasoning had been used at the beginning, it would not have been necessary to develop an intermediate z-set. The transform (5.8) could be written directly. However, the reader should try this, and note that it is not easy. The 3 dimensional thinking required is confusing, and prone to error. In most cases, it is safer and easier to develop such transforms in a series of simple steps.

Sometimes, a rotation takes place about an axis that is not one of the coordinate axes given in the problem. In that case, (as will be seen in the example problem, below), an intermediate set is set up specifically to orient the rotation about one of its coordinate axis. To do this, it is necessary to take the cross product of two existing vectors to generate one of the coordinate axes in the new, rotated set. For this reason, we should first review this product (see also, Chapter 1, Section 1.2).

The "vector product," or "cross product" of two vectors produces a *vector* which is orthogonal to both of the vectors crossed. In contrast, it will be recalled that the dot product of two vectors produces a scalar. The magnitude of the new vector is the product of the input vector magnitudes times the sine of the angle between them. For example, consider two vectors, $\mathbf{u} = \{u_1, u_2, u_3\}$, and $\mathbf{v} = \{v_1, v_2, v_3\}$ in a coordinate system, x. Their cross product is a vector, whose elements can be found by the first row "expansion" of the following determinant. This "expansion" is quite special, however, ***involving the unit vectors as the first row elements***. In this (fabricated) way, the result is a three-dimensional vector rather than a scalar.

$$\mathbf{u}_x \times \mathbf{v}_x = \mathbf{u}_x \text{ "cross" } \mathbf{v}_x \Rightarrow \begin{vmatrix} 1_x & 2_x & 3_x \\ u_1 & u_2 & u_3 \\ v_1 & v_2 & v_3 \end{vmatrix} \Rightarrow \begin{Bmatrix} u_2 v_3 - u_3 v_2 \\ u_3 v_1 - u_1 v_3 \\ u_1 v_2 - u_2 v_1 \end{Bmatrix}. \tag{5.10}$$

The same result can be obtained by premultiplying $\mathbf{v}$ by a skew symmetric matrix made from the elements of $\mathbf{u}$, as given in (5.11), below:

$$\mathbf{u}_x \times \mathbf{v}_x = \mathbf{U}\mathbf{v} = \begin{bmatrix} 0 & -u_3 & u_2 \\ u_3 & 0 & -u_1 \\ -u_2 & u_1 & 0 \end{bmatrix} \begin{Bmatrix} v_1 \\ v_2 \\ v_3 \end{Bmatrix} = \begin{Bmatrix} u_2 v_3 & -u_3 v_2 \\ u_3 v_1 & -u_1 v_3 \\ u_1 v_2 & -u_2 v_1 \end{Bmatrix}. \tag{5.11}$$

Equation (5.11) can be "read in reverse:" A matrix-vector product in which the premultiplying matrix is skew symmetric can be interpreted as a vector cross product.

The resultant vector from (5.10) or (5.11) has to be orthogonal to both $\mathbf{v}_x$ and $\mathbf{u}_x$. It is a worthwhile exercise for the reader to prove that this is true.

Note that the product $(\mathbf{u} \times \mathbf{v})$ is different than $(\mathbf{v} \times \mathbf{u})$. Specifically, if rows 2 and 3 of the determinant in (5.10) are interchanged, the determinant expansion (5.10) will yield $(\mathbf{v} \times \mathbf{u})$. And, the elements will be of reversed sign. Then $(\mathbf{v} \times \mathbf{u})$ is the negative of $(\mathbf{u} \times \mathbf{v})$. Again, the righthand rule is handy: Curl your right fingers from $+\mathbf{u}$ to $+\mathbf{v}$ (the fingers being parallel to the plane of $\mathbf{u}$ and $\mathbf{v}$), the outstretched thumb will point in the positive direction of $(\mathbf{u} \times \mathbf{v})$.

With the unit vectors of a right-handed coordinate system, curl your fingers from $+1_x$ to $+2_x$ – note that the thumb points in the direction of $+3_x$. The order is, of course, important. For example, if one were to cross 2_x into 1_x, the result would point the 3_x axis in the wrong direction. The following equations summarize the correct results:

$$\begin{aligned} 1_x \times 2_x &= 3_x \\ 2_x \times 3_x &= 1_x \\ 3_x \times 1_x &= 2_x \ . \end{aligned} \tag{5.12}$$

5.3.2   ROTATION ABOUT A VECTOR (NOT A COORDINATE AXIS)

Consider two coordinate sets, x and y. Initially, they are superimposed, but the y-set is free to rotate, the x-set is fixed. Now, enter the vector, $\mathbf{r} = \{-3, -4, 5\}$, and "glue" its base to the origin of the y-set. At this point, the coordinates of $\mathbf{r}$ are the same in both the x-set, and the y-set. Now, looking down $\mathbf{r}$, from its tip toward the origin, rotate $\mathbf{r}$ through a positive (counterclockwise) angle θ. Note that the y-set must rotate as well; however, the rotation is not in any of the coordinate planes of this set.

Figure 5.4 shows the two superimposed x- and y-sets, and the $\mathbf{r}$ vector with a positive angular rotation indicated. The problem that will be discussed is the construction of the transform between the rotated y-set, and the fixed x-set.

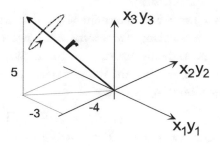

Figure 5.4:

First, define a coordinate set (say, w) one of whose coordinate planes lies in the plane of the rotation (then, one of its axes will be along the vector, $\mathbf{r}$). Its origin is fixed to that of the x-set (i.e., the w-set will not rotate). Somewhat arbitrarily, define the unit vector along $\mathbf{r}$ as the 3-axis of the w-set (i.e., w_3). Since the length of $\mathbf{r}$ is $r_m = \sqrt{9 + 16 + 25} = 5\sqrt{2}$:

$$3_w = \frac{1}{r_m} \left\{ \begin{array}{ccc} -3 & -4 & 5 \end{array} \right\}; \quad r_m = 5\sqrt{2}.$$

Now, construct the 1-2 plane of the w-set. The specific direction of each of these two axes is quite arbitrary, but, they certainly must be orthogonal to the 3_w axis. If we cross 3_x into 3_w the result will be perpendicular to both 3_x and 3_w and it will point in the general direction of 1_x (not necessary, but easier to visualize). Normalized, it'll be the 2_w axis:

$$3_x \times 3_w = \left\{ \begin{array}{ccc} \dfrac{4}{r_m} & \dfrac{-3}{r_m} & 0 \end{array} \right\} \text{ normalized } = 2_w = \frac{1}{5} \left\{ \begin{array}{ccc} 4 & -3 & 0 \end{array} \right\}.$$

Now, following the relations (5.12), cross 2_w into 3_w, to define 1_w. This cross product will yield the unit vector 1_w directly (already normalized):

$$1_w = \frac{1}{r_m} \left\{ \begin{array}{ccc} -3 & -4 & -5 \end{array} \right\}.$$

The 3 vectors just defined as $1_w, 2_w,$ and 3_w, define, in turn, the transform matrix, between the x-set and the w-set:

$$\mathbf{w} = \mathbf{T}'\mathbf{x}; \quad \text{where} \quad \mathbf{T}' = \begin{bmatrix} \dfrac{-3}{r_m} & \dfrac{-4}{r_m} & \dfrac{-5}{r_m} \\ \dfrac{4}{5} & \dfrac{-3}{5} & 0 \\ \dfrac{-3}{r_m} & \dfrac{-4}{r_m} & \dfrac{5}{r_m} \end{bmatrix}. \tag{5.13}$$

That is, the 1st row of $\mathbf{T}'$ is $\mathbf{1}_w$, the second row is $\mathbf{2}_w$, etc. To provide confidence that we have the transform in the right order, put $\{-3, -4, \ 5\}$ (the coordinates of $\mathbf{r}$ in the x set) into (5.13). These coordinates will transform through (5.13) to a vector in the w-set with a w_3 component equal to r_m, and the w_1 and w_2 components equal to zero.

The inverse transform, $\mathbf{x} = \mathbf{Tw}$, is also determined by simply transposing the matrix $\mathbf{T}'$.

The 1-2 plane of the w-set is the plane of rotation. Note, however, that the w-set is not rotated. Instead, we will define a new z-set, originally superimposed upon the w-set, but then rotated through the required θ angle. The transform between the w and z sets can be written directly, because it is the same as that defined in Equation (5.4), above:

$$\mathbf{z} = \mathbf{Qw} \ \Rightarrow \ \mathbf{z} = \begin{bmatrix} \cos\theta & -\sin\theta & 0 \\ \sin\theta & \cos\theta & 0 \\ 0 & 0 & 1 \end{bmatrix} \mathbf{w} \ . \tag{5.14}$$

Now that the transforms (5.13) and (5.14) are known, we can proceed with the solution to the problem. Originally, before the rotation, the y-set and x-set are superimposed. Therefore, Equation (5.13) holds for the y-set as well, and since before rotation the w- and z-sets are superimposed:

$$\mathbf{y} = \mathbf{Tw} = \mathbf{Tz} \quad \text{(before rotation)} \ . \tag{5.15}$$

After the rotation, (5.13) still relates the x-set to the w-set because neither of them moves. More importantly, (5.15) can still be used to relate the z-set to the y-set *after* rotation, because *they move together:*

$$\mathbf{y} = \mathbf{Tz} \quad \text{(after rotation)} \ . \tag{5.16}$$

Plugging the definition of (5.14) into (5.16):

$$\mathbf{y} = \mathbf{TQw} \tag{5.17}$$

and since, from (5.13), $\mathbf{w} = \mathbf{T}'\mathbf{x}$, the final transform is:

$$\mathbf{y} = \mathbf{TQT}'\mathbf{x} \tag{5.18}$$

and its inverse is (obviously):

$$\mathbf{x} = \mathbf{TQ}'\mathbf{T}'\mathbf{y} \ . \tag{5.19}$$

We have already seen that the transform of a vector, say $\mathbf{x}$, is done through the premultiplication of $\mathbf{x}$ by some matrix, $\mathbf{T}$ ($\mathbf{y} = \mathbf{Tx}$). Now, (5.19) implies that the rotational matrix, $\mathbf{Q}$, is transformed by both pre- and post multiplication (i.e., $\mathbf{TQT}'$). And this is, indeed the general case – *matrices are transformed by pre- and post multiplication by the transforming matrices*. This transformation of $\mathbf{Q}$ produces the rotation given in $\mathbf{Q}$, as observed in the x- and y-sets, respectively.

In (5.18), if we call the overall transform matrix $\mathbf{W}$, then $\mathbf{W} = \mathbf{TQT}'$. The matrix $\mathbf{W}$ is the "transform" of $\mathbf{Q}$. The transforming matrix, $\mathbf{T}$, is orthogonal. In this case, as will be discussed in a later article, $\mathbf{W}$ and $\mathbf{Q}$ are said to be related by a "congruent" transform.

Section 5.4, below, discusses the transformation of matrices.

5.3.3 ROTATION ABOUT ALL THREE COORDINATE AXES

In this section, we will develop a transform which includes rotation about all of the coordinate axes (in three dimensions). The 3 angles of rotation will be denoted θ_1, θ_2, and θ_3. These have been referred to as the "Eulerian" rotations, for it was Euler who showed that it is always possible to go from any initial orientation of coordinates, to any final orientation, by rotations about the three axes of the coordinate set – in a specific order. In the development, below, we will choose the order 3, 2, 1, somewhat arbitrarily. The angles will be referred to as "pitch," "roll," and "yaw," as if the axes lie within an airframe, with the positive $\mathbf{x}_2$ axis pointing "ahead," and the positive $\mathbf{x}_1$ axis pointing out the right wing. The angles, θ_i, are defined as the rotations about their respective axes, $\mathbf{x}_i$.

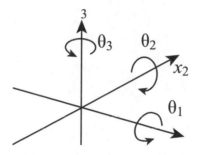

Figure 5.5: Rotation about all 3 axes.

For clarity in the equations to follow, define $C_i = \cos\theta_i$, and $S_i = \sin\theta_i$. Shown below are the transforms around each axis, corresponding to the diagram below.

Pitch (Rotation About $\mathbf{x}_1$)

$$\mathbf{x} = \mathbf{T}_1\mathbf{y} = \begin{bmatrix} 1 & 0 & 0 \\ 0 & C_1 & -S_1 \\ 0 & S_1 & C_1 \end{bmatrix}. \qquad (5.20)$$

Note in the diagram that the positive $\mathbf{x}_1$ axis is out of the paper.

The "airplane coordinates" are the y-set. The fuselage still lies along the $\mathbf{1}_2$ axis, but it is the y_2-axis. For example, a vector $\{0, 1, 0\}$ (along the axis of the aircraft — in the y-set) will have the coordinates $\{0, C_1, S_1\}$ in the x-set — showing a pitch upward.

Roll (Rotation About x_2)

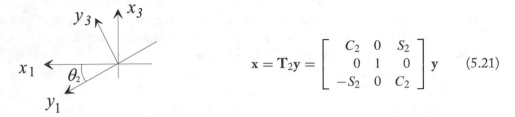

$$\mathbf{x} = \mathbf{T}_2\mathbf{y} = \begin{bmatrix} C_2 & 0 & S_2 \\ 0 & 1 & 0 \\ -S_2 & 0 & C_2 \end{bmatrix} \mathbf{y} \qquad (5.21)$$

Again, the airplane coordinates are the y-set. The positive x_2 axis is up, out of the paper.

Yaw (Rotation About x_3)

$$\mathbf{x} = \mathbf{T}_3\mathbf{y} = \begin{bmatrix} C_3 & -S_3 & 0 \\ S_3 & C_3 & 0 \\ 0 & 0 & 1 \end{bmatrix} \mathbf{y} \qquad (5.22)$$

If rotations are taken in 3, 2, 1 order, then $\mathbf{x} = \mathbf{T}_3\mathbf{T}_2\mathbf{T}_1\mathbf{y} = \mathbf{Ty}$, where T is given in (5.23):

$$\mathbf{T} = \begin{bmatrix} C_2C_3 & S_1S_2C_3 - C_1S_3 & S_1S_3 + C_1S_2C_3 \\ C_2S_3 & C_1C_3 + S_1S_2S_3 & C_1S_2S_3 - S_1C_3 \\ -S_2 & S_1C_2 & C_1C_2 \end{bmatrix}. \qquad (5.23)$$

It is to be noted, here, that the order of this product is important in that the final result is different for any different order. For example, if an aircraft rolls 90 degrees, and then pitches "up" by 90 degrees, the result is quite different than if it had pitched up 90 degrees, and then rolled. In the order given here, yaw is first, then roll, then pitch.

To make equations easier to read, the "shorthand," $C_j = \cos\theta_j$" and $S_j = \sin\theta_j$, is used above. This kind of shorthand will be used throughout this book.

5.3.4   SOLAR ANGLES

A solar panel converts the radiant energy from the sun to an electrical output. The output is proportional to the area of the panel exposed to the sun's rays (the "effective area"). The diagram below shows a single square foot of the panel surface. The lower half (plain view) shows this square area from above; the upper half shows an edge-view of the same area. If the sun is directly above that surface, the entire square foot is exposed as in the lower half, but when the sun's rays are at an angle, one of the dimensions of the area reduces (compare the length d (= 1 ft) to the length d' in the diagram, above). The effective area is proportional to the ratio of these dimensions. In numerical terms, that ratio, C_f, is equal to the ***trigonometric cosine of the "Angle of Incidence," i***, between the

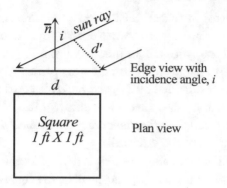

sun ray and the panel normal, $\bar{\mathbf{n}}$. To constrain i to angles between plus and minus 90°, the "**sun vector**" is perceived as the vector from the panel toward the sun (the negative of that shown).

In order to calculate C_f, a unit "sun vector" and the unit "panel vector" must be calculated. The dot product of these two *unit* vectors yields the required cosine of the angle of incidence. Both of these vectors must be defined in the same coordinate set. That set might be defined at the surface of the solar panel or elsewhere (possibly at earth center). Because the transforms between sets will be orthogonal, any convenient set will produce the same results (i.e., angle measurement is preserved).

There are two rotations involved. First, the earth orbits about the sun. A coordinate set at the earth center, the **o**-set, can be used to describe this motion, and define the sun vector. Second, the earth's rotation about its axis requires a second set (the **e**-set), one of whose coordinate axes collinear with the earth's axis.

The o-set: Arbitrarily, make the $\mathbf{o_3}$ axis orthogonal to the orbit plane with $+\mathbf{o_3}$ pointing to celestial north, the $\mathbf{o_1 o_2}$ plane in the orbital plane, and the $\mathbf{o_1}$ axis directed toward the sun. The coordinates of the sun vector in this set are then $\{1, 0, 0\}$. See Figure 5.6.

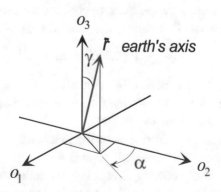

Figure 5.6: Earth Orbit in the o_1, o_2 plane.

The **o**-set is inertial (fixed in space), ***with the orbit rotation simulated by varying angle*** α. When α is 0, it is the March (spring) equinox, when $\alpha = 90°$ the earth axis is tilting directly toward the sun along $\mathbf{o}_1$ — the summer solstice (about June 20). From Figure 5.6, the $\mathbf{o}_3$ coordinate of the earth's axis is *cos(γ)*, written $C\gamma$. Its projection on the $\mathbf{o}_1\mathbf{o}_2$ plane is $S\gamma$. Then *sin(α)* is equal to the $\mathbf{o}_1$ component of the unit vector $\mathbf{r}$ divided by $S\gamma$, and *cos(α)* is equal to the $\mathbf{o}_2$ component divided by $S\gamma$. Then the unit vector earth axis has the components $\{S\gamma\,S\alpha,\ S\gamma\,C\alpha,\ C\gamma\}$. Note that $S\gamma\,S\alpha \equiv$ $\sin(\gamma)\sin(\alpha)$; as before, ***the trigonometric functions are given by their first character, capitalized***. The angle γ is the (constant) 23.5° tilt of the earth axis.

The e-set: Rotation of the earth about its axis is defined in the *e*-set, $\{\mathbf{e}_1, \mathbf{e}_2, \mathbf{e}_3\}$. Choose $\mathbf{e}_3$ to be collinear with the earth axis; then its $\mathbf{e}_1\mathbf{e}_2$ plane will be the equatorial plane. The $\mathbf{e}_3$ unit vector has the same **o**-set coordinates defined above: $\{S\gamma\,S\alpha,\ S\gamma\,C\alpha,\ C\gamma\}$.

Now, cross $\mathbf{e}_3 \times \mathbf{o}_3$ to define $\mathbf{e}_1$. The result, $\{S\gamma\,C\alpha,\ -S\gamma\,S\alpha,\ 0\}$, is a vector orthogonal to $\mathbf{e}_3$ and so must lie in the equatorial plane as required. It must be normalized to unit length, yielding the $\mathbf{e}_1$ coordinates in the **o**-set: $\{C\alpha,\ -S\alpha,\ 0\}$. Since $\mathbf{e}_1$ is also orthogonal to $\mathbf{o}_3$, it is in the earth orbit plane as well as the equatorial plane.

Finally, the $\mathbf{e}_2$ axis is defined by crossing $\mathbf{e}_3 \times \mathbf{e}_1 = \{C\gamma\,S\alpha,\ C\gamma\,C\alpha,\ -S\gamma\}$, a unit vector. This completes the definition of the **e**-set in terms of the **o**-set coordinates. Using the results of Equation (5.5), the transform relating these sets is

$$\mathbf{e} = \mathbf{T}_1\mathbf{o} = \begin{bmatrix} C\alpha & -S\alpha & 0 \\ C\gamma\,S\alpha & C\gamma\,C\alpha & -S\gamma \\ S\gamma\,S\alpha & S\gamma\,C\alpha & C\gamma \end{bmatrix}\mathbf{o}. \tag{5.24}$$

Note that the three vectors just defined are used as the rows of the transform matrix $\mathbf{T}_1$. Also, the **e**-set is defined solely by γ and α. The value of α is (0–360°) depending on a "day number," chosen (0–364). On day 0, $\alpha = 0$, on day 92 α is approximately 90°.

Since the sun vector (say, $\bar{\mathbf{s}}$) has the coordinates $\{1, 0, 0\}$ in the **o**-set, the first column of $\mathbf{T}_1$ gives the coordinates of the sun vector in the **e**-set: $\{C\alpha,\ C\gamma\,S\alpha,\ S\gamma\,S\alpha\}$.

Sun Latitude: The $\mathbf{e}_3$ sun vector coordinate, $S\gamma\,S\,\alpha$, is the cosine of the angle between the $\mathbf{e}_3$ axis and the sun vector (the $\mathbf{o}_1$ axis). This defines the "sun latitude," φ_s:

$$\varphi_s = \frac{\pi}{2} - \arccos(\sin\alpha\,\sin\gamma)\,.$$

Since γ is constant, 23.5°, φ_s is a function of α. When $\alpha = 0$, $\varphi = 0$; as α increases to 180, φ_s increases to 23.5°, then drops back to zero. During the winter months in the Northern Hemisphere, φ_s becomes negative, as α increases from 180 to 360.

During a day, the earth rotates 360° while moving in its orbit less than a degree. Then during this 24-hour period, consider the earth orbit position as fixed (i.e., α constant), making φ_s constant, ***and the same for all longitudes***. Then the longitude of the sun collector (the panel) is arbitrary.

Figure 5.7, shows the panel longitude at 0°, in line with the e_1 axis at "solar noon." Movement of the vector $\bar{\mathbf{s}}$ simulates time—the passing of the sun across the sky. Values of $\theta_s > 0$ corresponds to times before noon, $\theta_s < 0$ afternoon times.

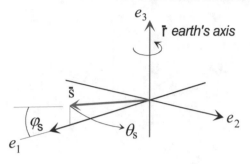

Figure 5.7: Earth rotation simulated by moving $\bar{s}$ through an angle θs.

In this **e**-set, the projection of $\bar{s}$ onto the $\mathbf{e}_1\mathbf{e}_2$ plane has the coordinates $\{C\theta, S\theta, 0\}$. Then, the coordinates of the sun vector in the set are $\{C\theta C\varphi_s, S\theta C\varphi_s, S\varphi_s\}$.

The x-set: An additional coordinate set is required in which to define the "panel vector" (the normal to the solar panel surface). Refer to Figure 5.2 used in the construction of earth-centered coordinates. In this case, the **x**-set is at the solar panel, the e-set is earth centered. Equation (5.8) can be used directly, changing only the names of the coordinate sets, and setting $\theta = 0$. As in Figure 5.2, the angle φ is the latitude of the panel.

$$\mathbf{e} = \mathbf{T}_2\mathbf{x} = \begin{bmatrix} C\varphi & 0 & -S\varphi \\ 0 & 1 & 0 \\ S\varphi & 0 & C\varphi \end{bmatrix}\mathbf{x}; \text{ or } \mathbf{x} = \begin{bmatrix} C\varphi & 0 & S\varphi \\ 0 & 1 & 0 \\ -S\varphi & 0 & C\varphi \end{bmatrix}\mathbf{e} \text{ and} \tag{5.25}$$

$$\mathbf{s}_x = \begin{bmatrix} C\varphi & 0 & S\varphi \\ 0 & 1 & 0 \\ -S\varphi & 0 & C\varphi \end{bmatrix}\begin{Bmatrix} C\theta_s C\varphi_s \\ S\theta_s C\varphi_s \\ S\varphi_s \end{Bmatrix} = \begin{Bmatrix} C\varphi C\theta_s C\varphi_s + S\varphi S\varphi_s \\ S\theta_s C\varphi_s \\ C\varphi S\varphi_s - S\varphi C\theta_s C\varphi_s \end{Bmatrix}. \tag{5.26}$$

Where

φ is the latitude of the sun panel.

φ_s is the "sun latitude," $\varphi_s = \frac{\pi}{2} - \arccos(S\gamma\, S\alpha)$

θ_s is the sun movement simulating earth rotation (see Figure 5.7).

On any given day, determined by the value of α, the only variable in this equation is θ_s. The latitude of the panel is, of course, constant; the sun latitude is assumed constant. The next succeeding day is set by incrementing α by 360/365.25 degrees.

Panel Vector

The **x**-set has its $\mathbf{x}_1$ axis pointing straight upward along a radius of the earth, its $\mathbf{x}_2\mathbf{x}_3$ plane is tangent to the earth surface (see the y-set in Figure 5.2). The $+\mathbf{x}_2$ axis points east, $+\mathbf{x}_3$ north.

Figure 5.8 is very similar to Figure 5.7. The panel normal, $\bar{p}$, is defined in terms of its "azimuth and elevation" — the angles θ_p and φ_p, respectively. If the panel were laying on the ground the

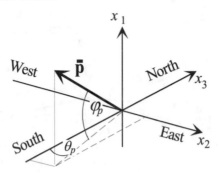

Figure 5.8: Solar Panel normal, $\bar{\mathbf{p}}$.

normal would be collinear with $\mathbf{x}_1$. Now, just move the panel vector to the desired angles θ_p and φ_p. In this diagram, the projection of the normal onto the $\mathbf{x}_2\mathbf{x}_3$ plane has the length $\cos\varphi_p$, and the $\mathbf{x}_1$ component is $\sin\varphi_p$. The panel vector, then, is:

$$\mathbf{p}_x = \{S\varphi_p, \quad S\theta_p C\varphi_p, \quad -C\varphi_p C\theta_p\}. \tag{5.27}$$

With both the sun vector and the panel vector defined, the cosine factor, C_f, is $\mathbf{p}_x \bullet \mathbf{s}_x$.

In residential applications, the two panel angles are often dictated by the roof of the building, its pitch angle and its orientation from south. In industrial applications (on a flat roof) the panel is movable and able to "track" the sun.

Appendix D contains a discussion of the use of these Equations (5.26) and (5.27) in determining "Solar energy geometric effects."

5.3.5   IMAGE ROTATION IN COMPUTER GRAPHICS

Computer graphics work has excellent use for the matrix $\mathbf{T}$ in (5.23). Consider a graphic (picture) consisting of an "assemblage of points," P_n, in a three-dimensional space. The position of each point is known in the y-set by its three coordinates. Certain of the points are to be connected on the monitor by (usually straight) lines forming the image seen by the user. The computer must "remember" not only the 3-coordinate positions of the points, but also which ones are connected. These points, together with their interconnections, may represent the (transparent) drawing of a machine part, or an entire machine.

The computer user often must be shown different views of the object being represented. So the graphics program must provide means by which the points appear to rotate about any of the three axes through the object. Of course, the display can only draw two coordinates onto the plane of the screen, but, the user must be given the perspective of three dimensions. The screen coordinates are clearly inertial (fixed). They can be chosen as any two of the 3 x-set coordinates – say x_1, and x_2. Usually, the $+ x_1$ axis is from left to right along the top of the screen, and $+ x_2$ is from top toward the bottom of the display. The non-inertial y-set is located at the centroid of the object, and probably at the center of the screen. In this case, then, the y-set is offset from the upper left corner of the screen, to its center, by the amounts h_0 (horizontal offset), and v_0 (vertical offset). These offsets are 1/2 the horizontal and vertical pixels of resolution of the screen.

When the command to rotate the object is given, the program uses equations like (5.23) to reposition the points and project them into computer screen coordinates. When the image is next displayed, the same points (in their new positions) are interconnected by lines, and the image will appear to have rotated by the given angles.

If the image is required to appear to move, dynamically, the rotations must then be taken in incremental fashion. At each increment, the image must be erased, then rotated again, and redisplayed – rapidly enough to give the impression of rotational motion at the screen. If the drawing is complicated, there will be many points, P_n. Since a vector multiplication is required for each point, plus reconnection of the points by lines, it can be seen that the computer must have a very large main memory, and be capable of high speed arithmetic ("floating point") operations. It has only been in recent years that such computers have been generally available.

Computer graphics software has become very complex. The above discussion omits all of the drawing part, the interaction with the user — virtually all of the very difficult problems. But, the transform matrix (5.23) is one of the many tools that make sophisticated graphics possible.

5.4   CONGRUENT AND SIMILARITY MATRIX TRANSFORMS

Earlier paragraphs have shown that a vector — a mathematical, and possibly physical, entity — can be viewed from different frames of reference, different coordinate sets. There is no particular significance to any given "frame," and we can easily erect a different one to afford a better perspective. This is especially true for orthogonal reference frames which retain the vector length. The vector transforms as easily as a single matrix-times-vector product.

The same can be said of a matrix, and functions of matrices. A matrix may be viewed from a given reference set, or it can be transformed, along with the vectors upon which it may be operating, to a new set affording a more convenient view. It is of interest to see how a matrix is transformed.

Consider again the vector equation $\mathbf{Ax} = \mathbf{b}$. The coordinates in which $\mathbf{A}$, $\mathbf{x}$, and $\mathbf{b}$ are described are quite arbitrary. Then, it may become necessary to transform these vectors using a (general) matrix $\mathbf{P}$. The transform need not be an orthogonal one, so consider that $\mathbf{P}$ is a nonsingular matrix whose inverse is $\mathbf{P}^{-1}$. Using $\mathbf{P}$, we obtain:

$$\mathbf{x} = \mathbf{P\bar{x}} \quad \text{and} \quad \mathbf{b} = \mathbf{P\bar{b}} \tag{5.28}$$

in which $\bar{\mathbf{x}}$ refers to the transformed vector $\mathbf{x}$, and $\bar{\mathbf{b}}$ refers to the transformed $\mathbf{b}$. Of course, our main interest is in the original matrix equation, and how $\mathbf{b}$ is obtained from $\mathbf{x}$. Upon substitution of the transform into the original equation $\mathbf{Ax} = \mathbf{b}$:

$$\mathbf{AP\bar{x}} = \mathbf{P\bar{b}} \quad \text{or} \quad \mathbf{P}^{-1}\mathbf{AP}\,\bar{\mathbf{x}} = \bar{\mathbf{b}}\,. \tag{5.29}$$

In the second Equation of (5.29) the matrix $\mathbf{A}$ is transformed to the new coordinates by combined pre- and post-multiplication. The transform of $\mathbf{A}$ is:

$$\bar{\mathbf{A}} = \mathbf{P}^{-1}\mathbf{A}\mathbf{P}. \tag{5.30}$$

The two matrices, $\mathbf{A}$ and $\bar{\mathbf{A}}$ are said to be "similar" matrices, and the transform is called a "similarity transform." Since $\mathbf{P}$ and its inverse have reciprocal determinants, then (5.30) shows that $\mathbf{A}$ and $\bar{\mathbf{A}}$ have the same determinant (i.e., $|\mathbf{A}| = |\bar{\mathbf{A}}|$).

Now it will be shown that algebraic functions of $\mathbf{A}$ are transformed in the manner of (5.30), and thus, these functions are invariant under similarity transforms.

Matrix Product

The product is transformed

$$\bar{\mathbf{A}}\bar{\mathbf{B}} \Rightarrow \mathbf{P}^{-1}\mathbf{A}\mathbf{P}\mathbf{P}^{-1}\mathbf{B}\mathbf{P} = \mathbf{P}^{-1}(\mathbf{A}\mathbf{B})\mathbf{P}.$$

Matrix Addition/Subtraction

$$\bar{\mathbf{A}} \pm \bar{\mathbf{B}} \Rightarrow \mathbf{P}^{-1}\mathbf{A}\mathbf{P} \pm \mathbf{P}^{-1}\mathbf{B}\mathbf{P} = \mathbf{P}^{-1}(\mathbf{A} \pm \mathbf{B})\mathbf{P}.$$

Matrix Inversion

Given that $\bar{\mathbf{A}} = \mathbf{P}^{-1}\mathbf{A}\mathbf{P},$ then by the inversion of a product rule: $\bar{\mathbf{A}}^{-1} = \mathbf{P}^{-1}\mathbf{A}^{-1}\mathbf{P}.$

Then, all these operations transform just as $\mathbf{A}$ itself transforms — these operations remain invariant under similarity transformation.

Matrix Transposition

This case is somewhat different.

Given $\bar{\mathbf{A}} = \mathbf{P}^{-1}\mathbf{A}\mathbf{P},$ by transposition of a product: $\bar{\mathbf{A}}' = \mathbf{P}'\mathbf{A}'[\mathbf{P}^{-1}]'.$ This is not the same as the transformation of $\mathbf{A}$ unless $\mathbf{P}$ is orthogonal. If the matrix, $\mathbf{P}$, is not orthogonal then the operation of transposition is *not* invariant under transformation.

Three out of four isn't bad. Functions of matrices which involve addition/subtraction, multiplication, and inversion, remain invariant under similarity transformation:

$$f(\mathbf{A}, \ \mathbf{B}, \ \mathbf{C}, \ , \ , \ \mathbf{A}^{-1}, \ \mathbf{B}^{-1}, \ \mathbf{C}^{-1}) \ \Leftrightarrow \ f(\bar{\mathbf{A}}, \ \bar{\mathbf{B}}, \ \bar{\mathbf{C}}, \ , \ , \ \bar{\mathbf{A}}^{-1}, \ \bar{\mathbf{B}}^{-1}, \ \bar{\mathbf{C}}^{-1}).$$

That is, a given function of matrices "implies" the same function of the same matrices, transformed to some new coordinate system by a similarity transform, as long as the function includes just those operations which passed the above test. For example, a given polynomial in $\mathbf{A}$:

$$c_0\mathbf{A}^n + c_1\mathbf{A}^{n-1} + c_2\mathbf{A}^{n-2} + \cdots + c_{n-1}\mathbf{A} + c_n\mathbf{I} = 0$$

implies the same polynomial, with the same coefficients, in the transformed matrix $\bar{\mathbf{A}}$.

If the transforming matrix is orthogonal, the transform is called "congruent," and as described earlier, the invariant functions will include transposition. Further, if $\mathbf{A} - \mathbf{A}' = \mathbf{0}$, the matrix is symmetric. Since the subtraction is invariant under congruent transformation then symmetric matrices remain symmetric under such transformation.

5.5   DIFFERENTIATION OF MATRICES, ANGULAR VELOCITY

The objectives of this section will be to define the derivative of a matrix whose elements are variable functions, and then to use this definition in the development of the angular velocity matrix. Of course, angular velocity is a vector quantity. It was shown earlier that the vector cross product can be affected by the product of a 3X3 skew-symmetric matrix elements times a 3X1 vector. In fact, this is just how the angular velocity vector emerges in this development.

Suppose the elements of the matrix $\mathbf{A}$ are functions of a scalar variable, t. Then:

$$\mathbf{A}(t) = [a_{ij}(t)] .$$

Now, if t is incremented by dt, note that each element of A is incremented – that is

$$\mathbf{A}(t + dt) = [a_{ij}(t + dt)] .$$

Then, if the original $\mathbf{A}$ matrix is subtracted, the result divided by dt, and the limit taken as dt approaches zero, we see that the overall result is

$$\frac{d}{dt}[\mathbf{A}(t)] = \left[\frac{da_{ij}(t)}{dt}\right] . \tag{5.31}$$

That is, the differentiation of $\mathbf{A}$ is accomplished by differentiating each element of $\mathbf{A}$. Now, considering the variable to be time, t, we denote the time derivative as

$$\frac{d}{dt}\mathbf{A}(t) \equiv \dot{\mathbf{A}} \equiv \mathbf{A}_t . \tag{5.32}$$

Notice the unusual notation $\mathbf{A}_t$ for the derivative of $\mathbf{A}$. We can define the following derivatives:

$$\frac{d}{dt}[\mathbf{A} + \mathbf{B}] = \dot{\mathbf{A}} + \dot{\mathbf{B}} = \mathbf{A}_t + \mathbf{B}_t \text{ and} \tag{5.33}$$

$$\frac{d}{dt}[\mathbf{AB}] = \dot{\mathbf{A}}\mathbf{B} + \dot{\mathbf{A}}\mathbf{B} = \mathbf{A}_t\mathbf{B} + \mathbf{A}\mathbf{B}_t . \tag{5.34}$$

The results (5.33) and (5.34) are just like their scalar counterparts. However, in (5.34), the original product order, $\mathbf{AB}$, must be maintained in the derivative of the product. Of course, if more than two matrices are involved in the product then

$$[\mathbf{ABC}]_t = \mathbf{A}_t[\mathbf{BC}] + \mathbf{A}[\mathbf{BC}]_t = \mathbf{A}_t\mathbf{BC} + \mathbf{AB}_t\mathbf{C} + \mathbf{ABC}_t$$

and again the order of the product is maintained.

The derivative of $\mathbf{A}^{-1}$ can be found by noting that

$$\mathbf{A}\mathbf{A}^{-1} = \mathbf{I} \quad \text{then}$$
$$\dot{\mathbf{A}}\mathbf{A}^{-1} + \mathbf{A}\dot{\mathbf{A}}^{-1} = \mathbf{0}$$
$$\dot{\mathbf{A}}^{-1} = -\mathbf{A}^{-1}\dot{\mathbf{A}}\mathbf{A}^{-1}.$$

5.5.1 VELOCITY OF A POINT ON A WHEEL

A point, p, rides on the periphery of a wheel (or disk), as shown in Figure 5.9. The axis of the wheel is attached to a shaft (in the plane of the paper) which is also capable of rotation.

As in previous examples, intermediate coordinate sets are used, with each one describing one angular displacement (and velocity) about one of its axes. In this case, the angles are θ_2 and θ_3.

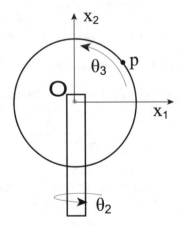

Figure 5.9:

An inertial (fixed) x-set is set up at the point "O" in the figure, with axes as shown (the $+x_3$ axis is up, out of the paper). A y-set is constructed, also at point O, which rotates about its y_2 axis (collinear with x_2). This rotation angle is denoted θ_2. Lastly, a z-set is constructed at point O, which rotates with angle θ_3 about the y_3, z_3 axes.

As observed in the z-set, the point p is fixed, with coordinates $\{r_p, 0, 0\}$; and note that the point p does remain at a constant distance from point O—equal to the radius of the disk. That is, all the motion is angular rotation.

To find the velocity of the point p, the vector $\mathbf{r}_x$ must first be found. Its time derivative is the velocity of p. To find $\mathbf{r}_x$, vector $\mathbf{r}_z$ is transformed from the z-set to the x-set. The two transforms are $\mathbf{T}_2$ and $\mathbf{T}_3$.

$$\mathbf{x} = \mathbf{T}_2\mathbf{y} \tag{5.35}$$
$$\mathbf{y} = \mathbf{T}_3\mathbf{z} \quad\text{then}\tag{5.36}$$
$$\mathbf{x} = \mathbf{T}_2\mathbf{T}_3\mathbf{z} \quad\text{and}\quad \mathbf{z} = \mathbf{T}_3'\mathbf{T}_2'\mathbf{x} \tag{5.37}$$

where

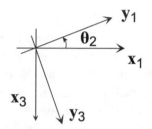

$$\mathbf{T}_2 = \begin{bmatrix} C\theta_2 & 0 & S\theta_2 \\ 0 & 1 & 0 \\ -S\theta_2 & 0 & C\theta_2 \end{bmatrix}, \tag{5.38}$$

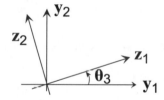

$$\mathbf{T}_3 = \begin{bmatrix} C\theta_3 & -S\theta_3 & 0 \\ S\theta_3 & C\theta_3 & 0 \\ 0 & 0 & 1 \end{bmatrix} \tag{5.39}$$

In (5.38) and (5.39), "C" is to be read as "cos," and "S" as "sin;" for example: $S\theta_2 = \sin\theta_2$.

Define the vector from the center of rotation to the point p as $\mathbf{r}$. Then $\mathbf{r}_z$ is the vector $\mathbf{r}$ as seen in the z-set, $\mathbf{r}_x$ is the same vector, seen in the inertial x-set. In order to derive the velocity of p, we must differentiate $\mathbf{r}_x$ – in the inertial x-set system that can "see" all the motion. From (5.37):

$$\mathbf{r}_x = \mathbf{T}_2\mathbf{T}_3\mathbf{r}_z \;.$$

The vector $\mathbf{r}_z$ is simply $\{r, 0, 0\}$. Then, defining $\mathbf{v}$ as the velocity of point p:

$$\mathbf{v}_x = \dot{\mathbf{r}}_x = \left[\dot{\mathbf{T}}_2\mathbf{T}_3 + \mathbf{T}_2\dot{\mathbf{T}}_3\right]\mathbf{r}_z \;. \tag{5.40}$$

In (5.40), we can eliminate $\mathbf{r}_z$ through the use of (5.37):

$$\mathbf{v}_x = \dot{\mathbf{r}}_x = \left[\dot{\mathbf{T}}_2\mathbf{T}_3 + \mathbf{T}_2\dot{\mathbf{T}}_3\right]\mathbf{T}_3'\mathbf{T}_2'\mathbf{r}_x = \left[\dot{\mathbf{T}}_2\mathbf{T}_2' + \mathbf{T}_2(\dot{\mathbf{T}}_3\mathbf{T}_3')\mathbf{T}_2'\right]\mathbf{r}_x = \mathbf{W}_x\mathbf{r}_x \;. \tag{5.41}$$

In (5.41) the two important products are:

$$\dot{\mathbf{T}}_2\mathbf{T}_2' = \dot{\theta}_2 \begin{bmatrix} -S\theta_2 & 0 & C\theta_2 \\ 0 & 0 & 0 \\ -C\theta_2 & 0 & -S\theta_2 \end{bmatrix} \begin{bmatrix} C\theta_2 & 0 & -S\theta_2 \\ 0 & 1 & 0 \\ S\theta_2 & 0 & C\theta_2 \end{bmatrix} = \begin{bmatrix} 0 & 0 & \dot{\theta}_2 \\ 0 & 0 & 0 \\ -\dot{\theta}_2 & 0 & 0 \end{bmatrix} \quad \text{and}$$

$$\dot{\mathbf{T}}_3\mathbf{T}_3' = \dot{\theta}_3 \begin{bmatrix} -S\theta_3 & -C\theta_3 & 0 \\ C\theta_3 & -S\theta_3 & 0 \\ 0 & 0 & 0 \end{bmatrix} \begin{bmatrix} C\theta_3 & S\theta_3 & 0 \\ -S\theta_3 & C\theta_3 & 0 \\ 0 & 0 & 1 \end{bmatrix} = \begin{bmatrix} 0 & -\dot{\theta}_3 & 0 \\ \dot{\theta}_3 & 0 & 0 \\ 0 & 0 & 0 \end{bmatrix}$$

where, again, S means sine (e.g., $S\theta_2 = \sin\theta_2$), and C means cosine.

It should be clear that the elements of angular velocity are emerging in the products of these $\dot{\mathbf{T}}_j\mathbf{T}_j'$ matrices. That is, if $\mathbf{T}_j$ is the transform matrix defining rotation about the jth (inertial) axis, then $\dot{\mathbf{T}}_j\mathbf{T}_j'$ provides the jth component of angular velocity. Also, in (5.41), note that the components of rotation about the 3-axis must be transformed back to the inertial x-set, while the rotation about the 2-axis is already described in the x-set, and need not be transformed. Note again that the transform af a matrix is accomplished by pre- and postmultiplying matrices. Specifically, in the $\mathbf{W}_x$ matrix, the components, $\dot{\mathbf{T}}_3\mathbf{T}_3'$, must be transformed, while those from $\dot{\mathbf{T}}_2\mathbf{T}_2'$ do not.

$$\mathbf{W}_x = \dot{\mathbf{T}}_2\mathbf{T}_2' + \mathbf{T}_2(\dot{\mathbf{T}}_3\mathbf{T}_3')\mathbf{T}_2' . \tag{5.42}$$

Angular velocity matrices which "emerge" in this way are always "skew symmetric." That is

$$\mathbf{W} = -\mathbf{W}' = \begin{bmatrix} 0 & -\omega_3 & \omega_2 \\ \omega_3 & 0 & -\omega_1 \\ -\omega_2 & \omega_1 & 0 \end{bmatrix} . \tag{5.43}$$

In the general case, with the transform $\mathbf{T} = \mathbf{T}_1\mathbf{T}_2\mathbf{T}_3$, (rotation about all three coordinate axes) the angular velocity matrix would be:

$$\mathbf{W}_x = \dot{\mathbf{T}}_1\mathbf{T}_1' + \mathbf{T}_1(\dot{\mathbf{T}}_2\mathbf{T}_2')\mathbf{T}_1' + \mathbf{T}_1\mathbf{T}_2(\dot{\mathbf{T}}_3\mathbf{T}_3')\mathbf{T}_2'\mathbf{T}_1' .$$

And, again note the transformation of the 2-axis and 3-axis angular velocity components.

In the example problem of Figure 5.9, multiplying the terms out in (5.42)

$$\mathbf{W}_x = \dot{\mathbf{T}}_2\mathbf{T}_2' + \mathbf{T}_2(\dot{\mathbf{T}}_3\mathbf{T}_3')\mathbf{T}_2' = \begin{bmatrix} 0 & -C\theta_2\dot{\theta}_3 & \dot{\theta}_2 \\ C\theta_2\dot{\theta}_3 & 0 & -S\theta_2\dot{\theta}_3 \\ -\dot{\theta}_2 & S\theta_2\dot{\theta}_3 & 0 \end{bmatrix} . \tag{5.44}$$

Therefore, the angular velocity (vector quantity) for the problem is:

$$\boldsymbol{\omega}_x = \begin{Bmatrix} \omega_1 \\ \omega_2 \\ \omega_3 \end{Bmatrix} = \begin{Bmatrix} \dot{\theta}_3 \sin\theta_2 \\ \dot{\theta}_2 \\ \dot{\theta}_3 \cos\theta_2 \end{Bmatrix} . \tag{5.45}$$

In hindsight, the angular velocity, $\boldsymbol{\omega}$, could be calculated in vector form. The z-set "sees" no rotation, the y-set "sees" the vector $\boldsymbol{\omega}_3 = \{0, 0, \dot{\theta}_3\}$ about its 3-axis, and the x-set "sees" $\varpi_2 = \{0, \dot{\theta}_2, 0\}$. Then instead of transforming matrices, the simpler vector would do:

$$\boldsymbol{\omega}_x = \left\{ \begin{array}{c} 0 \\ \dot{\theta}_2 \\ 0 \end{array} \right\} + \mathbf{T}_2 \left\{ \begin{array}{c} 0 \\ 0 \\ \dot{\theta}_3 \end{array} \right\} = \left\{ \begin{array}{c} 0 \\ \dot{\theta}_2 \\ 0 \end{array} \right\} + \left[\begin{array}{ccc} C\theta_2 & 0 & S\theta_2 \\ 0 & 1 & 0 \\ -S\theta_2 & 0 & C\theta_2 \end{array} \right] \left\{ \begin{array}{c} 0 \\ 0 \\ \dot{\theta}_3 \end{array} \right\} \tag{5.46}$$

which clearly has the same result. In the general case, with the transform $\mathbf{T} = \mathbf{T}_1\mathbf{T}_2\mathbf{T}_3$, the angular velocity vector would be:

$$\boldsymbol{\omega} = \boldsymbol{\omega}_1 + \mathbf{T}_1\boldsymbol{\omega}_2 + \mathbf{T}_1\mathbf{T}_2\boldsymbol{\omega}_3 . \tag{5.47}$$

The terms $\boldsymbol{\omega}_j$ are vectors with non-zero element values only at the jth element. The angular velocity matrix, $\mathbf{W}$, can then be written by simply putting the elements from (5.47) into their proper places in a skew-symmetric matrix.

Returning then to (5.41), the velocity of the point p is

$$\mathbf{v}_x = \mathbf{W}\mathbf{r}_x = (\boldsymbol{\omega}_x \times \mathbf{r}_x) . \tag{5.48}$$

The velocity of $\mathbf{p}$ is equal to the cross product of the total angular velocity times the vector, $\mathbf{r}$, both expressed in the inertial x-set. This result is certainly obvious. But, the importance of the development is the introduction of angular velocity as a skew symmetric matrix quantity that emerges in the form $\dot{\mathbf{T}}\mathbf{T}'$. Furthermore, the development leaves no uncertainty as to the correct vector quantities to be cross multiplied; and for this reason it is more than just an introduction. The next section will continue with the same matrix and vector quantities.

5.6   DYNAMICS OF A PARTICLE

In the study of classical mechanics, the velocity and acceleration of a particle in motion are developed as vector entities. The development is troublesome because part of the motion is described in a moving coordinate system. In the classic vector development some of the terms in these equations mysteriously appear as "correction terms." Using the insight gained through matrix manipulation, and specifically, the angular velocity matrix, we will develop the equations directly, and watch the "correction terms" as they appear.

In Figure 5.10, the position of the point p is determined by the vector $\mathbf{r}$ in a non-inertial y-set. The position of the y-set is determined by the vector $\mathbf{R}$ and angular motion between the coordinate sets is measured in the transformation matrix, $\mathbf{T}$. We will determine the absolute velocity and acceleration of the point as vector equations, and identify, in a matrix sense, each of the terms.

From the figure:

$$\boldsymbol{\rho}x = \mathbf{R}_x + \mathbf{r}_x \tag{5.49}$$

and the subscripts, x, are the reminder that to derive a true ("absolute") velocity we must differentiate in the x-set. The transform between coordinate sets is $\mathbf{x} = \mathbf{T}\mathbf{y}$. Specifically, note that $\mathbf{r}$ is known in

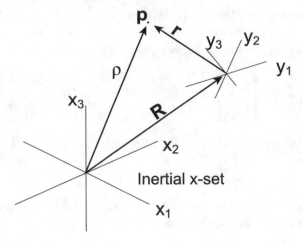

Figure 5.10: Particle Dynamics.

the y-set and must be transformed as $\mathbf{r}_x = \mathbf{T}\mathbf{r}_y$ in (5.49). Then, by differentiation:

$$\dot{\rho}_x = \dot{\mathbf{R}}_x + \mathbf{T}\dot{\mathbf{r}}_y + \dot{\mathbf{T}}\mathbf{r}_y$$

$$\dot{\rho}_x = \dot{\mathbf{R}}_x + \mathbf{T}\dot{\mathbf{r}}_y + \dot{\mathbf{T}}\mathbf{T}'\mathbf{r}_x = \dot{\mathbf{R}}_x + \mathbf{T}\dot{\mathbf{r}}_y + (\boldsymbol{\omega}_x \times \mathbf{r}_x)\,. \tag{5.50}$$

$$\dot{\rho}_x = \dot{\mathbf{R}}_x + \{\dot{\mathbf{r}}_y\}_x + (\boldsymbol{\omega} \times \mathbf{r}_x) \tag{5.51}$$

which is the absolute velocity of the point p, expressed in the x-set. The $(\boldsymbol{\varpi} \times \mathbf{r})$ term emerges the same way that it did in the last section: $\dot{\mathbf{T}}\mathbf{T}'$ is $\mathbf{W}$, and note that $\mathbf{W}$ is $\mathbf{W}_x$. The notation $\{\dot{\mathbf{r}}_y\}_x$ is used to emphasize that the derivative of $\mathbf{r}_y$ is taken, and the results then transformed to the x-set (not the derivative of $\mathbf{r}_x$). The derivative of $\mathbf{r}_y$ is usually referred to as the "apparent velocity." It is the velocity that would be measured without any knowledge that the y-set is not inertial. Note that this matrix development is straightforward and leaves no question as to which coordinate set the vectors are to be defined in.

Often, it is desired to express the absolute velocity in the non-inertial set. This can be done by simply transforming $\dot{\rho}_x$ to the y-set. However, it is interesting to start back at the first of (5.50), and to consider the transform of the angular velocity:

$$\mathbf{W}_y = \mathbf{T}'\mathbf{W}_x\mathbf{T} \tag{5.52}$$

$$\text{but } \mathbf{W}_x = \dot{\mathbf{T}}\mathbf{T}'$$

$$\text{then } \mathbf{W}_y = \mathbf{T}'\dot{\mathbf{T}}\mathbf{T}'\mathbf{T} = \mathbf{T}'\dot{\mathbf{T}}\,. \tag{5.53}$$

Now, returning to (5.50)

$$\{\dot{\rho}_x\}_y = \mathbf{T}'\dot{\mathbf{R}}_x + \dot{\mathbf{r}}_y + \mathbf{T}'\dot{\mathbf{T}}\mathbf{r}_y$$

$$\{\dot{\rho}_x\}_y = \{\dot{\mathbf{R}}_x\}_y + \dot{\mathbf{r}}_y + \mathbf{W}_y\mathbf{r}_y \tag{5.54}$$

$$\{\dot{\rho}_x\}_y = \{\dot{\mathbf{R}}_x\}_y + \dot{\mathbf{r}}_y + (\boldsymbol{\omega}_y \times \mathbf{r}_y)$$

which is the absolute velocity of p, expressed in the y-set. Note the similarity to (5.51).

Now, for the acceleration, we must differentiate the first (5.50) equation:

$$\ddot{\rho}_x = \ddot{\mathbf{R}}_x + \mathbf{T}\ddot{\mathbf{r}}_y + \dot{\mathbf{T}}\dot{\mathbf{r}}_y + \dot{\mathbf{T}}\dot{r}_y + \ddot{\mathbf{T}}\mathbf{r}_y$$
$$\ddot{\rho}_x = \ddot{\mathbf{R}}_x + \mathbf{T}\ddot{r}_y + 2\dot{\mathbf{T}}\mathbf{T}'\{\dot{\mathbf{r}}_y\}_x + \ddot{\mathbf{T}}\mathbf{T}'\mathbf{r}_x \qquad (5.55)$$
$$\ddot{\rho}_x = \ddot{\mathbf{R}}_x + \{\ddot{\mathbf{r}}_y\}_x + 2[\mathbf{W}_x]\{\dot{\mathbf{r}}_y\}_x + \ddot{\mathbf{T}}\mathbf{T}'\mathbf{r}_x \ .$$

In (5.55), note the new correction term consisting of the angular velocity crossed into the "apparent velocity" transformed to the x-set. But, to interpret further, consider the definition $\dot{\mathbf{T}}\mathbf{T}' = \mathbf{W}_x$. By differentiation:

$$\frac{d}{dt}[\dot{\mathbf{T}}\mathbf{T}'] = \ddot{\mathbf{T}}\mathbf{T}' + \dot{\mathbf{T}}\dot{\mathbf{T}}' = \dot{\mathbf{W}}_x; \quad \text{then} \quad \ddot{\mathbf{T}}\mathbf{T}' = \dot{\mathbf{W}}_x - \dot{\mathbf{T}}\dot{\mathbf{T}}' \ .$$

Since the angular velocity matrix is skew symmetric:

$$\dot{\mathbf{T}}\mathbf{T}' = \mathbf{W}_x = -\mathbf{W}'_x = -\mathbf{T}\dot{\mathbf{T}}'$$
$$\text{then} \quad \mathbf{W}_x^2 = [\dot{\mathbf{T}}\mathbf{T}'][-\mathbf{T}\dot{\mathbf{T}}'] = -\dot{\mathbf{T}}\dot{\mathbf{T}}'$$
$$\text{finally} \quad \ddot{\mathbf{T}}\mathbf{T}' = \dot{\mathbf{W}}_x + \mathbf{W}_x^2 \ .$$

Then, taking this back to (5.55):

$$\ddot{\rho}_x = \ddot{\mathbf{R}}_x + \{\ddot{\mathbf{r}}_y\}_x + 2[\mathbf{W}_x]\{\dot{\mathbf{r}}_y\}_x + \ddot{\mathbf{T}}\mathbf{T}'\mathbf{r}_x$$
$$\ddot{\rho}_x = \ddot{\mathbf{R}}_x + \{\ddot{\mathbf{r}}_y\}_x + 2[\mathbf{W}_x]\{\dot{\mathbf{r}}_y\}_x + [\dot{\mathbf{W}}_x + \mathbf{W}_x^2]\mathbf{r}_x \qquad (5.56)$$
$$\ddot{\boldsymbol{\rho}}_x = \ddot{\mathbf{R}}_x + \{\ddot{\mathbf{r}}_y\}_x + 2(\boldsymbol{\omega}_x \times \{\dot{\mathbf{r}}_y\}_x) + (\dot{\boldsymbol{\omega}}_x \times \mathbf{r}_x) + \boldsymbol{\omega}_x \times (\boldsymbol{\omega}_x \times \mathbf{r}_x) \qquad (5.57)$$

both (5.56) and (5.57) show the final result for the absolute acceleration. The vector form, (5.57), is the form most often seen. Note that $\mathbf{W}^2\mathbf{r} = \mathbf{W}\mathbf{W}\mathbf{r}$ is simply the cross product of a cross product, shown as the final term of (5.57). The absolute acceleration, then, has three cross product "correction terms." This also shows that when parts of the total motion of the particle are described in a non-inertial coordinate set, the equations of motion can become somewhat complicated.

This absolute acceleration is transformed to the y-set in the same way that velocity is transformed. This time, however, we must transform $\mathbf{W}^2$.

$$\mathbf{T}'\mathbf{W}_x^2\mathbf{T} = [\mathbf{T}'\mathbf{W}_x\mathbf{T}][\mathbf{T}'\mathbf{W}_x\mathbf{T}] = \mathbf{W}_y^2 \qquad (5.58)$$

which shows that the square of $\mathbf{W}_x$ (or, in fact, any integer power of $\mathbf{W}_x$) transforms just like $\mathbf{W}_x$. Then, basically we must transform the equation:

$$\ddot{\rho}_x = \ddot{\mathbf{R}}_x + \mathbf{T}\ddot{\mathbf{r}}_y + 2[\mathbf{W}_x]\{\dot{\mathbf{r}}_y\}_x + [\dot{\mathbf{W}}_x + \mathbf{W}_x^2]\mathbf{r}_x \ .$$

The transformation is accomplished by premultiplying the above equation by $\mathbf{T}'$. Then:

$$\{\ddot{\rho}_x\}_y = \{\ddot{\mathbf{R}}_x\}_y + \ddot{\mathbf{r}}_y + 2\mathbf{T}'\mathbf{W}_x\mathbf{T}\dot{\mathbf{r}}_y + \mathbf{T}'[\dot{\mathbf{W}}_x + \mathbf{W}_x^2]\mathbf{T}\mathbf{r}_y$$
$$\{\ddot{\rho}_x\}_y = \{\ddot{\mathbf{R}}_x\}_y + \ddot{\mathbf{r}}_y + 2[\mathbf{W}_y]\{\dot{\mathbf{r}}_y\} + [\dot{\mathbf{W}}_y + \mathbf{W}_y^2]\mathbf{r}_y \qquad (5.59)$$
$$\{\ddot{\boldsymbol{\rho}}_x\}_y = \{\ddot{\mathbf{R}}_x\}_y + \ddot{\mathbf{r}}_y + 2(\boldsymbol{\omega}_y \times \dot{\mathbf{r}}_y) + (\dot{\boldsymbol{\omega}}_y \times \mathbf{r}_y) + \boldsymbol{\omega}_y \times (\boldsymbol{\omega}_y \times \mathbf{r}_y) \ . \qquad (5.60)$$

$\ddot{\rho}$	The absolute acceleration of the particle, as found in an inertial coordinate system, although the quantity can be expressed in (transformed to) any set.
$\ddot{\mathbf{R}}$	The absolute acceleration of the origin of the non-inertial set relative to the inertial set.
$\ddot{\mathbf{r}}_y$	The apparent acceleration of the point p, as measured in the non-inertial coordinate set.
$2(\boldsymbol{\omega} \times \dot{\mathbf{r}})$	The compound acceleration of Coriolis; a correction term that must be applied whenever there is angular motion and apparent velocity, simultaneously.
$(\dot{\omega} \times \mathbf{r})$	The correction term that relates the acceleration of the point to a change in the angular velocity.
$\omega \times (\omega \times \mathbf{r})$	The well known centripetal acceleration, in the amount of $\mathbf{W}^2$ times the radius, $\mathbf{r}$.

And, note again the similarity of these Equations to (5.56) and (5.57). Because of this similarity, they can be discussed in general terms – being specific about the coordinate set only when it is important (for example, in the discussion of apparent acceleration).

5.7   RIGID BODY DYNAMICS

The analysis of rigid body dynamics follows from that of a single particle in that the body is perceived as an aggregate of particles. The dynamics of one chosen particle is examined, and then a summation is made to include all such particles.

In the diagram, a rigid body is indicated by the wavy, closed, line. A chosen, ith, particle is located in an inertial set by the vector ρ. The center, O′ of a non-inertial z-set is located by the vector $\mathbf{d}$. Within the z-set, the vector $\mathbf{r}_i$ locates the particle. As in the previous section, both sets are

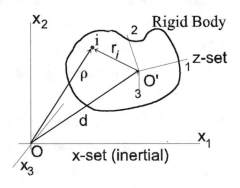

Figure 5.11:

required because the z-set is often the one in which the particle is observed, but the x-set is required in which to do the differentiation necessary in the use of Newton's second law. Since the ith is just one of the particles, the vectors $\mathbf{r}$ and ρ must be given subscripts:

$$\rho_i = \mathbf{d} + \mathbf{r}_i \ . \tag{5.61}$$

Using Newton's second law for the ith particle:

$$\mathbf{F}_i = m_i \ddot{\rho}_i \ . \tag{5.62}$$

Where

$\mathbf{F}_i$ is the total force applied to the particle
m_i is the mass of the particle, and
$\ddot{\rho}_i$ is its absolute acceleration.

The force on the particle is the result of both internal forces, $\mathbf{f}_{ij}$, from the adjoining particles (two subscripts), and the external applied force, $\mathbf{f}_i$ (one subscript). Then $\mathbf{F}_i = \mathbf{f}_{ij} + \mathbf{f}_i$.

Making the substitutions for $\mathbf{F}_i$ and for $\ddot{\rho}_i$, the equation of equilibrium is obtained by summing over all the particles in the system (rigid body):

$$\sum_i \mathbf{f}_{ij} + \sum_i \mathbf{f}_i = \sum_i m_i (\ddot{\mathbf{d}} + \ddot{\mathbf{r}}_i) \ . \tag{5.63}$$

Since the particles do not move relative to one another within the rigid body, each $\mathbf{f}_{ij}$ must be accompanied by an equal but opposite force, $\mathbf{f}_{ji}$. Then the sum of forces $\mathbf{f}_{ij}$ must be zero. The sum of $\mathbf{f}_i$ is simply the external force vector, $\mathbf{f}$ on the body. Also, in (5.63), the vector $\mathbf{d}$ is independent of which particle is chosen, so the sum is just that of the particle masses:

$$\mathbf{f} = m\ddot{\mathbf{d}} + \sum_i m_i \ddot{\mathbf{r}}_i \ \text{ where } m \text{ is the total mass of the rigid body.} \tag{5.64}$$

An important simplification results if the mass points are located in relation to the center of gravity of the body. This is accomplished in the figure by redefining $\mathbf{r}_i$:

$$\mathbf{r}_i = \mathbf{r}_c + (\mathbf{r}_i)_c \ . \tag{5.65}$$

In (5.65), $\mathbf{r}_c$ is a fixed vector from the origin O′ to the center of gravity, cg. The new vector $(\mathbf{r}_i)_c$ emanates from the cg and terminates at the ith mass point. Then

$$\sum_i m_i \ddot{\mathbf{r}}_i = \sum_i m_i \ddot{\mathbf{r}}_c + \sum_i m_i (\ddot{\mathbf{r}}_i)_c \ . \tag{5.66}$$

By definition of the center of gravity, the last term in (5.66) is zero. Thus,

$$\mathbf{f} = m\ddot{\mathbf{d}} + m\ddot{\mathbf{r}}_c = m(\ddot{\mathbf{d}} + \ddot{\mathbf{r}}_c) \ . \tag{5.67}$$

The motion of translation can be determined by treating the rigid body as a single particle, with the total mass located at its center of gravity.

5.7.1 ROTATION OF A RIGID BODY

The analysis of the rotation of the rigid body is more complex than that of its translational motion. However, it is also based on Newton's Laws. It will be shown that an external "torque" produces a change in "angular momentum" in the same way that the external force produces a change in linear momentum.

First, two important assertions are discussed, that will provide "physical picture" of a rigid body rotating with angular velocity $\boldsymbol{\omega}$.

1. All lines within the body rotate at the angular velocity equal to ω. This first point is intuitive, since (in the figure), it is clear that rotations of OA and OB are both equal to ω.

2. The complete angular velocity, ω, of the rigid body, can be visualized as occurring about any arbitrarily chosen point. This assertion is not obvious.

The figure below shows an arbitrarily chosen line, AB within the rigid body. At the instant shown, the center of rotation is at the point O. These three points form the triangle OAB. The velocities of A and B are

$$V_A = a\omega; \quad V_B = b\omega.$$

The component of velocity along the line AB must be the same for both points A and B (because the body is rigid).

Then $V_A \sin\alpha = V_B \sin\beta$, or $a\omega_a \sin\alpha = b\omega_b \sin\beta$.

But using the law of sines: $\dfrac{a}{\sin\beta} = \dfrac{b}{\sin\alpha}$, note that

$$a \sin\alpha = b \sin\beta .$$

Then $\omega_A = \omega_B$. That is, the arbitrarily chosen line, AB rotates at the angular velocity, ω: ***All lines within the rigid body rotate at the same angular velocity,*** Assertion (1).

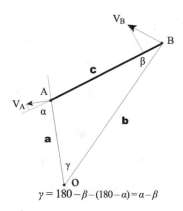

$$\gamma = 180 - \beta - (180 - \alpha) = \alpha - \beta$$

Now, consider the point A as the "apparent center of rotation," and the velocity of B about A. The velocity of B relative to A equals $c\omega_{AB} = V_B \cos\beta - V_A \cos\alpha = b\omega\cos\beta - a\omega\cos\alpha$.

Again from the law of sines, $c = a\dfrac{\sin\gamma}{\sin\beta}$, and $b = a\dfrac{\sin\alpha}{\sin\beta}$. Then

$$a\omega_{BA}\frac{\sin\gamma}{\sin\beta} = a\omega\frac{\sin\alpha}{\sin\beta}\cos\beta - a\omega\cos\alpha; \text{ and note that } \gamma = \alpha - \beta.$$

$$a\omega_{AB}\sin(\alpha - \beta) = a\omega(\sin\alpha\cos\beta - \cos\alpha\sin\beta) = a\omega\sin(\alpha - \beta).$$

Then $\omega_{AB} = \omega$. Thus, (Assertion 2), point B rotates about A with the total angular velocity of the rigid body.

The total rotational motion of a rigid body can be considered to occur about any convenient point. The complete motion of the body is then the sum of the translational motion of this point plus the rotation around it. In the usual case, if the point chosen is the cg, the equations of both translation and rotation are simplified.

5.7.2 MOMENT OF MOMENTUM

In Figure 5.11, the momentum of the ith particle is given by $m_i\mathbf{v}_i$. Newton's second law states that the time rate of change of this momentum is equal to the net force acting upon it. Equation (5.62) is rewritten:

$$\mathbf{F}_i = \frac{d}{dt}(m_i\mathbf{v}_i). \tag{5.68}$$

The momentum, $m_i\mathbf{v}_i$, of the ith particle (Figure 5.11) produces a moment about the point O', defined as its ***moment of momentum.***" Its value is determined as the cross product:

$$\text{moment of momentum} \equiv \mathbf{h}_i = \mathbf{r}_i \times m_i\mathbf{v}_i. \tag{5.69}$$

The "*Angular momentum*" (or "moment of momentum") of the rigid body is the sum of the moments of all the particles within the body

$$\mathbf{h} = \sum_i m_i\mathbf{r}_i \times \mathbf{v}_i.$$

Note that the moment of momentum/angular momentum, **h,** is a vector quantity. The velocity is $\mathbf{v}_i = \dot{\rho}_i = \dot{\mathbf{d}} + \dot{\mathbf{r}}_i$. Then

$$\mathbf{h} = \sum_i m_i\mathbf{r}_i \times \dot{\mathbf{d}} + \sum_i m_i\mathbf{r}_i \times \dot{\mathbf{r}}_i \tag{5.70}$$

and this is the general expression for angular momentum, in terms of the inertial coordinate set. The first term in (5.70) will vanish if:

1. The origin of the inertial set is at the cg of the rigid body, $\sum_i m_i\mathbf{r}_i = 0$; or

2. The origin of the non-inertial set is fixed, $\dot{\mathbf{d}} = \mathbf{0}$.

In either of these cases, this term vanishes. Further, since the motion is rotational, $\dot{\mathbf{r}}_i = \boldsymbol{\omega} \times \mathbf{r}_i$

$$\mathbf{h} = \sum_i m_i \mathbf{r}_i \times \boldsymbol{\omega} \times \mathbf{r}_i \ . \tag{5.71}$$

By expressing (5.71) in matrix terms (and noting that $\boldsymbol{\omega} \times \mathbf{r} = -\mathbf{r} \times \boldsymbol{\omega}$):

$$\mathbf{h} = \sum_i m_i \mathbf{R}_i \mathbf{W} \mathbf{r}_i = -\sum_i m_i \mathbf{R}_i^2 \boldsymbol{\omega} \tag{5.72}$$

in which $\mathbf{R}_i$ is the skew-symmetric matrix of the $\mathbf{r}_i$ coordinates, and $\mathbf{W}$ is the skew-symmetric matrix of the ω coordinates.

In (5.68), note that the cross of $\mathbf{r}_i$ into $\mathbf{F}_i$ produces a torque, $\mathbf{t} = \mathbf{r}_i \times \mathbf{F}_i = \mathbf{r}_i \times (\mathbf{f}_{ij} + \mathbf{f}_i)$. For the same reason that the internal forces cancel when summed over all particles, their contribution to torque also cancels. The result is that the torque is simply the moment of the external forces applied to the rigid body. Then the torque required to produce a change in the angular momentum of a rigid body is

$$\mathbf{t} = \mathbf{r} \times \mathbf{f} = \frac{d}{dt}\mathbf{h}, \text{ see footnote}^1$$

$$\mathbf{t} = \frac{d}{dt}\left[\sum_i m_i \mathbf{R}_i \mathbf{W} \mathbf{r}_i\right] = -\frac{d}{dt}\left[\sum_i m_i \mathbf{R}_i^2 \boldsymbol{\omega}\right] \ . \tag{5.73}$$

Although (5.73) correctly expresses the torque in terms of angular momentum, it is not in a form that is useful.

5.7.3   THE INERTIA MATRIX

The problem in (5.73) is with $\mathbf{R}_i$, the skew symmetric matrix formed from the coordinates of the vector, $\mathbf{r}_i$.

$$\mathbf{r}_i = \{r_1, \ r_2, \ r_3\}_i \ \Rightarrow \ \mathbf{R}_i = \begin{bmatrix} 0 & -r_3 & r_2 \\ r_3 & 0 & -r_1 \\ -r_2 & r_1 & 0 \end{bmatrix}_{(i)} . \tag{5.74}$$

The subscript, i, has been omitted from the terms within $\mathbf{R}_i$, but it must be remembered that there is a different $\mathbf{R}_i$ for each particle. *The problem, however, is that the r_i components vary as the rigid body turns relative to the inertial axes*. This can be remedied by expressing these terms in the non-inertial set — at the expense of a somewhat more complicated angular velocity, whose direction and magnitude may change with the motion. It will be worth it.

The transform between the inertial x set and the moving z set is the orthogonal matrix, $\mathbf{T}$. For the vectors involved we write $\mathbf{v}_x = \mathbf{T}\mathbf{v}_z$, and note that the Matrices, (R and W) transform as

---

[1]Although the symbol, $\mathbf{t}$, is used to denote torque, there should be no confusion with "T," which is used to define a 3X3 transform matrix. The elements of $\mathbf{T}$ will not be shown in bold type.

T′MT. Then, the angular momentum is:

$$\mathbf{Th}_z = \sum_i m_i (\mathbf{R}_i \mathbf{W})_x \mathbf{T}(\mathbf{r}_i)_z = -\sum_i m_i (\mathbf{R}_i^2)_x \mathbf{T}\boldsymbol{\omega}_z$$

$$\mathbf{h}_z = \sum_i m_i (\mathbf{R}_i \mathbf{W})_z (\mathbf{r}_i)_z = -\sum_i m_i (\mathbf{R}_i^2)_z \boldsymbol{\omega}_z \ .$$

(5.75)

Now the components of each particle are constant since the non inertial set is fixed in the body; and the term $-\sum_i m_i \mathbf{R}_i^2$ is a physical characteristic of the body itself. It is defined as the "**inertia matrix**" of the body. Since the symbol "**I**" denotes the unit matrix, the inertia matrix will be assigned the letter "**J**."

$$\mathbf{J} = -\sum_i \mathbf{m}_i \mathbf{R}_i^2 = \sum_i \mathbf{m}_i \begin{bmatrix} r_2^2 + r_3^2 & -r_1 r_2 & -r_1 r_3 \\ -r_2 r_1 & r_1^2 + r_3^2 & -r_2 r_3 \\ -r_3 r_1 & -r_3 r_2 & r_1^2 + r_2^2 \end{bmatrix} .$$

(5.76)

Note that $\mathbf{J}$ is necessarily defined in the z set, with the z axes fixed in the rigid body. Its terms arise because of a moment arm between the velocity of the particle and a given axis. The main diagonal terms are called "**moments of inertia**." In these terms the moment arm is the same as the radius of the velocity vector, giving rise to squared "r" factors. The off-diagonal terms are called " **products of inertia**," in which the moment arm is different than the radius of the velocity vector.

Physical Picture of the Inertia Matrix
The diagram shows a single mass point, **m**, rotating about the x_1 axis. The mass is located within the non-inertial set by $\mathbf{r} = \{r_1, r_2, r_3\}$. Its velocity is $\mathbf{v} = \{0, -v\cos\theta, v\sin\theta\}$.

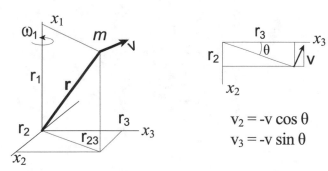

Mass m rotating about x_1 axis

Since the velocity vector is given by $-R\omega_1$:

$$\boldsymbol{\omega}_1 \times \mathbf{r} = -\mathbf{R}\boldsymbol{\omega}_1 = \begin{bmatrix} 0 & r_3 & -r_2 \\ -r_3 & 0 & r_1 \\ r_2 & -r_1 & 0 \end{bmatrix} \begin{Bmatrix} \omega_1 \\ 0 \\ 0 \end{Bmatrix} = \begin{Bmatrix} \mathbf{v}_1 = 0 \\ v_2 \\ v_3 \end{Bmatrix} = \begin{Bmatrix} 0 \\ -r_3\omega_1 \\ r_2\omega_1 \end{Bmatrix}$$

$$\text{Angular momentum is} \quad \mathbf{h} = \mathbf{r} \times (\boldsymbol{\omega} \times \mathbf{r}) = -\mathbf{R}^2\boldsymbol{\omega} = \left\{ \begin{array}{c} \omega_1\left(r_2^2 + r_3^2\right) \\ -\omega_1 r_1 r_2 \\ -\omega_1 r_1 r_3 \end{array} \right\}$$

$$\omega_1(r_2^2 + r_3^2) \quad \rightarrow \quad \text{velocity component} \ = \ \omega_1\sqrt{r_2^2 + r_3^2}, \quad \text{moment arm} = \sqrt{r_2^2 + r_3^2} \ ;$$
$$-\omega_1 r_1 r_2 \quad \rightarrow \quad \text{velocity component} \ = \ \omega_1 r_2, \qquad\quad \text{moment arm} = r_1 \ ;$$
$$-\omega_1 r_1 r_3 \quad \rightarrow \quad \text{velocity component} \ = \ \omega_1 r_3, \qquad\quad \text{moment arm} = \ r_1 \ .$$

Note that r_j components vary as the point mass rotates. For this reason, a non-inertial set whose axes perform the rotation(s) is always used.

Every particle contributes to the inertia matrix. As the particles are summed, each brings both moments of inertia. The products of inertia might cancel, while the moments of inertia can only add, being inherently positive.

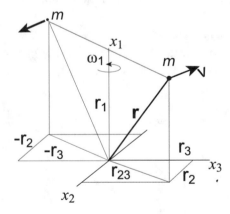

In the diagram above, the two mass points are arranged symmetrically, and the product terms cancel—the r_2 and r_3 coordinates are equal and opposite in sign. To achieve this result in the rigid body, *the non-inertial set is set along the axes of symmetry*.

Inertia Matrix of the Rigid Body
In the limit, the particles become infinitesimal, but infinite in number, and the summations become integrals in (5.77) producing the inertia matrix of (5.78).

In the following, the vector $\mathbf{r}_i = \{r_1, r_2, r_3\}$ is expressed in the z-set $(\mathbf{z}_1, \mathbf{z}_2, \mathbf{z}_3)$. The elemental mass points, $\boldsymbol{dm}$, are equal to the mass density, σ, times an elemental volume, $\boldsymbol{dV}$:

$$I_{11} = \sum_i m_i(r_2^2 + r_3^2) \Rightarrow \int_V \sigma(z_2^2 + z_3^2)\boldsymbol{dV}$$

$$I_{22} = \sum_i m_i(r_1^2 + r_3^2) \Rightarrow \int_V \sigma(z_1^2 + z_3^2)\boldsymbol{dV}$$

$$I_{33} = \sum_i m_i(r_1^2 + r_2^2) \Rightarrow \int_V \sigma(z_1^2 + z_2^2)\boldsymbol{dV}$$

$$I_{12} = \sum_i m_i r_1 r_2 \Rightarrow \int_V \sigma z_1 z_2 \boldsymbol{dV} \qquad (5.77)$$

$$I_{13} = \sum_i m_i r_1 r_3 \Rightarrow \int_V \sigma z_1 z_3 \boldsymbol{dV}$$

$$I_{23} = \sum_i m_i r_2 r_3 \Rightarrow \int_V \sigma z_2 z_3 \boldsymbol{dV}.$$

And the inertia matrix is written:

$$\mathbf{J} = \begin{bmatrix} I_{11} & -I_{12} & -I_{13} \\ -I_{21} & I_{22} & -I_{23} \\ -I_{31} & -I_{32} & I_{33} \end{bmatrix}. \qquad (5.78)$$

The elements of $\mathbf{J}$ are given in upper case—against the rules of this work. But, it is simply too common for the inertia terms to be named this way. The rules must bend, and there is no confusion with the unit matrix.

As mentioned above, it is advantageous to set the non-inertial axes along axes of symmetry to get rid of the off-diagonal terms in (5.78). This is usually done visually, but the matrix in Equation (5.78) can **always** be reduced to diagonal form by the eigenvalue methods discussed in Chapter 6. Thus, every rigid body has axes of symmetry. However, in practice it is rarely worth the effort to diagonalize $\mathbf{J}$.

5.7.4 THE TORQUE EQUATION

The torque required is given as the time rate of change of angular momentum, $\mathbf{t}_x = \dfrac{d}{dt}\mathbf{h}_x$. $\mathbf{t}_x$ can be expressed in terms of $\mathbf{t}_z$, by the equation $\mathbf{t}_x = \mathbf{T}\mathbf{t}_z$. Then

$$\mathbf{t}_x = \frac{d}{dt}\mathbf{T}\mathbf{h}_z = \dot{\mathbf{T}}\mathbf{h}_z + \mathbf{T}\dot{\mathbf{h}}_z. \qquad (5.79)$$

Transforming $\mathbf{t}_x$ to the z-set

$$\mathbf{t}_z = \mathbf{T}'\mathbf{t}_x = \dot{\mathbf{T}}'\mathbf{T}\mathbf{h}_z + \dot{\mathbf{h}}_z. \qquad (5.80)$$

The matrix $\dot{\mathbf{T}}\mathbf{T}'$ has previously been defined as $\mathbf{W}_x$. If this matrix is transformed to the z set, the result is $\mathbf{T}''\dot{\mathbf{T}} = \mathbf{W}_z$. Then, finally:

$$t_z = \mathbf{W}_z \mathbf{h}_z + \dot{\mathbf{h}}_z \; ;$$
$$t_z = \mathbf{W}_z \mathbf{J}\omega_z + \mathbf{J}\dot{\omega}_z \; . \tag{5.81}$$

Equations (5.81) have been developed directly from (5.71). Then, they assume that the center of the z-set is either at the center of gravity of the body, or that the center is at a stationary point (actually the point is only required to be non accelerating).

5.8   EXAMPLES

The following simple example illustrates the concepts of momentum and torque.

Two small "mass points" are attached to a weightless rod of length $2a$. The rod is tilted at an angle θ from the horizontal, and rotates with an angular velocity, ω, about a vertical axis at its center, marked O in the diagram. Is a torque required, and if so, what is its magnitude?

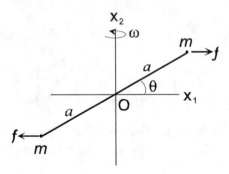

As soon as the centrifugal forces, f, are added to the diagram, it is clear that an external balancing torque is required. Each force is in the amount of $m\omega^2 a\cos\theta$. These forces produce a (total) moment about the negative x_3 axis of $2m\omega^2 a^2 \cos\theta \sin\theta$. To maintain the motion a torque of this same amount is required, about the *positive* x_3 axis.

This torque can be found by using the equation, $\mathbf{t} = \mathbf{WJ}\varpi$ where

$$\varpi = \{0, \; \omega, \; 0\}; \; \mathbf{R} = \begin{bmatrix} 0 & 0 & r_2 \\ 0 & 0 & -r_1 \\ -r_2 & r_1 & 0 \end{bmatrix}; \; \mathbf{R}^2 = 2\begin{bmatrix} -r_2^2 & r_1 r_2 & 0 \\ r_1 r_2 & -r_1^2 & 0 \\ 0 & 0 & -(r_1^2 + r_2^2) \end{bmatrix}.$$

The $\mathbf{R}$ matrix for each mass point is the negative of the other; but, the $\mathbf{R}^2$ matrices are identical, and add — which accounts for the "2," above. Note the product terms.

$$\mathbf{J}\varpi = -2m\mathbf{R}^2\varpi = 2m\begin{bmatrix} r_2^2 & -r_1 r_2 & 0 \\ -r_1 r_2 & r_1^2 & 0 \\ 0 & 0 & r_1^2 + r_2^2 \end{bmatrix}\begin{Bmatrix} 0 \\ \omega \\ 0 \end{Bmatrix} = 2m\omega\begin{Bmatrix} -r_1 r_2 \\ r_1^2 \\ 0 \end{Bmatrix}.$$

Then $\mathbf{t} = 2m\omega \begin{bmatrix} 0 & 0 & \omega \\ 0 & 0 & 0 \\ -\omega & 0 & 0 \end{bmatrix} \begin{Bmatrix} -r_1 r_2 \\ r_1^2 \\ 0 \end{Bmatrix} = 2m\omega^2 \begin{Bmatrix} 0 \\ 0 \\ r_1 r_2 \end{Bmatrix} = 2m\omega^2 \begin{Bmatrix} 0 \\ 0 \\ a^2 \cos\theta \sin\theta \end{Bmatrix}.$

Since $r_1 = a\cos\theta$ and $r_2 = a\sin\theta$, the external applied torque is the same as predicted above. This torque would have to be applied by the mechanism that holds the rod at the center, O.

Rotating Plate

The square plate, dimensions, a by a, has moments of inertia I_{11} and I_{22} about the x_1 and x_2 axes, shown in the diagram. It is set into rotational motion about the x_2 axis at the rate of ω r/sec. What are the torques involved?

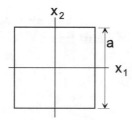

There are obviously no product of inertia terms, $\mathbf{J}$ is diagonal, with elements I_{11}, I_{22}, and 0

$$\mathbf{t} = \mathbf{WJ}\varpi = \begin{bmatrix} 0 & 0 & \omega \\ 0 & 0 & 0 \\ -\omega & 0 & 0 \end{bmatrix} \begin{bmatrix} I_{11} & 0 & 0 \\ 0 & I_{22} & 0 \\ 0 & 0 & I_{33} \end{bmatrix} \begin{Bmatrix} 0 \\ \omega \\ 0 \end{Bmatrix} = \begin{bmatrix} 0 & 0 & \omega \\ 0 & 0 & 0 \\ -\omega & 0 & 0 \end{bmatrix} \begin{Bmatrix} 0 \\ I_{11}\omega \\ 0 \end{Bmatrix} = \{0\}.$$

This is expected.

Now, incline the rotational axis at an angle θ. The question: Is the torque still zero? The x_2 component of the angular motion produces the same analysis, and result as the previous problem. The x_1 component also has the same analysis and result. So, surely this could be regarded as proof that these problems are the same, even though it intuitively seems that the plate should be "out of balance."

In the equation below, define $S \equiv \sin\theta$; and $C \equiv \sin\theta$

$$\mathbf{t} = \mathbf{WJ}\varpi = \begin{bmatrix} 0 & 0 & \omega C \\ 0 & 0 & -\omega S \\ -\omega C & \omega S & 0 \end{bmatrix} \begin{bmatrix} I_{11} & 0 & 0 \\ 0 & I_{22} & 0 \\ 0 & 0 & I_{33} \end{bmatrix} \begin{Bmatrix} \omega S \\ \omega C \\ 0 \end{Bmatrix} = \begin{Bmatrix} 0 \\ 0 \\ \omega^2 SC(I_{22} - I_{11}) \end{Bmatrix} = \{0\}.$$

There is no external torque required because $I_{11} = I_{22}$.

The Spinning Top

A top is diagrammed below, shown within a non-inertial coordinate set. Its center of gravity, cg, is at a distance a from its apex, its weight is mg. The moment of inertia about its vertical z_1 axis is I_{11};

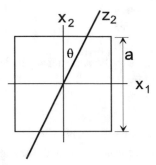

the moments of inertia I_{22} and I_{33}, about these respective, axes are equal because of symmetry. For the same reason, there are no product of inertia terms.

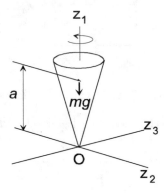

The top is caused to spin with its apex on a horizontal (x_2, x_3) plane at point O. The x_1 axis is vertical, the x-set is inertial. The apex is not held at O, but, there is just enough friction to hold it in place without slipping.

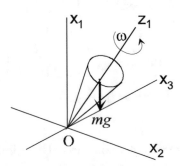

The top spins at the rate ω_1 about its centroidal, z_1, axis which makes an angle θ with x_1. In addition to its spin, the axis of the top will "precess" (rotate) about the x_1 axis, and 'nutate" toward, or away from vertical (i.e., the "nutation" rate is defined as the time derivative of angle θ).

The study of the motion of the top is a popular subject in the literature. We will only set up the problem and determine the equation of motion (the torque equation), in order to illustrate the matrices involved.

The precession rate is to be measured by a rotation of an intermediate coordinate y-set, initially collinear with the x-set, but free to rotate about its y_1 axis, with the precession of the top. The axis of the top will be fixed in the $y1_1$ y_2 plane, an arbitrary choice. This will define an angle φ whose time derivative will be the precession rate.

$$\mathbf{x} = \mathbf{T}_1 \mathbf{y} = \begin{bmatrix} 1 & 0 & 0 \\ 0 & \cos\varphi & -\sin\varphi \\ 0 & \sin\varphi & \cos\varphi \end{bmatrix} \mathbf{y} . \quad (5.82)$$

The precession rate will be $\dfrac{d\varphi}{dt} = \dot\varphi$ and note that the y-set experiences this rotation.

Now, define the rest of the z-set, whose z_1 axis is collinear with the axis of the top. As noted above, it tips at an angle θ with the x_1 and y_1 axes. Using the figure at the left

$$\mathbf{y} = \mathbf{T}_2 \mathbf{z} = \begin{bmatrix} \cos\theta & -\sin\theta & 0 \\ \sin\theta & \cos\theta & 0 \\ 0 & 0 & 1 \end{bmatrix} \mathbf{z} . \quad (5.83)$$

It is not necessary to define a set that spins with the top. All the terms, angular velocity, momentum, the force, mg, etc., will be the same in the z-set as they would be in a set which rotates with the top. In particular, the inertia matrix will be the same, because the top is symmetric about its axis.

Then the transform between the z and x sets is simply $\mathbf{x} = \mathbf{T}_1\mathbf{T}_2\mathbf{z}$.

In the constructing of the torque equation, differentiaion must be done in the inertial x-set:

$$\mathbf{t}_x = \frac{d}{dt}\mathbf{h}_x = \frac{d}{dt}(\mathbf{T}_1\mathbf{T}_2\mathbf{h}_z) = \dot{\mathbf{T}}_1\mathbf{T}_2\mathbf{h}_z + \mathbf{T}_1\dot{\mathbf{T}}_2\mathbf{h}_z + \mathbf{T}_1\mathbf{T}_2\dot{\mathbf{h}}_z .$$

Transforming this torque to the z-set,

$$t_z = \mathbf{T}'_2\mathbf{T}'_1\mathbf{t}_x = \mathbf{T}'_2(\mathbf{T}'_1\dot{\mathbf{T}}_1)\mathbf{T}_2\mathbf{h}_z + \mathbf{T}'_2\dot{\mathbf{T}}_2\mathbf{h}_z + \dot{\mathbf{h}}_z .$$ (5.84)

The matrix $\mathbf{T}'_1\dot{\mathbf{T}}_1$ is the precession rate about the x_1 axis: $\{\dot{\varphi},\ 0,\ 0\}$

$$\mathbf{T}'_1\dot{\mathbf{T}}_1 = \begin{bmatrix} 0 & 0 & 0 \\ 0 & 0 & -\dot{\varphi} \\ 0 & \dot{\varphi} & 0 \end{bmatrix} ;\ \text{transforms to } \mathbf{T}'_2(\mathbf{T}'_1\dot{\mathbf{T}}_1)\mathbf{T}_2 = \begin{bmatrix} 0 & 0 & -\dot{\varphi}\sin\theta \\ 0 & 0 & -\dot{\varphi}\cos\theta \\ \dot{\varphi}\sin\theta & \dot{\varphi}\cos\theta & 0 \end{bmatrix} .$$

The matrix $\mathbf{T}'_2\dot{\mathbf{T}}_2$ describes the tip (nutation) about the y_3 axis, $\mathbf{T}'_2\dot{\mathbf{T}}_2 = \begin{bmatrix} 0 & -\dot{\theta} & 0 \\ \dot{\theta} & 0 & 0 \\ 0 & 0 & 0 \end{bmatrix} .$

When summed together these matrices form the $\mathbf{W}_z$ matrix

$$\mathbf{W_z} = \mathbf{T}'_2\mathbf{T}'_1\dot{\mathbf{T}}_1\mathbf{T}_2 + \mathbf{T}'_2\dot{\mathbf{T}}_2 = \begin{bmatrix} 0 & -\dot{\theta} & -\dot{\varphi}\sin\theta \\ \dot{\theta} & 0 & -\dot{\varphi}\cos\theta \\ \dot{\varphi}\sin\theta & \dot{\varphi}\cos\theta & 0 \end{bmatrix} .$$ (5.85)

The related vector representation is $\mathbf{w}_z = \{\dot{\varphi}\cos\theta,\ -\dot{\varphi}\sin\theta,\ \dot{\theta}\ \}$. *Note that this does not include the rotation of the top*, which is $\{\omega_1, 0, 0\}$ in the z-set.

The three dimensional torque equation is given in (5.86). Note that the bold $\boldsymbol{\omega}_z$, and its derivative, are vectors. The elements of $\boldsymbol{\omega}_z$ are given in (5.87), including the scalar spin, ω_1. *This total angular velocity must be used in* (5.86).

The angular momentum of the top is the product of the inertia matrix and the total angular velocity vector. Its rate of change is the inertia matrix times the derivative of total angular velocity. The final equation is

$$t_z = \mathbf{W}_z\mathbf{h}_z + \dot{\mathbf{h}}_z = \mathbf{W}_z\mathbf{J}\boldsymbol{\omega} + \mathbf{J}\dot{\boldsymbol{\omega}}$$ (5.86)

where $\mathbf{W}_z$ is defined in (5.85), and:

$$\mathbf{J} = \begin{bmatrix} \mathbf{I}_{11} & 0 & 0 \\ 0 & \mathbf{I}_{22} & 0 \\ 0 & 0 & \mathbf{I}_{33} \end{bmatrix} ;\ \mathbf{I}_{33} = \mathbf{I}_{22};\ \boldsymbol{\omega} = \left\{ \begin{array}{c} \dot{\varphi}\cos\theta + \omega \\ -\dot{\varphi}\sin\theta \\ \dot{\theta} \end{array} \right\} .$$ (5.87)

The expansion of (5.86) into its three coordinate elements in three non-linear differential equations whose solutions are a numerical analysis problem. However, there are simplifications that are solvable. For example, the initial setup, above, implies that the only external torque is the moment of the weight of the top about the z_3 axis, $\mathbf{t} = \{0, 0, mga\sin\theta\}$.

In this case, over a short period, the precession and spin rates, and the angle θ are assumed constant. The expansion of (5.86) then yields non-zero values only about z_3. The three variables are the, θ, ω_1, and $\dot{\varphi}$, in an algebraic equation.

5.9 EXERCISES

5.1. An Airplane is to fly, direct, from a point A, 74 degrees west longitude and 41 degrees (north) latitude (roughly on the east coast of the US), to a point B, 122 degrees west longitude and 41 degrees latitude (roughly on the west coast). Assume a spherical earth, with a radius equal to 4000mi.

(a) Construct a coordinate y-set at the point A, as in Equation (5.8) of the text.

(b) What is the great circle distance between the two points?

(c) If the airplane simply flies west, along the 41 degree latitude, what is that distance?

(d) After takeoff, what is the correct heading to fly the great circle path?

5.2. A two-dimensional transform matrix is $\mathbf{T} = \begin{bmatrix} \cos\dfrac{\pi}{5} & \sin\dfrac{\pi}{5} \\ -\sin\dfrac{\pi}{5} & \cos\dfrac{\pi}{5} \end{bmatrix}$. Find $\mathbf{T}^{10}$.

5.3. Given;

$$\mathbf{R} = \begin{bmatrix} 0 & -r_3 & r_2 \\ r_3 & 0 & -r_1 \\ -r_2 & r_1 & 0 \end{bmatrix}; \mathbf{W} = \begin{bmatrix} 0 & -\omega_3 & \omega_2 \\ \omega_3 & 0 & -\omega_1 \\ -\omega_2 & \omega_1 & 0 \end{bmatrix}; \mathbf{r} = \begin{Bmatrix} r_1 \\ r_2 \\ r_3 \end{Bmatrix}; \varpi = \begin{Bmatrix} \omega_1 \\ \omega_2 \\ \omega_3 \end{Bmatrix}.$$

(a) Is $\mathbf{RWr}$ equal to $\mathbf{r} \times (\omega \times \mathbf{r})$ or equal to $(\mathbf{r} \times \omega) \times \mathbf{r}$?

(b) Is $\mathbf{RWr} = \mathbf{R}^2\omega$?

(c) Is $\mathbf{W}^2\mathbf{r}$ equal to $\omega \times (\omega \times \mathbf{r})$ or equal to $(\omega \times \omega) \times \mathbf{r}$?

5.4. It has been shown that no external torques are required to maintain the

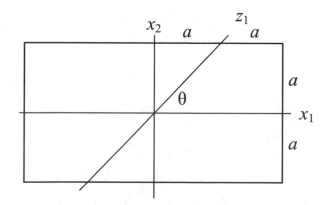

rotation of the square flat plate about an inclined axis. The $4a$ by $2a$ plate in the diagram is to rotate at constant angular velocity about z_1, inclined at $45°$. Find the torques required, if any. The plate has mass "M," and its moment of inertia about x_1 is $I_{11} = \dfrac{Ma^2}{3}$.

5.5. The spinning top, discussed in section 5.8, is to be put into the state of "Steady precession," in which $\dot{\varphi}$ and ω (the precessing and spin rates) remain constant, and the nutation rate is zero (θ remains constant).

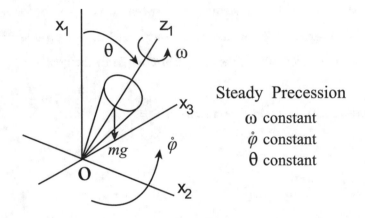

Steady Precession

ω constant

$\dot{\varphi}$ constant

θ constant

Determine the rate of change of angular momentum, as a function of these constants, that balances the single external torque produced by the weight of the top ($mag\sin\theta$).

CHAPTER 6

Matrix Eigenvalue Analysis

6.1 INTRODUCTION

Matrix analysis is particularly interesting because of the insight that it brings to so many areas of engineering. With the advent of the modern computer, much of the numerical labor is at least transferred into the fascinating realm of programming.

Perhaps the single most interesting matrix analysis is that which will now be discussed. It has fundamental bearing on the solution to many differential equations governing vibration problems, and the analysis of electrical networks. The eigenvalue problem is basically concerned with the transformation of vectors and matrices in a most advantageous fashion.

6.2 THE EIGENVALUE PROBLEM

The beginning is simple enough: Concerning the transform

$$\mathbf{A}\mathbf{x} = \mathbf{y} \tag{6.1}$$

where $\mathbf{A}$ is a general, real, *square* matrix, we ask whether or not an (input) $\mathbf{x}$ vector can be found such that the (output) $\mathbf{y}$ vector is proportional to $\mathbf{x}$. That is:

$$\mathbf{A}\mathbf{x} = \lambda\mathbf{x} . \tag{6.2}$$

The constant λ is the (scalar) factor of proportionality. We can bring $\lambda\mathbf{x}$ to the left side of (6.2):

$$[\mathbf{A} - \lambda\mathbf{I}]\mathbf{x} = \mathbf{A}(\lambda)\mathbf{x} = \mathbf{0} . \tag{6.3}$$

In (6.3), the notation $[\mathbf{A} - \lambda\mathbf{I}]$ is used rather than the more familiar $(\mathbf{A} - \lambda\mathbf{I})$ in order to emphasize that the quantity within the "[..]" is a square matrix. $\mathbf{A}$ is nXn, $\mathbf{x}$ is nX1, $\mathbf{I}$ is the nXn unit matrix; so the right side zero is an nX1 null column. The matrix $[\mathbf{A} - \lambda\mathbf{I}]$ is often referred to as $\mathbf{A}(\lambda)$, the "Lambda Matrix," or "Characteristic Matrix."

When $\mathbf{A}$ is not symmetric, the "companion" Equation (6.4) must also be considered:

$$\mathbf{z}[\mathbf{A} - \lambda\mathbf{I}] = \mathbf{0} \tag{6.4}$$

where, now, $\mathbf{z}$ is 1Xn (a row vector), and the $\mathbf{0}$ is a null 1Xn row. As will be seen, these two equations are "bound together," and will be solved together.

From Chapter 4, the homogeneous sets (6.3) and (6.4) have nontrivial solution iff the matrix $[\mathbf{A} - \lambda\mathbf{I}]$ is singular. ***Furthermore, in this treatment of the problem, we will require that the rank of the matrix $[\mathbf{A} - \lambda\mathbf{I}]$ be n-1.*** This condition is met by most engineering problems of interest.

6.2.1   THE CHARACTERISTIC EQUATION AND EIGENVALUES

In order for $[\mathbf{A} - \lambda\mathbf{I}]$ to be singular, the determinant must vanish:

$$\left|\mathbf{A} - \lambda\mathbf{I}\right| = (-1)^n \begin{vmatrix} a_{11} - \lambda & a_{12} & \cdots & a_{1n} \\ a_{21} & a_{22} - \lambda & \cdots & \cdots \\ \cdots & \cdots & a_{22} - \lambda & \cdots \\ a_{n1} & a_{n2} & \cdots & a_{nn} - \lambda \end{vmatrix} = 0. \tag{6.5}$$

The expansion of the determinant in (6.5) clearly will result in a polynomial of degree n. The multiplier $(-1)^n$ is used simply to cause the coefficient of λ^n to be positive (and, the determinant (6.5) would be more accurately written as $|\lambda\mathbf{I}-\mathbf{A}|$). Thus,

$$f(\lambda) = \lambda^n + c_1\lambda^{n-1} + \ldots c_{n-1}\lambda + c_n = 0 . \tag{6.6}$$

$f(\lambda)$ is called the "characteristic equation" and the polynomial is called the "characteristic polynomial" related to the matrix $\mathbf{A}$. The coefficients, c_k, are all functions of the $[a_{ij}]$ elements, and the coefficient c_n is equal to the determinant of $\mathbf{A}$ (and the product of the λ values). Now, represent the polynomial in (6.6) in its factored form:

$$f(\lambda) = (\lambda - \lambda_1)(\lambda - \lambda_2)(\lambda - \lambda_3)\cdots(\lambda - \lambda_n) = 0 \tag{6.7}$$

and it is clear that for each $\lambda = \lambda_j$, $f(\lambda_j) = 0$. These roots of the characteristic equation are called the "**eigenvalues**," or "**characteristic values**" of $\mathbf{A}$. Since (6.6) and (6.5) represent the same equation, these λ_j values also cause the determinant in (6.5) to vanish. If the λ_j eigenvalues are all distinct (i.e., no two roots the same), then the above constraint that the rank of the determinant be $n - 1$ will be met. Except for a short discussion concerning what happens when multiple roots occur, this chapter will assume distinct roots. In the general case, in which $\mathbf{A}$ is not symmetric, the eigenvalues may be complex numbers. While this fact is not much of a conceptual difficulty, it does pose calculation problems.

Now, consider the case $\lambda = \lambda_1$. The determinant $|\mathbf{A}(\lambda_1)|$ is zero and there will be exactly one solution to each of the Equations (6.3) and (6.4) above. These solutions, being associated with the eigenvalue, λ_1, are known as "**eigenvectors**," or "**characteristic vectors**." Equation (6.3) will yield a *column* vector, and (6.4) will yield a *row* vector. These vectors will "emerge" together.

The adjoint of $[\mathbf{A} - \lambda_1\mathbf{I}]$ will be of unit rank. All its rows (columns) will be proportional to each other (some, but not all, of the rows (columns) of the adjoint may be null). Denote the row eigenvector as $\mathbf{u}_1$, and the column eigenvector as $\mathbf{v}_1$. The adjoint of $[\mathbf{A} - \lambda_1\mathbf{I}]$ can be written

$$\mathbf{A}^a(\lambda_1) = [\mathbf{A} - \lambda_1\mathbf{I}]_{\text{adj}} = k\{\mathbf{v}_1\}[\mathbf{u}_1] . \tag{6.8}$$

The vector product given in (6.8) is nX1X1Xn = nXn. It is certainly not the dot product of $\mathbf{u}_1$ and $\mathbf{v}_1$. (In fact, every matrix of unit rank can be written as this type of single column times single row. That is essentially the definition of rank = 1.)

Any column of the adjoint of (6.8) solves $[\mathbf{A} - \lambda_1\mathbf{I}]\mathbf{x} = \mathbf{0}$. Yet, there can be only one solution. Thus, all the adjoint columns must be the same—i.e., proportional, differing only in magnitude. That is, *only the eigenvector's direction is obtained*.

Similarly, every (non-zero) row of the adjoint solves $\mathbf{z}[\mathbf{A} - \lambda_1\mathbf{I}] = \mathbf{0}$ (note, again that this is a row equation). Then, for any given eigenvalue, (6.8) yields exactly one row, and one column eigenvector. And, since the vector magnitudes are arbitrary, we can always choose scaling such that the dot product of $\mathbf{u}_1$ and $\mathbf{v}_1$ is equal to +1 (unity).

In the same way, all n eigenvectors are obtained from their respective adjoint matrices. And, each time they are scaled to +1. It will now be shown that $\mathbf{v}_j$ is orthogonal to $\mathbf{u}_i$, for the subscript i not equal to j. Write (**v** for column vectors, and **u** for rows)

$$\begin{aligned}\mathbf{A}\mathbf{v}_j &= \lambda_j\mathbf{v}_j \;;\quad \text{Note that } \mathbf{v}_j \text{ is (nX1)}\\ \mathbf{u}_i\mathbf{A} &= \lambda_i\mathbf{u}_i \;;\quad \text{Note that } \mathbf{u}_i \text{ is (1Xn) .}\end{aligned} \qquad (6.9)$$

Premultiply the first of (6.9) by $[\mathbf{u}_i]$ and postmultiply the second by $\{\mathbf{v}_j\}$. The left sides of both equations will then be identical. If the two are subtracted, the result is

$$\mathbf{u}_i \bullet \mathbf{v}_j(\lambda_i - \lambda_j) = 0 . \qquad (6.10)$$

Since the eigenvalues are distinct (by hypothesis), then $(\lambda_i - \lambda_j)$ cannot be zero. Thus, the dot product $\mathbf{u}_i\bullet\mathbf{v}_j$ must be zero, proving that the two eigenvectors are orthogonal. But, the original choice of i and j was arbitrary. So, the assertion of orthogonality must be true for any choice. Then, if all the row eigenvectors are collected into the square matrix, **U**, and the column eigenvectors collected into **V** (and remembering that $\mathbf{u}_i\bullet\mathbf{v}_i$ can be normalized to +1):

$$\mathbf{UV} = \mathbf{VU} = \mathbf{I}, \quad \text{(the unit matrix) .} \qquad (6.11)$$

Also, the entire eigenvalue problem can be displayed in the following 2 equations:

$$\begin{aligned}\mathbf{AV} &= \mathbf{V}\Lambda\\ \mathbf{UA} &= \Lambda\mathbf{U}\end{aligned} \qquad (6.12)$$

In (6.12), the matrix, Λ, is a diagonal matrix whose main diagonal elements are the eigenvalues – arranged, carefully, in the same order in which the eigenvectors are placed in the matrices **U** and **V**. Now, choose the first of the Equations (6.12), and premultiply by **U**. Using (6.11), the result is:

$$\mathbf{UAV} = \Lambda . \qquad (6.13)$$

That is, the **A** matrix is transformed by **U** and **V** into its "eigenvalue matrix," Λ.

6.2.2 SYNTHESIS OF A BY ITS EIGENVALUES AND EIGENVECTORS

Premultiplying the second of Equations (6.12) by **V** reveals an interesting, and useful result.

$$\mathbf{A} = \mathbf{V}\Lambda\mathbf{U} .$$

The eigenvalue analysis "resolves" matrix $\mathbf{A}$ into its component eigenvalues and vectors. This becomes more evident if it is remembered that premultiplying $\mathbf{U}$ by Λ has the effect of multiplying every element of *row* $\mathbf{u}_j$ by its corresponding λ_j. Now, simply visualize this and also *partition* $\mathbf{V}$ *by columns and the* $\Lambda\mathbf{U}$ *product by rows*:

$$\mathbf{A} = \sum_j^n \lambda_j \{\mathbf{v}_j\}[\mathbf{u}_j] . \tag{6.14}$$

$\mathbf{A}$ is shown as a sum of n matrices; $\lambda_j\{\mathbf{v}_j\}[\mathbf{u}_j]$, each nXn, and each composed only of corresponding eigenvalues and eigenvectors. It is instructive to postmultiply the eigenvector $\mathbf{v}_k$ on both sides of (6.14)

$$\mathbf{A}\mathbf{v}_k = \sum_j^n \lambda_j \{\mathbf{v}_j\}[\mathbf{u}_j]\{\mathbf{v}_k\} .$$

Now, all the products $\mathbf{u}_j \bullet \mathbf{v}_k$ vanish (the eigenvectors are orthogonal), except the $\mathbf{u}_k \bullet \mathbf{v}_k$ one (which is normalized to +1). Then

$$\mathbf{A}\mathbf{v}_k = \lambda_k \mathbf{v}_k$$

which is the same as Equation (6.2), with the appropriate subscripts.

In (6.14), if any one of the nXn matrices in the summation were to be subtracted away, a new matrix, say $\mathbf{B}$, would result. $\mathbf{B}$ would have all the same eigenvalues and vectors that $\mathbf{A}$ possesses – except the one subtracted away. This fact is useful in "matrix iteration" (not yet discussed here), in which iterative techniques are used to obtain eigenvalues and vectors, one at a time. When one set is found, its effects can be subtracted away, to move on to iterate for the next. See the article on **matrix iteration** in Section 6.7.2 of this chapter.

6.2.3   EXAMPLE ANALYSIS OF A NONSYMMETRIC 3X3

To illustrate the eigenvalue problem numerically, consider the following 3X3:

$$\mathbf{A} = \begin{bmatrix} 25 & -44 & 18 \\ 12 & -21 & 8 \\ -3 & 6 & -4 \end{bmatrix} .$$

This matrix is a particularly simple one numerically. But, its analysis will nevertheless illustrate the eigenvalue problem. Eigenvalues will be denoted using λ, and the characteristic equation is the expansion of the determinant:

$$f(\lambda) = (-1)^3 \begin{vmatrix} 25 - \lambda & -44 & 18 \\ 12 & -21 - \lambda & 8 \\ -3 & 6 & -4 - \lambda \end{vmatrix} = 0 . \tag{6.15}$$

Since $(-1)^3 = -1$, negate the first row:

$$|(\lambda - 25) \qquad + 44 \qquad - 18| ,$$

and expand by first minors of the first row,

$$f(\lambda) = (\lambda - 25)[(-21 - \lambda)(-4 - \lambda) - 48] - 44[12(-4 - \lambda) + 24]$$
$$- 18[72 + 3(-21 - \lambda)] = 0$$

which reduces algebraically to:

$$f(\lambda) = \lambda^3 - 7\lambda - 6 = 0 . \tag{6.16}$$

Notice that in this case there is no λ^2 term (i.e., its coefficient is zero), and that the *"trace"* of A (the sum of its diagonal elements) is also zero. In fact, the (negative) trace of A is always equal to its coefficient in its characteristic polynomial.

By inspection, -1 is a root of (6.16). Dividing by $(\lambda + 1)$, and factoring the quadratic:

$$f(\lambda) = (\lambda + 1)(\lambda^2 - \lambda - 6) = (\lambda + 1)(\lambda + 2)(\lambda - 3) . \tag{6.17}$$

The three roots, -1, -2, and 3, are the three **eigenvalues** of A. For each eigenvalue there will be **two eigenvectors**—a row eigenvector, and a column eigenvector.

With $\lambda_1 = -1$, and denoting $[A - \lambda_1 I]$, as $A(\lambda_1)$:

$$A(\lambda_1) = \begin{bmatrix} 26 & -44 & 18 \\ 12 & -20 & 8 \\ -3 & 6 & -3 \end{bmatrix}; \qquad \begin{array}{l} 26x_1 - 44x_2 + 18x_3 = 0 \\ 12x_1 - 20x_2 + 8x_3 = 0 \\ -3x_1 + 6x_2 - 3x_3 = 0 \end{array} .$$

The solution of the linear equation set at the above/right determines the eigenvector, v_1. Since $A(\lambda_1)$ is known to be singular, this set must have a non-trivial solution. One way to do this is to set x_3 arbitrarily (say, $x_3 = 1$), delete the third equation, and solve the remaining two variables:

$$26x_1 - 44x_2 = -18$$
$$12x_1 - 20x_2 = -8$$

From which it is found that $x_1 = x_2 = x_3 = 1$.

$|A(\lambda_1)|$ is equal to zero. However, the adjoint matrix must have at least one non zero row and column (the rank of $A(\lambda_1)$ is $n - 1$). Then, by calculating the adjoint, both row and columns are found:

$$A(\lambda_1) = \begin{bmatrix} 26 & -44 & 18 \\ 12 & -20 & 8 \\ -3 & 6 & -3 \end{bmatrix}; \qquad A^a(\lambda_1) = \begin{bmatrix} 12 & -24 & 8 \\ 12 & -24 & 8 \\ 12 & -24 & 8 \end{bmatrix} . \tag{6.18}$$

From this adjoint, any row can be chosen as the row vector, and any column can be chosen as the column vector, for example $[12, -24, 8]$ for the row vector, and $\{12, 12, 12\}$ for the column. However, the eigenvectors emerge in direction only, then any multiples of these vectors are also eigenvectors. Then:

$$u_1 = [3, -6, 2] \quad \text{and} \quad v_1 = \{1, 1, 1\}$$

where $\mathbf{u}_1$ denotes the row vector, and $\mathbf{v}_1$ denotes the column. Since $\mathbf{u}_1 \bullet \mathbf{v}_1$ product must be 1, normalize by multiplying $\mathbf{u}_1$ by -1.

$$\mathbf{u}_1 = [-3, 6, -2] \quad \text{and} \quad \mathbf{v}_1 = \{1, 1, 1\}$$

Now, if these two vectors are truly eigenvectors they must solve

$$(\mathbf{A} - \lambda_1 \mathbf{I})\mathbf{v}_1 = \mathbf{0}; \quad \& \quad \mathbf{u}_1(\mathbf{A} - \lambda_1 \mathbf{I}) = \mathbf{0}.$$

And they do:

$$\begin{bmatrix} -3 & 6 & -2 \end{bmatrix} \begin{bmatrix} 26 & -44 & 18 \\ 12 & -20 & 8 \\ -3 & 6 & -3 \end{bmatrix} = \begin{bmatrix} 0 & 0 & 0 \end{bmatrix} \quad \text{and}$$

$$\begin{bmatrix} 26 & -44 & 18 \\ 12 & -20 & 8 \\ -3 & 6 & -3 \end{bmatrix} \begin{Bmatrix} 1 \\ 1 \\ 1 \end{Bmatrix} = \begin{Bmatrix} 0 \\ 0 \\ 0 \end{Bmatrix}.$$

With $\lambda_2 = -2$

$$\mathbf{A}(\lambda_2) = \begin{bmatrix} 27 & -44 & 18 \\ 12 & -19 & 8 \\ -3 & 6 & -2 \end{bmatrix}; \quad \mathbf{A}^a(\lambda_2) = \begin{bmatrix} 10 & 20 & -10 \\ 0 & 0 & 0 \\ 15 & -30 & 15 \end{bmatrix}. \tag{6.19}$$

In the same manner as before,

$$\mathbf{u}_2 = [-1, 2, -1] \quad \text{and} \quad \mathbf{v}_2 = \{2, 0, -3\}$$

Note that the adjoint has a zero row, which must not be chosen as an eigenvector. This is simple by sight, but if the computer is choosing eigenvectors, it must be taught to avoid such things.

With $\lambda_3 = 3$:

$$\mathbf{A}(\lambda_3) = \begin{bmatrix} 22 & -44 & 18 \\ 12 & -24 & 8 \\ -3 & 6 & -7 \end{bmatrix}; \quad \mathbf{A}^a(\lambda_3) = \begin{bmatrix} 120 & -200 & 80 \\ 60 & -100 & 40 \\ 0 & 0 & 0 \end{bmatrix}. \tag{6.20}$$

$$\mathbf{u}_3 = [3, -5, 2] \quad \text{and} \quad \mathbf{v}_3 = \{2, 1, 0\}$$

Now that the eigenvectors have all been chosen, and normalized $\mathbf{u}_k \bullet \mathbf{v}_k = 1$, the 3X3 $\mathbf{U}$ and $\mathbf{V}$ matrices are:

$$\mathbf{U} = \begin{bmatrix} -3 & 6 & -2 \\ -1 & 2 & -1 \\ 3 & -5 & 2 \end{bmatrix}; \quad \text{and} \quad \mathbf{V} = \begin{bmatrix} 1 & 2 & 2 \\ 1 & 0 & 1 \\ 1 & -3 & 0 \end{bmatrix}. \tag{6.21}$$

These matrices are inverses, i.e., $\mathbf{UV} = \mathbf{VU} = \mathbf{I}$, and

$$\mathbf{AV} = \mathbf{V}\Lambda \quad \text{(See Equations (6.12))}$$

where Λ is the (3X3) diagonal eigenvalue matrix. Now, postmultiply by $\mathbf{U}$:

$$\mathbf{AVU} = \mathbf{A} = \mathbf{V\Lambda U} . \tag{6.22}$$

Which shows the synthesis of $\mathbf{A}$ by its eigenvalues and eigenvectors. In this example:

$$\mathbf{V\Lambda U} = \begin{bmatrix} 1 & 2 & 2 \\ 1 & 0 & 1 \\ 1 & -3 & 0 \end{bmatrix} \begin{bmatrix} -1 & 0 & 0 \\ 0 & -2 & 0 \\ 0 & 0 & 3 \end{bmatrix} \begin{bmatrix} -3 & 6 & -2 \\ -1 & 2 & -1 \\ 3 & -5 & 2 \end{bmatrix} = \begin{bmatrix} 25 & -44 & 18 \\ 12 & -21 & 8 \\ -3 & 6 & -4 \end{bmatrix} = \mathbf{A} .$$

An important result. Alternatively, the matrices $\mathbf{U}$ and $\mathbf{V}$ "diagonalize" the original matrix:

$$\mathbf{UAV} = \Lambda \tag{6.23}$$

$$\mathbf{UAV} = \begin{bmatrix} -3 & 6 & -2 \\ -1 & 2 & -1 \\ 3 & -5 & 2 \end{bmatrix} \begin{bmatrix} 25 & -44 & 18 \\ 12 & -21 & 8 \\ -3 & 6 & -4 \end{bmatrix} \begin{bmatrix} 1 & 2 & 2 \\ 1 & 0 & 1 \\ 1 & -3 & 0 \end{bmatrix} = \begin{bmatrix} -1 & 0 & 0 \\ 0 & -2 & 0 \\ 0 & 0 & 3 \end{bmatrix} .$$

Now, to illustrate the point about the synthesis of $\mathbf{A}$ via its eigenvalues and vectors, from the matrix $\mathbf{A}$, subtract the 3X3 $= \lambda_1\{\mathbf{v}_1\}[\mathbf{u}_1]$:

$$\mathbf{B} = \begin{bmatrix} 25 & -44 & 18 \\ 12 & -21 & 8 \\ -3 & 6 & -4 \end{bmatrix} - \begin{bmatrix} -3 & 6 & -2 \\ -3 & 6 & -2 \\ -3 & 6 & -2 \end{bmatrix} = \begin{bmatrix} 22 & -38 & 16 \\ 9 & -15 & 6 \\ -6 & 12 & -6 \end{bmatrix} . \tag{6.24}$$

An analysis of $\mathbf{B}$ shows that it still possesses the eigenvalues -2, and 3, but, in place of $\lambda_1 = -1$, its λ_1 is zero ($\mathbf{B}$ is singular). Interestingly, all its eigenvectors are the same—even $\mathbf{u}_1$ and $\mathbf{v}_1$. However, $\mathbf{u}_1$ and $\mathbf{v}_1$ can play no part in the synthesis of $\mathbf{B}$, because these are multiplied by zero.

6.2.4   EIGENVALUE ANALYSIS OF SYMMETRIC MATRICES

In the general (non-symmetric) case, 2 equations are required to define the eigenvalue problem (Equations (6.3) and (6.4)). When the given matrix is symmetric, a simplification occurs. If (6.3) is transposed ($\lambda\mathbf{I}$ is diagonal) the result is $\mathbf{x}'[\mathbf{A}' - \lambda\,\mathbf{I}] = \mathbf{0}$. But, in this case $\mathbf{A}' = \mathbf{A}$, and the result is that the row vector is simply the transposed column vector.

For any eigenvalue, λ_i the adjoint matrix $[\mathbf{A} - \lambda_i\mathbf{I}]_{\mathrm{adj}}$ is also a symmetric matrix, proportional to the product of $\mathbf{v}_i\mathbf{v}_i'$. Any nonzero row or column can be chosen.

The orthogonality of these eigenvectors is shown in the following way. For any two vectors, write:

$$\begin{aligned} \mathbf{Av}_i &= \lambda_i\mathbf{v}_i \\ \mathbf{Av}_j &= \lambda_j\mathbf{v}_j . \end{aligned} \tag{6.25}$$

Now, premultiply the first of these by $\mathbf{v}_j$ and the second by $\mathbf{v}_i$.

$$\begin{aligned} \mathbf{v}_j'\mathbf{Av}_i &= \lambda_i\mathbf{v}_j'\mathbf{v}_i \\ \mathbf{v}_i'\mathbf{Av}_j &= \lambda_j\mathbf{v}_i'\mathbf{v}_j . \end{aligned} \tag{6.26}$$

If the second of these is transposed, the left sides become identical, because $\mathbf{A}$ is symmetric. Then, when the two are subtracted, as before, the eigenvectors must be orthogonal (again assuming distinct eigenvalues). This orthogonality can be expressed in terms of all the eigenvectors, as

$$\mathbf{V'V} = \mathbf{I} \text{ (compare with (6.11))} . \tag{6.27}$$

And the entire eigenvalue problem can be displayed in the single equation,

$$\mathbf{AV} = \mathbf{V}\Lambda \text{ (compare with (6.12))} . \tag{6.28}$$

The diagonalization of $\mathbf{A}$ is shown by premultiplying by $\mathbf{V'}$

$$\mathbf{V'AV} = \mathbf{V'V}\Lambda = \Lambda \text{ (compare with (6.13))} . \tag{6.29}$$

The synthesis of $\mathbf{A}$ is given by postmultiplying: $\mathbf{AVV'} = \mathbf{A} = \mathbf{V}\Lambda\mathbf{V'}$ and (6.14) becomes:

$$\mathbf{A} = \sum_{j}^{n} \lambda_j \{\mathbf{v}_j\}[\mathbf{v}_j] . \tag{6.30}$$

Again note that the vector product shown here is nX1X1Xn, resulting in nXn matrices.

6.3   GEOMETRY OF THE EIGENVALUE PROBLEM

The dot product of a vector $\mathbf{x}$, times itself, is equal to the sum of squares of its elements. If this sum is equated to unity, we have

$$\mathbf{x'x} = x_1^2 + x_2^2 + \cdots + x_n^2 = 1 .$$

In two or three dimensions, the above equation is identified as that of a circle, or sphere, of unit radius. By analogy, the n dimensional case, written above, is called an n-dimensional sphere.

The dot product of $\mathbf{x}$ into the vector $\Lambda\mathbf{x}$, where Λ is a diagonal matrix, is $\mathbf{x'}\Lambda\mathbf{x}$. Equated to unity:

$$\mathbf{x'}\Lambda\mathbf{x} = \lambda_1 x_1^2 + \lambda_2 x_2^2 + \cdots + \lambda_n x_n^2 = 1 . \tag{6.31}$$

Depending upon the sign of the λ values, the above equation in three dimensions would be an ellipsoid or hyperboloid. For our purposes, it is most beneficial to visualize an ellipsoid. In the accompanying figure, note that the coordinate axes are aligned along the principal axes of the ellipsoid. But, in (6.31), if we affect an arbitrary orthogonal coordinate transform, $\mathbf{x} = \mathbf{Tq}$. Then:

$$\mathbf{q'T'}\Lambda\mathbf{Tq} = \mathbf{q'Aq} = 1 \tag{6.32}$$

and note that $\mathbf{A}$ is a symmetric (not diagonal) matrix whose eigenvalues are in Λ and whose eigenvectors are in the transform matrix $\mathbf{T}$.

Chapter 5, shows that such a transform amounts to a series of rotations about the axes of a rectangular coordinate system—apparently, in this case, *rotating the axes away from the principal axes*

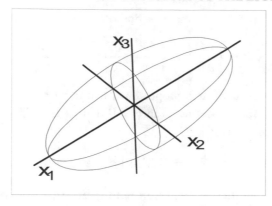

of the ellipsoidal surface. Again from Chapter 5, all vectors and angles remain invariant under such a transform. Therefore, the surface itself does not change, just the coordinate perspective through which it is viewed.

In practical situations the coordinate axes are rarely aligned along the principal axes. Instead, the "quadratic form" is derived as the dot product of a vector $\mathbf{x}$ multiplied by its transform $\mathbf{Ax}$—as in (6.33), below. In Chapter 4, Section 4.3, the "quadratic form" was introduced. The equation of the ellipsoid described here is just such a form:

$$F = \mathbf{x}'\mathbf{Ax} = 1 \tag{6.33}$$

a scalar, defined by the *symmetric matrix*, $\mathbf{A}$. In general the form is the equation of an n-dimensional ellipsoid, whose principal axes do not lie along the axes of the coordinate set. When the form F is expanded, it includes "cross product terms," involving $x_i x_j$ in addition to the squared terms found in (6.31). The problem is to affect a coordinate transform, such that F appears with squared terms only.

Figure 6.1 shows a simple 2-dimensional case. $\mathbf{x}$ is simply the vector drawn from coordinate center to any arbitrarily chosen point. At that same point, the normal to the surface is identified as

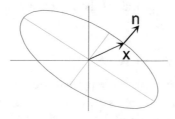

Figure 6.1:

the vector $\mathbf{n}$. Analytic geometry tells us that $\mathbf{n}$ is proportional to the *column vector* ∇F:

$$\nabla F = \left\{ \frac{\partial F}{\partial x_i} \right\} = \left\{ \frac{\partial F}{\partial x_1}, \ \frac{\partial F}{\partial x_2}, \ \cdots, \ \frac{\partial F}{\partial x_n} \right\} . \tag{6.34}$$

That is, the ith direction cosine of $\mathbf{n}$ is proportional to the partial of F with respect to the ith coordinate. Assembling all these together as in (6.34) derives the ∇F column vector. But, again from Appendix A, Equation (A.12):

$$\nabla F = \nabla(\mathbf{x}'\mathbf{A}\mathbf{x}) = 2\mathbf{A}\mathbf{x} . \tag{6.35}$$

In general, the normal, $\mathbf{n}$, is different from $\mathbf{x}$ in both direction and magnitude, as in the figure. However, note that along the principal axes of the ellipse the vector $\mathbf{x}$, itself, is normal to the surface. Then, at these points, the vectors $\mathbf{x}$ and $\mathbf{n}$ are collinear, and are proportional:

$$\mathbf{n} = 2\mathbf{A}\mathbf{x} = \lambda\mathbf{x} . \tag{6.36}$$

Note that Equation (6.36) *is simply the statement of the eigenvalue problem* (with the constant 2 absorbed into the proportionality factor, λ).

The eigenvalue analysis leads to a solution for n characteristic numbers (eigenvalues), λ and their n eigenvectors, $\mathbf{v}$. Assembling these quantities into matrix form:

$$\begin{aligned} \Lambda &= [\lambda_j \delta_{ij}]; \ \text{The diagonal matrix of eigenvalues} \\ \mathbf{V} &= [\mathbf{v}_j]; \ \text{The orthogonal matrix of eigenvectors} . \end{aligned} \tag{6.37}$$

The given matrix $\mathbf{A}$ is symmetric, so $\mathbf{V}$ is orthogonal (the rows of $\mathbf{V}$ are the row eigenvectors). Now, define the new coordinates as the q-set, where $\mathbf{x} = \mathbf{V}\mathbf{q}$. The quadratic form is

$$F = \mathbf{x}'\mathbf{A}\mathbf{x} = \mathbf{q}'\mathbf{V}'\mathbf{A}\mathbf{V}\mathbf{q} = \mathbf{q}'\Lambda\mathbf{q} = \lambda_1 q_1^2 + \lambda_2 q_2^2 + \cdots + \lambda_n q_n^2 .$$

Because $\mathbf{V}'\mathbf{A}\mathbf{V}$ transforms $\mathbf{A}$ to the diagonal matrix, Λ, F is now composed of squared terms only, the familiar form of the ellipsoid from analytic geometry. Such a transform preserves both magnitude and angle, the square roots of the reciprocals of the eigenvalues are equal to the lengths of the semi major axes.

6.3.1 NON-SYMMETRIC MATRICES

In the general, nonsymmetric eigenvalue problem, there are two sets of eigenvectors, that are identified as $\mathbf{u}_i$ (the row set), and $\mathbf{v}_i$ (the column set). When the full complement of vectors is gathered together, these occupy the rows and columns, respectively, of $\mathbf{U}$, and $\mathbf{V}$. However, neither $\mathbf{U}$ nor $\mathbf{V}$ is orthogonal. Neither of these, then, represent rectangular coordinate sets. Instead, they represent base vectors in two "skewed" systems in which the $\mathbf{u}$ (as well as the $\mathbf{v}$) axes are at oblique angles (within each set).

But, $\mathbf{U}$ and $\mathbf{V}$ are inverses. Then, $\mathbf{u}_i \bullet \mathbf{v}_j = \delta_{ij}$. That is, the $\mathbf{u}$ axes are orthogonal to the $\mathbf{v}$. *In the oblique (nonsymmetric) case, it takes two sets of coordinates to take part in a coordinate transform* and, the diagonalization of a quadratic form (now called a "bilinear form").

In the rectangular set, a given vector, $\mathbf{r} = \{r_1, r_2, \cdots, r_n\}$, is represented by the r_j set of numbers, each of which is determined by taking the dot product of $\mathbf{r}$ with the jth base vector. A transform of $\mathbf{r}$ to a new orthogonal set is given by $\mathbf{Vr}$, where $\mathbf{V}$ is an orthogonal transform matrix relating the new unit vector axes to the old ones.

But, in an oblique system, this convenience is absent. Any vector, say $\mathbf{r}$, has two sets of coordinate values:

$$\mathbf{r} = \left\{ \begin{array}{cccc} r_1\mathbf{v}_1 & r_2\mathbf{v}_2 & \cdots & r_n\mathbf{v}_n \\ \rho_1\mathbf{u}_1 & \rho_2\mathbf{u}_2 & \cdots & \rho_n\mathbf{u}_n \end{array} \right\}$$

where $\mathbf{u}_i$ and $\mathbf{v}_i$ represent unit vectors. That is (for example), the scalar r_1 is the coordinate of $\mathbf{r}$ in the direction of the unit vector $\mathbf{v}_1$, while ρ_1 is the coordinate of $\mathbf{r}$ in the direction of the unit vector $\mathbf{u}_1$. Furthermore, in order to determine r_1 we must take the dot product of $\mathbf{r}$, not with the unit vector $\mathbf{v}_1$, but, with the unit vector, $\mathbf{u}_1$. Similarly, given another vector $\mathbf{y}$:

$$\mathbf{y} = \left\{ \begin{array}{ccc} y_1, & \cdots, & y_n \text{ in the } \mathbf{v}\text{-set} \\ \psi_1, & \cdots, & \psi_n \text{ in the } \mathbf{u}\text{-set} \end{array} \right\}$$

(the unit vectors are omitted) the dot product of the two vectors is given by

$$\mathbf{r} \bullet \mathbf{y} = r_1\psi_1 + r_2\psi_2 + \cdots + r_n\psi_n = \rho_1 y_1 + \rho_2 y_2 + \cdots + \rho_n y_n \,.$$

Dot products must be taken between coordinates of the two sets. The products involving terms like $r_i y_i$, or $\rho_i \psi_i$, have absolutely no meaning. With this in mind, we write the form:

$$\xi' \mathbf{A} \mathbf{x} = 1 \ (\mathbf{A} \text{ nonsymmetric}) \,.$$

The principal axes of the form are still those that are normal to the surface, and the form is once again expressed in the defining equation $\mathbf{A}\mathbf{x} = \lambda\mathbf{x}$. However, ξ and $\mathbf{x}$ are the two different representations of the same entity. Another expression is required in order to derive "the other half" of the normal to the surface. That is, the transposed set

$$(\xi' \mathbf{A} \mathbf{x})' = \mathbf{x}' \mathbf{A}' \xi = 1 \,.$$

This time the defining equation is $\mathbf{A}'\xi = \lambda\xi$ which will bring out the companion set coordinates of the vector which is proportional to the normal. Note here, that ξ is represented as a column. In our development of the eigenvalue problem, this same equation is written as a **row equation** because it was important to identify this half of the problem, "the row half." That is:

$$\mathbf{A}'\xi = \lambda\xi \ \Leftrightarrow \ \xi\mathbf{A} = \lambda\xi$$

are the same equation, but the second form, with $\xi\mathbf{A}$ makes it clear that ξ is a row vector.

The eigenvalue analysis finds that two sets of n vectors emerge (as in Section 6.2), and they are mutually orthogonal. In the geometric sense, there is only one set of principal axes, and these are orthogonal. But the analysis of them is required to take place within the two oblique, mutually orthogonal systems.

Once this is accomplished, a transform is made to the principal axes, which are an orthogonal set. The extra complication is removed. Thus, with the transforms $\xi = z'U$ and $x = Vz$.

$$F = \xi Ax = z'UAVz = z'\Lambda z = 1 .$$

And the transformation of F is complete.

Yet, not all the complications can be avoided. ***In the symmetric case, the eigenvalues are always real, and a full complement of eigenvectors can always be determined***. In the general case this is not true. Both eigenvalues and eigenvectors may be complex numbers; the matrices U, V, and Λ, then complex. When eigenvalues are repeated in the symmetric case, it simply means that ellipses become circles, providing another degree of choice in choosing rectangular axes.

In the oblique case, when repeated eigenvalues occur, it might be that some of the oblique axes collapse into one, and it cannot be guaranteed that a full set of eigenvectors can be found. Thus, the eigenvalue problem and its geometric representation is far easier when the quadratic form is originally given in terms of a symmetric matrix A.

Matrices that arise in engineering problems are often symmetric, and the associated quadratic form has physical as well as geometric significance. For this reason the symmetric eigenvalue problem is particularly important. However, it is also true that eigenvalue analysis is often required of nonsymmetric matrices, with complex roots and vectors. For example, the kinetic and potential energies in vibrating systems are described by quadratic forms. However, when energy dissipation terms are involved, the system is dynamically described by a nonsymmetric matrix, with complex eigenvalues and eigenvectors. Such systems will be discussed in the following chapter.

6.3.2 MATRIX WITH A DOUBLE ROOT

When a non symmetric matrix, A, is found to have a repeated root, there is the question of whether or not the matrix is defective—does A possess a full complement of eigenvectors?

As an example consider $\quad A = \begin{bmatrix} 0 & -2 & -2 \\ 1 & 3 & 1 \\ 0 & 0 & 2 \end{bmatrix}$

whose characteristic polynomial is $(\lambda - 1)(\lambda - 2)(\lambda - 2)$.

For the eigenvalue $\lambda = 1$ the adjoint of $A(\lambda 1)$ is the product $[A - 2I][A - 2I]$, from which a row and a column vector emerge. For the double root, $\lambda = 2$, the matrix $[A - I][A - 2I]$ is ***null*** (no row or column can be chosen as an eigenvector). So, look at $[\mathbf{A} - 2\mathbf{I}]\mathbf{x}$

$$[\mathbf{A} - 2\mathbf{I}]\mathbf{x} = \mathbf{A}(\lambda_2)\mathbf{x} = \begin{bmatrix} 2 & 2 & 2 \\ -1 & -1 & -1 \\ 0 & 0 & 0 \end{bmatrix} \mathbf{x} \,.$$

This matrix clearly has rank = 1. **Two** independent vectors can be found that are orthogonal to the rows/columns of $\mathbf{A}(\lambda_2)$. Thus, $\mathbf{A}$ is **not defective**—it has *all three eigenvectors*. If $\mathbf{A}(\lambda_2)$ had rank = 2, only one eigenvector could be found, and $\mathbf{A}$ would be defective. This $\mathbf{A}$ matrix satisfies an equation $f(\mathbf{A})$ that is of lower rank than the Cayley-Hamilton equation, namely $[\mathbf{A} - 2\mathbf{I}][\mathbf{A} - 2\mathbf{I}] = \mathbf{0}$. This leaves just enough room for the definition of the necessary eigenvectors.

Now, consider the matrix below. Its characteristic equation is $f(\lambda) = (\lambda + 1)^2(\lambda + 2) = 0$; a double root $\lambda = -1$.

$$\mathbf{A} = \begin{bmatrix} 0 & 1 & 0 \\ 0 & 0 & 1 \\ -2 & -5 & -4 \end{bmatrix}$$

In this case,

$$[\mathbf{A} - \lambda_1\mathbf{I}] = \begin{bmatrix} 1 & 1 & 0 \\ 0 & 1 & 1 \\ -2 & -5 & -3 \end{bmatrix} \quad \text{whose rank} > 1, \text{ and } [\mathbf{A} + \mathbf{I}][\mathbf{A} + 2\mathbf{I}] \text{ is } \textit{\textbf{not null}}.$$

Then, only one eigenvector can be found, and the matrix is defective.

6.4   THE EIGENVECTORS AND ORTHOGONALITY

The importance of orthogonality, and just what it means, cannot be overemphasized. The fact that eigenvectors come in orthogonal sets makes them very special—they are the stuff solutions are made of. The matrix, itself, is synthesized by its eigenvectors. Equation (6.14) rewritten, here:

$$\mathbf{A} = \sum_j^n \lambda_j \{\mathbf{v}_j\}[\mathbf{u}_j]; \qquad \mathbf{u}_i \bullet \mathbf{v}_j = \delta_{ij} \,.$$

The solution to equation sets involves some sort of "diagonalizing" (reduction) of the matrix so that a solution for one of the variables can be made without interference from the others (i.e., decoupling the original equations). Note the simplification that occurs if the eigenvalues of the matrix are known in advance:

Given $\mathbf{Ax} = \mathbf{c}$, just transform the $\mathbf{x}$ vector by $\mathbf{x} = \mathbf{Vz}$. Then:

$$\mathbf{Ax} = \mathbf{c} \ \Rightarrow \ \mathbf{AVz} = \mathbf{c}$$
$$\mathbf{UAVz} = \mathbf{\Lambda z} = \mathbf{Uc} \,.$$

Now, the equations are decoupled in the variables, z. Each equation can be solved individually, the Λ matrix is inverted by simply taking the reciprocals of its diagonal elements. Then just transform back to the x-set, $\mathbf{z} = \mathbf{Ux}$:

$$\mathbf{z} = \Lambda^{-1}\mathbf{Uc}$$
$$\mathbf{Ux} = \Lambda^{-1}\mathbf{Uc}$$
$$\mathbf{VUx} = \mathbf{x} = \mathbf{V}\Lambda^{-1}\mathbf{Uc} \,.$$

Granted, in the general case this approach is not practical, because it is at least as difficult to obtain the eigenvalue analysis as it is to invert the original matrix. However, the point here is to illustrate the *"power"* of the orthogonal eigenvector set. Furthermore, in the next chapter this approach is used, and is practical in the case of differential equation sets.

6.4.1   INVERSE OF THE CHARACTERISTIC MATRIX

The inverse of the characteristic matrix is found in the same manner as above. It will be shown here because of its importance to the solution of differential equation sets in Chapter 7.

The solution to $[\mathbf{A} - \lambda\mathbf{I}]\mathbf{x} = \mathbf{d}$ is required—tantamount to the inversion of $[\mathbf{A} - \lambda\mathbf{I}]$. Orthogonality of the eigenvectors is required; thus the eigenvalue analysis of $\mathbf{A}$ (i.e., the matrices $\mathbf{U}, \mathbf{V}$ and Λ) must be known. The solution will be shown here for the non-symmetric case.

The vector $\mathbf{x}$ is first transformed via $\mathbf{x} = \mathbf{Vz}$ (then $\mathbf{z} = \mathbf{Ux}$)

$$[\mathbf{A} - \lambda\mathbf{I}]\mathbf{Vz} = \mathbf{d} \text{ and then premultiply by } \mathbf{U}:$$
$$\mathbf{U}[\mathbf{A} - \lambda\mathbf{I}]\mathbf{Vz} = \mathbf{Ud} = [\mathbf{UAV} - \mathbf{UIV}\lambda]\mathbf{z} = \mathbf{Ud}$$
$$= [\Lambda - \lambda\mathbf{I}]\mathbf{z} = \mathbf{Ud} \,.$$

But, the matrix $[\Lambda - \lambda\mathbf{I}]$ is easy to invert—it is a diagonal matrix. So $\mathbf{z} = [\Lambda - \lambda\mathbf{I}]^{-1}\mathbf{Ud}$. Now, $\mathbf{x}$ is determined by the inverse transform $\mathbf{z} = \mathbf{Ux}$

$$\mathbf{Ux} = [\Lambda - \lambda\mathbf{I}]^{-1}\mathbf{Ud}, \;\; \rightarrow \;\; \mathbf{x} = \mathbf{V}[\Lambda - \lambda\mathbf{I}^{-1}\mathbf{Ud} \,.$$

Then, the inverse of the characteristic matrix is

$$[\mathbf{A} - \lambda\mathbf{I}]^{-1} = \mathbf{V}[\Lambda - \lambda\mathbf{I}]^{-1}\mathbf{U} \;\; \textbf{Note:} \; \lambda \neq \lambda_k \,. \tag{6.38}$$

This equation can be interpreted in the manner of (6.14):

$$[\mathbf{A} - \lambda\mathbf{I}]^{-1} = \sum_{j}^{n} \frac{\{\mathbf{v}_j\}[\mathbf{u}_j]}{\lambda - \lambda_j} \,; \lambda \neq \lambda_j \,. \tag{6.39}$$

The fundamentally important concept of orthogonality is not just found in matrix analysis. It carries over from orthogonality of vector sets into orthogonality of continuous functions within a given range. Our first exposure to the concept is in the determination of Fourier series coefficients.

An excellent example of the way that an orthogonal set of eigenvectors is used to build the solution to a problem is given in the following paragraphs. It then shows the "evolution" of the matrix/vector solution into the continuous solution of the vibrating string problem.

6.4.2   VIBRATING STRING PROBLEM

A tightly stretched string of lengh L and mass M, vibrates freely following an initial deformation. The problem is to determine the equations of the vibration at points along the string as functions of time.

The matrix approach, summarized here, divides the continuous string into n parts of mass m, and concentrating it into a single point at the center (of the part). Points, m_k, are located horizontally by x_k; the deflection of the string at that point is measured by y_k. A load, P, is applied at the kth point (loads can be applied only at these points), and a "free-body diagram" at that point determines the displacement, $y_k(x_k)$, as a function of the load, the tension, T, the position of the point, x_k, and the point at which the load is applied.

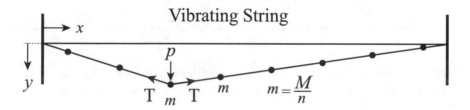

Summarizing for all points, a load vector, $\mathbf{p}$, is formed, and the resulting matrix equation relating the displacements of the loads is $\mathbf{y} = \mathbf{Wp}$.

The elements of the vector, $\mathbf{y}$, are the displacements at the sequential points. The $\mathbf{p}$ vector gives the load at these points, and the elements, w_{ij}, of the *symmetric* matrix $\mathbf{W}$, are the deflections at x_i, due to unit loads at x_j. $\mathbf{W}$ is referred to as the "influence matrix."

Appendix C develops the following set of second order differential equations:

$$\mathbf{y}(t) = -\frac{LM}{T}\mathbf{W}\ddot{\mathbf{y}}(t)\; ;\; \text{with } w_{ij} = \frac{1}{n^3}(i - \tfrac{1}{2})(n - j + \tfrac{1}{2})\; ;\; \text{for } i \leq j\; .$$

The solution to this equation is a weighted sum of the eigenvectors of $\mathbf{W}$:

$$\mathbf{y}(t) = \sum_{r=1}^{n} \mathbf{v}_r(a_r \cos \omega_r t + b_r \sin \omega_r t) \tag{6.40}$$

and the orthogonality of these vectors is used to determine the coefficients a_r and b_r. Specifically, note what happens when the solution equation is multiplied by $\mathbf{v}'_s$. Since the vector set is an orthogonal one, only one term in the series survives, "decoupling" the a_s (or b_s) coefficient.

Note that the vectors are "spatial" in the sense that they describe a possible spatial shape of deflections along the string. They are not time-variable (although they are multiplied by time variable functions). These spatial-template shapes are called "normal modes," and they can be plotted along an abscissa in the x dimension; such as that, below.

The graph shown here plots the first four eigenvectors, with the string divided into 12 parts. The black rectangles represent the mass points along the string. It is evident that these modes are in the shape of sinusoids, and are an orthogonal set (easiest to see this are numbers 1 and 2).

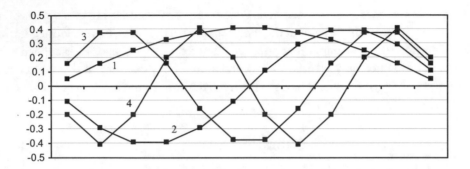

The continuous function approach: As the number of divisions of the string increases toward infinity, the vector function, $\mathbf{y}(t)$, becomes a continuous function $y(x, t)$. There comes a point in its solution when

$$y(x, t) = \sum_{n=1}^{\infty} \sin\left(\frac{n\pi x}{L}\right)\left(A_n \cos \frac{n\pi at}{L} + B_n \sin \frac{n\pi at}{L}\right). \tag{6.41}$$

The similarity between this and the matrix approach is striking! Compare (6.40) to (6.41). In this case, the solution is an infinite **summation of (continuous) sinusoidal functions**, that are an orthogonal set, over the interval from 0 to L:

$$\text{Note that } \int_0^L \sin \frac{n\pi x}{L} \sin \frac{k\pi x}{L} = \begin{cases} 0, & k \neq n \\ \frac{L}{2}, & k = n \end{cases}.$$

Then, to determine the coefficients, A_n and B_n, (6.39) is multiplied by $\sin \frac{n\pi x}{L}$ and integrated over the interval—very much like taking the dot product of two modes of the vector $\mathbf{y}(t)$.

6.5   THE CAYLEY-HAMILTON THEOREM

Intertwined with the eigenvalue analysis is a most amazing, and famous result, independently found by Cayley and Hamilton. It has no parallel in conventional algebra. Briefly, this theorem states that **any square matrix identically satisfies its own characteristic equation**. The most direct way to develop this theorem is given by Lanczos [3] as follows:

The equation $[\mathbf{A} - \lambda_1\mathbf{I}]\mathbf{x} = \mathbf{0}$ is solved by $c_1\mathbf{v}_1$, c_1 arbitrary. The equation:

$$[\mathbf{A} - \lambda_1\mathbf{I}][\mathbf{A} - \lambda_2\mathbf{I}]\mathbf{x} = \mathbf{0}$$

is solved by $\mathbf{x} = c_1\mathbf{v}_1 + c_2\mathbf{v}_2$ (i.e., any linear combination of $\mathbf{v}_1$ and $\mathbf{v}_2$):

$$[\mathbf{A} - \lambda_1\mathbf{I}][\mathbf{A} - \lambda_2\mathbf{I}](c_1\mathbf{v}_1 + c_2\mathbf{v}_2) = [\mathbf{A} - \lambda_1\mathbf{I}][\mathbf{A} - \lambda_2\mathbf{I}]c_1\mathbf{v}_1 + [\mathbf{A} - \lambda_1\mathbf{I}][\mathbf{A} - \lambda_2\mathbf{I}]c_2\mathbf{v}_2 .$$

The second term is obviously zero. The first term:

$$[\mathbf{A} - \lambda_1\mathbf{I}][\mathbf{A} - \lambda_2\mathbf{I}]c_1\mathbf{v}_1 = c_1[\mathbf{A} - \lambda_1\mathbf{I}](\mathbf{A}\mathbf{v}_1 - \lambda_2\mathbf{v}_1)$$
$$= c_1[\mathbf{A} - \lambda_1\mathbf{I}](\lambda_1\mathbf{v}_1 - \lambda_2\mathbf{v}_1) .$$

Which is also obviously zero. Using this same reasoning, adding one more term at each step, we see that

$$(\mathbf{A} - \lambda_1\mathbf{I})(\mathbf{A} - \lambda_2\mathbf{I})(\mathbf{A} - \lambda_3\mathbf{I}) \cdots (\mathbf{A} - \lambda_n\mathbf{I})\mathbf{x} = \mathbf{0}$$

is satisfied by any linear combination of all the eigenvectors. But, these vectors are linearly independent, and they fill the n-space. Thus, any vector at all can be represented by such a linear combination. *Then every n-dimensional vector satisfies this equation.* The only way that this could occur is that the equation is an identity.

$$(\mathbf{A} - \lambda_1\mathbf{I})(\mathbf{A} - \lambda_2\mathbf{I})(\mathbf{A} - \lambda_3\mathbf{I}) \cdots (\mathbf{A} - \lambda_n\mathbf{I}) \equiv \mathbf{0} \quad \textit{identically} . \tag{6.42}$$

It must be noted here that this proof assumes that the matrix $\mathbf{A}$ has a full set of eigenvectors. For some "defective" matrices (that are non-symmetric, and have repeated eigenvalues), a full set does not exist. However, it has been proven, via a limiting process, that even in the defective case, the Cayley-Hamilton theorem is still true. It should also be mentioned that not all non-symmetric matrices, even with repeated roots, are "defective."

From Section 6.2 above, given the matrix, $\mathbf{A}$, the determinant of $[\mathbf{A} - \lambda\mathbf{I}]$ expands to the characteristic equation:

$$f(\lambda) = c_0\lambda^n + c_1\lambda^{n-1} + \cdots + c_{n-1}\lambda + c_n = 0 . \tag{(6.6)rewrite}$$

The Cayley Hamilton theorem states that:

$$f(\mathbf{A}) = c_0\mathbf{A}^n + c_1\mathbf{A}^{n-1} + \cdots + c_{n-1}\mathbf{A} + c_n\mathbf{I} \equiv [\mathbf{0}] . \tag{6.43}$$

An amazing and powerful theorem. For example, by multiplying through by $\mathbf{A}^{-1}$

$$c_0\mathbf{A}^{n-1} + c_1\mathbf{A}^{n-2} + \cdots + c_{n-1}\mathbf{I} + c_n\mathbf{A}^{-1} = [\mathbf{0}] .$$

Then

$$\mathbf{A}^{-1} = (c_0\mathbf{A}^{n-1} + c_1\mathbf{A}^{n-2} + \cdots + c_{n-1}\mathbf{I})/c_n . \tag{6.44}$$

By the same reasoning (6.43) shows that any power of $\mathbf{A}(n\mathrm{X}n)$ can be represented in terms of powers of $\mathbf{A}$ no greater than $n - 1$. For example, the $\mathbf{A}$ matrix shown here is that which was used in the previous eigenvalue analysis, the characteristic equation was:

$$f(\lambda) = \lambda^3 - 7\lambda - 6 = 0 .$$

Then

$$A^3 = 7A + 6I$$
$$A^4 = 7A^2 + 6A, \quad \text{and}$$
$$A^5 = 7A^3 + 6A^2 = 6A^2 + 49A + 42I .$$

Any power of this A is a function of the A matrix raised to *powers no greater than 2*, and the unit matrix. Using the given A matrix, try it!

$$A = \begin{bmatrix} 25 & -44 & 18 \\ 12 & -21 & 8 \\ -3 & 6 & -4 \end{bmatrix} .$$

The eigenvalue analysis and Cayley Hamilton theorem also provide the solution to the analysis of A^{-1}. Define B as A^{-1}. Now premultiply (6.43) by B^n

$$f(B) = c_0 I + c_1 B + \cdots + c_{n-1} B^{n-1} + c_n B^n = [0] . \tag{6.45}$$

Then the characteristic polynomial for B has the same coefficients as that for A, except in reverse order—and therefore its roots are the reciprocals of those of $f(A)$ (see the Appendix B, "Polynomials"). But are the eigenvectors of B the same as those of A? By definition, $BA = I$. Now assume that B has the same eigenvectors, and set $B = V\Lambda_b U$ ($A = V\Lambda_a U$ has already been shown to be true). The matrices Λ_a and Λ_b are the eigenvalue matrices of A and B.

$$BA = V\Lambda_b UA = (V\Lambda_b U)(V\Lambda_a U) = (V\Lambda_b)(\Lambda_a U) = VU = I . \tag{6.46}$$

Since this product does produce the unit matrix, and since the inverse of A must be unique, it follows that $B = V\Lambda_b U$ is that inverse—i.e., that the eigenvectors of B are the same as those of A, and the eigenvalues are the reciprocals.

The calculations involved in an eigenvalue analysis are at least as complex as those involved in the inversion process. Therefore, it is unlikely that an eigenvalue analysis would ever be done just to determine A^{-1}. Perhaps it might be useful when A is very nearly singular.

6.5.1   FUNCTIONS OF A SQUARE MATRIX

This discussion makes frequent use of the transforms defined by the eigenvalue problem and is therefore limited to square matrices, A, which have a full complement of eigenvectors.

Note that $A = V\Lambda U$, and $A^2 = (V\Lambda U)(V\Lambda U) = V\Lambda^2 U$. Extending this

$$A^n = V\Lambda^n U \tag{6.47}$$

which shows that the eigenvectors of A^n are the same as those for A, and the eigenvalues are the nth power of those of A. This same argument holds for any polynomial in A.

$$P(A) = c_0 A^n + c_1 A^{n-1} + \cdots + c_{n-1} A + c_n \tag{6.48}$$

by transforming $\mathbf{A} = \mathbf{V}\Lambda\mathbf{U}$, the polynomial, $P(\mathbf{A})$ is diagonalized to $P(\Lambda)$. Then $P(\mathbf{A})$ also has the eigenvectors of $\mathbf{A}$, and its jth eigenvalue equals $P(\lambda_j)$, where λ_j is the jth eigenvalue of $\mathbf{A}$. An example is shown above for $P(\mathbf{A}) = \mathbf{A}^5$. In addition, if (6.48) is postmultiplied by $\mathbf{v}_k$, and since $\mathbf{A}^k \mathbf{v}_k = \lambda^k \mathbf{v}_k$:

$$P(\mathbf{A})\mathbf{v}_k = P(\lambda_k)\mathbf{v}_k \ . \tag{6.49}$$

General Polynomial Fnctions

The algebraic effort involved in actually expressing $P(\mathbf{A})$ in terms of the lower degrees of $\mathbf{A}$ can be daunting. The following development will help a great deal. Note, here that the general polynomial is given as P (i.e., upper case), while the characteristic polynomial will be denoted p (lower case). Then, first divide P by p. The result will be a quotient, Q, and a remainder, R:

$$\frac{P(x)}{p(x)} = Q(x) + \frac{R(x)}{p(x)} \ ; \quad \text{Then } P(x) = p(x)Q(x) + R(x) \text{ and therefore:} \tag{6.50}$$

$$P(\mathbf{A}) = p(\mathbf{A})Q(\mathbf{A}) + R(\mathbf{A}) \ .$$

But, $p(\mathbf{A})$ is identically equal to zero, by the Cayley-Hamilton theorem. So, $P(\mathbf{A}) = R(\mathbf{A})$. A simple example is when x^5 is divided by $p(x)$, the characteristic polynomial for the example matrix at the beginning of Section 6.5. The remainder is $6x^2 + 49x + 42$, which is consistent with the $\mathbf{A}^5$ given above.

For a formidable-looking example, find $P(\mathbf{A})$, where $P(x)$ given below, and $\mathbf{A}$ is the same matrix

$$P(x) = x^6 - x^5 - 7x^4 + 31x^3 + 40x^2 - 19x + 5 \ .$$

The bulk of the work can be done before $\mathbf{A}$ is inserted. Just use synthetic division to divide $P(x)$ by the characteristic polynomial, $p(x) = x^3 - 7x - 6$, and then retain the remainder, $10x^2 - 5x + 17$.

$$P(\mathbf{A}) = 10\mathbf{A}^2 - 5\mathbf{A} + 17\mathbf{I} \ .$$

This method, is handy, easy to use for polynomial functions. It can be extended beyond matric functions to analytic functions. However, its extension involves the Lagrange polynomials (see Chapter 4, Section 4.5) and arrives at the same method that is to be discussed next.

6.5.2   SYLVESTER'S THEOREM

General functions of A. This method is directly related to the Lagrange interpolation method, and could possibly be deduced from it. To derive it, first define the characteristic polynomial as $p(\lambda)$, and then consider the polynomial, $p_k(\lambda)$:

$$p_k(\lambda) = \prod_{j \neq k}^{n} (\lambda - \lambda_j); \ \text{ for example } p_1(\lambda) = (\lambda - \lambda_2)(\lambda - \lambda_3) \cdots (\lambda - \lambda_n)$$

which contains all the factors in $p(\lambda)$ except $(\lambda - \lambda_k)$. For the $\mathbf{A}$ matrix shown, $p_1(\lambda) = (\lambda + 2)(\lambda - 3)$. Then:

$$p_k(\mathbf{A}) = \prod_{j \neq k}^{n} (\mathbf{A} - \lambda_j \mathbf{I}) . \tag{6.51}$$

For the $\mathbf{A}$ matrix given here, $p_1(\mathbf{A}) = (\mathbf{A} + 2\mathbf{I})(\mathbf{A} - 3\mathbf{I})$, and in the general (nXn) case, there will be $n - 1$ $(\mathbf{A} - \lambda \mathbf{I})$ terms. Each p_k will be a polynomial of degree $n - 1$.

$$\mathbf{A} = \begin{bmatrix} 25 & -44 & 18 \\ 12 & -21 & 8 \\ -3 & 6 & -4 \end{bmatrix}$$

Note that $p_k(\mathbf{A})\mathbf{v}_j = \mathbf{0}$, except when $j = k$, as was shown in the development of the Cayley-Hamilton theorem. And when $j = k$ (in the example 3X3)

$$p_1(\mathbf{A})\mathbf{v}_1 = (\mathbf{A} + \lambda_2 \mathbf{I})(\mathbf{A} - \lambda_3 \mathbf{I})\mathbf{v}_1 = (\lambda_1 - \lambda_2)(\lambda_1 - \lambda_3)\mathbf{v}_1 .$$

(To derive this, use the fact that $\mathbf{A}\mathbf{v}_1 = \lambda_1 \mathbf{v}_1$.)

In general: $p_k(\mathbf{A})\mathbf{v}_k = \mathbf{v}_k \prod_{j \neq k}(\lambda_k - \lambda_j) . \tag{6.52}$

Now, define the problem: Given a general polynomial, $P(\mathbf{A})$, determine a set of n coefficients c_k, such that

$$P(\mathbf{A}) = \sum_{k=1}^{n} c_k p_k(\mathbf{A}) . \tag{6.53}$$

Now, just postmultiply successively by $\mathbf{v}_j (j = 1, 2, , n)$. When $j = k$: and $P(\mathbf{A})\mathbf{v}_k = P(\lambda_k)\mathbf{v}_k$

$$P(\lambda_k)\mathbf{v}_k = c_k \prod_{j \neq k} (\lambda_k - \lambda_j)\mathbf{v}_k \quad \text{and solving for } c_k:$$

$$c_k = \frac{P(\lambda_k)}{\prod_{j \neq k} (\lambda_k - \lambda_j)} . \tag{6.54}$$

Plugging these constants back into (6.53), with the definitions of the $p_k(\mathbf{A})$ polynomials:

$$P(\mathbf{A}) = \sum_{k}^{n} P(\lambda_k) \frac{\prod_{j \neq k} (\mathbf{A} - \lambda_j \mathbf{I})}{\prod_{j \neq k} (\lambda_k - \lambda_j)} . \tag{6.55}$$

The ratios of product factors are often referred to as $Z_k(\mathbf{A})$, and (6.55) is written

$$P(\mathbf{A}) = \sum_{k}^{n} P(\lambda_k) Z_k(\mathbf{A}) . \tag{6.56}$$

The foregoing development assumes *distinct eigenvalues* . In (6.55) the numerator terms are the adjoint of the matrix $\mathbf{A}(\lambda) = [\mathbf{A} - \lambda\mathbf{I}]$, and the denominator is the derivative of $p(\lambda)$ evaluated at λ_k. The equation can be generalized, and rewritten as:

$$F(\mathbf{A}) = \sum_{k=1}^{n} F(\lambda_k)\frac{A^a(\lambda_k)}{p'(\lambda_k)} \tag{6.57}$$

where it is known as Sylvester's Theorem. Equations (6.56) and (6.57) are more different than they appear. The function F can be any analytic function, and $\mathbf{A}^a$ is the adjoint of $[\mathbf{A} - \lambda\mathbf{I}]$ whether or not it has repeated roots (and the function p also represents the lowest degree polynomial satisfied by $\mathbf{A}$). Thus, Sylvester's Theorem is more general than (6.56).

When the matrix, $\mathbf{A}$, has distinct eigenvalues Equations (6.55) and (6.57) are the same. That is

$$\mathbf{A}^a(\lambda_k) = \prod_{j\neq k}^{n} (\mathbf{A} - \lambda_j\mathbf{I}) \text{ and } p'(\lambda_k) = \prod_{j\neq k}^{n} (\lambda_k - \lambda_j)$$

and (6.56) will be extended into analytic functions which possess an infinite series expansion. The question of convergence will not be addressed. However, the series themselves converge, and the Cayley-Hamilton theorem says that any sub-series of terms can be written in terms of a polynomial of degree $n - 1$. Therefore, convergence will be assumed.

Then the matrix series $e^{\mathbf{A}} = \sum_{k=0}^{\infty} \frac{\mathbf{A}^k}{k!}$ is a valid equation. Suppose $\mathbf{A}$ is 3X3. Then

$$Z_1 = \frac{(\mathbf{\Lambda} - \lambda_2\mathbf{I})(\mathbf{\Lambda} - \lambda_3\mathbf{I})}{(\lambda_1 - \lambda_1)(\lambda_1 - \lambda_3)}; \; Z_2 = \frac{(\mathbf{A} - \lambda_1\mathbf{I})(\mathbf{A} - \lambda_3\mathbf{I})}{(\lambda_2 - \lambda_1)(\lambda_2 - \lambda_3)}; \text{ and } Z_3 = \frac{(\mathbf{A} - \lambda_2\mathbf{I})(\mathbf{A} - \lambda_3\mathbf{I})}{(\lambda_3 - \lambda_1)(\lambda_3 - \lambda_2)}$$

and

$$e^{\mathbf{A}} = e^{\lambda_1}Z_1 + e^{\lambda_2}Z_2 + e^{\lambda_3}Z_3 . \tag{6.58}$$

It will take a lot of algebraic manipulation to "condense" Equation (6.58) into a single matrix; but note that it's just algebra. The usual phrase here is "This will be left as an exercise for the student."

6.6   MECHANICS OF THE EIGENVALUE PROBLEM

Efficient eigenvalue analysis is a problem in numerical analysis—beyond the scope of this work. The steps described below are those that illustrate the problem and the matrix characteristics. They are

- Determine the characteristic equation (calculation of the polynomial coefficients).

- Factor the characteristic equation, to obtain the eigenvalues, λ_i.

- For each value, λ_i, find the corresponding eigenvectors.

In a later section, a more sophisticated method is presented, which cleverly transforms the given matrix into one whose eigenvalues and eigenvectors are easily calculated—even when these are complex numbers. Known as Danilevsky's method, it is far superior to these methods for realistic matrices. And, even so, there may be methods that are superior to Danilevsky's.

6.6.1   CALCULATING THE CHARACTERISTIC EQUATION COEFFICIENTS

Pipes[1] reports that Maxime Bôcher has shown that the coefficients are related to the "traces" (sum of the diagonal elements) of the powers of the input matrix, A. Let S_j denote the trace of the ith power of A:

$$S_1 = \text{Trace}[A] = \text{Tr}[A], \qquad S_2 = \text{Trace}[A^2], \ldots \quad S_n = \text{Trace}[A^n]$$

then the coefficients, c_k, of the characteristic Equation (6.6) are calculated successively, as follows:

$$
\begin{aligned}
c_0 &= 1 \\
c_1 &= -S_1 \\
c_2 &= -(c_1 S_1 + S_2)2; \qquad \text{and, in general:} \\
c_k &= -(c_{k-1} S_1 + c_{k-2} S_2) + \ldots + c_1 S_{k-1} + S_k)/k
\end{aligned}
\qquad (6.59)
$$

This relationship is easily programmed, providing an easy method for developing $p(\lambda)$. Also, the powers of the A matrix can be saved to be used later (in determining the adjoints, $A^a(\lambda_i)$).

6.6.2   FACTORING THE CHARACTERISTIC EQUATION

There are handbook methods for factoring polynomials up to degree 4. Although there will not be any examples herein resulting in $p(\lambda)$ of higher degree, Appendix B, "Polynomials," discusses polynomial arithmetic and outlines computer methods, including root determination, real or complex.

Finding the roots of a polynomial requires a computer; and the computer routines for polynomial manipulation are very simple. See Appendix B.

6.6.3   CALCULATION OF THE EIGENVECTORS

Using Gauss-Jordan Reduction

The matrix $[A - \lambda_j I]$ is singular, and in this discussion will be assumed to have rank $n - 1$. Then, the Gauss-Jordan is an excellent tool to derive the eigenvectors one at a time. The method is described in Section 3.3. A 4X4 will be used in illustration from the point at which the Gauss-Jordan reduction of $[A - \lambda_j I]$ terminates. If $A(\lambda_j)$ is complex, the reduction must be done in a complex arithmetic.

$$
\begin{Vmatrix}
1 & 0 & 0 & z_1 \\
0 & 1 & 0 & z_2 \\
0 & 0 & 1 & z_3 \\
0 & 0 & 0 & 0
\end{Vmatrix}
$$

The reduced matrix will appear as in the diagram. If λ_j is complex, then the "z" values shown here will be complex. There will be a complete row of zero values along the bottom, showing that a solution $\{x_1, x_2, x_3, x_4\}$ does exist, with the value for x_4 chosen arbitrarily, say k.

[1]See [4], page 90.

The complete solution is: $\mathbf{x}_1 = \left\{ \begin{array}{c} -z_1 \\ -z_2 \\ -z_3 \\ 1 \end{array} \right\} k.$

The reduction/solution will have to be repeated for each eigenvalue, and again for the transposed matrix to obtain the row eigenvectors.

Calculation of the Adjoint of $[\mathbf{A} - \lambda_j \mathbf{I}]$

This method has been shown in an earlier example. It derives both the row and column eigenvectors together.

From the Cayley-Hamilton theorem, denoting the characteristic equation as $p(\lambda) = 0$:

$$p(\mathbf{A}) = [\mathbf{A} - \lambda_1 \mathbf{I}][\mathbf{A} - \lambda_2 \mathbf{I}] \cdots [\mathbf{A} - \lambda_n \mathbf{I}] = [\mathbf{0}] .$$

Since the numbering of the eigenvalues is arbitrary, we can write the ith term first in the above, and then gather the rest of the product terms into a polynomial, p_i:

$$[\mathbf{A} - \lambda_i \mathbf{I}] p_i (\mathbf{A}) = [\mathbf{0}] \qquad (6.60)$$

where $p_i (\mathbf{A}) = \prod_{k \neq i}^{n} [\mathbf{A} - \lambda_k \mathbf{I}]; \quad (n - 1 \text{ product terms}).$

Since $\mathbf{A}(\lambda_i) \mathbf{A}^a (\lambda_i) = |\mathbf{A}(\lambda_i)| \mathbf{I} = [\mathbf{0}]$, and comparing this to (6.60), note that $p_i(\mathbf{A})$ is the adjoint of $\mathbf{A}(\lambda_i)$. $p_i(\mathbf{A})$ will not be null, as long as λ_i is distinct—not a repeated root of the characteristic equation. Therefore, $p_i(\mathbf{A})$ will be the source of the eigenvectors.

p_i can be found by synthetic division of $p(\lambda)$. If the synthetic division is done in complex arithmetic, then $p_i(\lambda)$ is found by the synthetic division of $p(\lambda)$ by $(\lambda - \lambda_i)$. If the division routine accepts only reals, and λ_i is complex, $(a + jb)$, then its conjugate is also a root and the divisor can be the quadratic, $\lambda^2 + 2a\lambda + a^2 + b^2$.

The result of this division must then be multiplied by $(\lambda - a + jb)$:

$$p_i (\lambda) = \frac{P(\lambda)}{\lambda^2 + 2a\lambda + a^2 + b^2} \times (\lambda - a + jb) \qquad (6.61)$$

When $p_i(\lambda)$ has been found, matrix multiplications are then needed to derive $p_i(\mathbf{A}) = \mathbf{A}^a(\lambda_i)$.

This method of determining the adjoint of $[\mathbf{A} - \lambda_i \mathbf{I}]$, containing the eigenvectors, has the advantage that operations with complex numbers are minimized. Only the final multiplication by $[\mathbf{A} - (a + jb)\mathbf{I}]$ involves complex arithmetic. The powers of the original matrix are available, having been calculated for defining the coefficients of the characteristic polynomial.

When the initial row, $\mathbf{x}$, and column, $\mathbf{z}$, vectors have been determined (in general complex numbers), they must be normalized—usually such that $\mathbf{x}_j \bullet \mathbf{z}_i = 1$—defining $\mathbf{v}_j$ and $\mathbf{u}_i$.

6.7   EXAMPLE EIGENVALUE ANALYSIS

6.7.1   EXAMPLE EIGENVALUE ANALYSIS; COMPLEX CASE

The methods of eigenvalue analysis discussed in Section 6.6 are valid for matrices whose eigenvalues and eigenvectors are complex. The following matrix analysis follows the outlined method, showing the complex results. The given matrix is:

$$\mathbf{A} = \begin{bmatrix} -4.0 & 3.0 & 3.0 \\ 5.0 & -2.0 & 2.0 \\ 0.0 & 6.0 & 1.0 \end{bmatrix} \tag{6.62}$$

The given $\mathbf{A}$ matrix is non-symmetric. Its elements are integer (shown in decimal form). Since it is of third order, expect at least one real root (to the characteristic equation); and if there are complex roots, these will emerge in complex conjugate pairs.

The traces of the powers of $\mathbf{A}$ are given, below. Bôchers Formulae are then used to find the coefficients of the characteristic polynomial, $p(x)$:

	Traces		Coefficients	
$[\mathbf{A}]$	-5.0		c_1	5.0
$[\mathbf{A}]^2$	75.0		c_2	-25.0
$[\mathbf{A}]^3$	-107.0		c_3	-131.0

Of course, $c_0 = 1$, and the characteristic equation reads:

$$f(\lambda) = \lambda^3 + 5\lambda^2 - 25\lambda - 131 = 0 . \tag{6.63}$$

The three roots of this polynomial are the eigenvalues (λ_i) of $\mathbf{A}$. They are:

$$\lambda_1 = 5.05929 + j0.00000$$
$$\lambda_2 = -5.02965 + j0.77174$$
$$\lambda_3 = -5.02965 - j0.77174 .$$

The termination point of the *Gauss-Jordan reduction* is shown below for λ_2, for both $\mathbf{A}$ and $\mathbf{A}'$:

	$[\lambda_2\mathbf{I} - \mathbf{A}]$			$[\lambda_2\mathbf{I} - \mathbf{A}']$	
1.0	0.0	-0.189071	1.0	0.0	2.261977
0.0	0.0	0.233048	0.0	0.0	-0.567977
0.0	1.0	1.004941	0.0	1.0	-0.378141
0.0	0.0	-0.128624	0.0	0.0	0.466093
0.0	0.0	0.0	0.0	0.0	0.0
0.0	0.0	0.0	0.0	0.0	0.0

In the above table, the matrix elements are complex, with the imaginary parts shown below the reals. In both cases, the 3rd element value can be chosen arbitrarily (choose $1 + j0$), and the column and row vectors are

$$\mathbf{x} = \left\{ \begin{array}{c} 0.189071 - j0.233048 \\ -1.0004941 + j0.128624 \\ 1.0 + j0.0 \end{array} \right\} ; \text{ and } \mathbf{z} = \left\{ \begin{array}{c} -2.261977 + j0.567977 \\ 0.378141 - j0.466095 \\ 1.0 + j0.0 \end{array} \right\}.$$

After normalization, $\mathbf{x}$ and $\mathbf{z}$ will become the eigenvectors $\mathbf{v}_2$ and $\mathbf{u}_2$. Further, $\mathbf{v}_3$ and $\mathbf{u}_3$ are just the complex conjugates of $\mathbf{v}_2$ and $\mathbf{u}_2$.

The *adjoint method* is illustrated by calculating $A_{adj}(\lambda_2)$. In this case the result of the division indicated in (6.61) is simply $(\lambda - \lambda_1)$, and $p_2(\lambda) = (\lambda - \lambda_1)(\lambda - \lambda_3)$. Then $A_{adj}(\lambda_2)$ is:

$$\left\| \begin{array}{ccc} 5.672121 & -0.0889421 & -3.088942 \\ -6.99143 & 2.315223 & 2.315223 \\ \\ -30.14824 & 5.612826 & 12.94071 \\ 3.858705 & -5.447948 & 1.543482 \\ \\ 30.000 & -6.177884 & -12.47612 \\ 0.0000 & 4.630447 & -3.132725 \end{array} \right\|$$

If the first column of this table is divided by 30.00 it will show agreement with the x column obtained by the Gauss-Jordan reduction. Note that this table yields row and column vectors for both the complex eigenvalues, because they are complex conjugates.

The Normalized Eigenvectors

The eigenvectors emerge "in direction only." Their magnitudes are arbitrary. As before, the row eigenvectors are the rows of the matrix $\mathbf{U}$; and the column eigenvectors are the columns of $\mathbf{V}$. Then, we will normalize these vectors such that $\mathbf{UV} = \mathbf{I}$, by dividing each element of both $\mathbf{u}_i$ and $\mathbf{v}_i$ by the square root of the unnormalized dot product. The resulting (complex) $\mathbf{U}$ and $\mathbf{V}$ matrices are given in the table below. *The imaginary parts are again shown directly below the reals*

V Matrix			U Matrix		
0.40334	−0.70538	−0.70538	0.40334	0.73079	0.65814
0.00000	−0.28089	0.28089	0.00000	0.00000	0.00000
0.49150	1.06836	1.06836	−0.70538	0.18861	0.26401
0.00000	2.32999	−2.32999	−0.28089	−0.05103	0.19047
0.72648	−0.75401	−0.75401	−0.70538	0.18861	0.26401
0.00000	−2.41504	2.41504	0.28089	0.05103	−0.19047

6.7.2   EIGENVALUES BY MATRIX ITERATION

If $\mathbf{A}$ is a square matrix with distinct eigenvalues, then any arbitrary vector, x_0, can be expressed as a linear combination of the eigenvectors of $\mathbf{A}$. Thus, $\mathbf{x}_0 = \sum_{k=1}^{n} \alpha_k \mathbf{v}_k$. If the vector is multiplied by $\mathbf{A}$, the result is $\mathbf{x}_1 = \mathbf{A}\mathbf{x}_0$. Using $\mathbf{A}\mathbf{x} = \lambda\mathbf{x}$, write:

$$\mathbf{x}_1 = \mathbf{A}\mathbf{x}_0 = \sum_{k=1}^{n} \alpha_k \mathbf{A}\mathbf{v}_k = \sum_{k=1}^{n} \alpha_k \lambda_k \mathbf{v}_k \ . \tag{6.64}$$

Now, if a multiple of $\mathbf{x}_1$ (say, $w\mathbf{x}_1$) is premultiplied by $\mathbf{A}$, the result is $\mathbf{x}_2 = w \sum_{k=1}^{n} \alpha_k \lambda_k^2 \mathbf{v}_k$. If this process is continued, the term in the summation which has the largest eigenvalue will predominate. That is, after r iterations, the rth power of the largest eigenvalue, λ_k, will be much greater than the others, and

$$x_r \approx \mu \alpha_k \lambda_k^r \mathbf{v}_k \tag{6.65}$$

showing that the iterative process converges toward the eigenvector multiplied by the eigenvalue times a proportionality factor. We can absorb the factor into the vector, and control the iteration by its convergence to the eigenvalue.

This suggests a method for determining the largest eigenvalue and eigenvector of $\mathbf{A}$. Pick the first eigenvector arbitrarily, say $\{1, 1, 1, \text{,, } 1\}$. Premultiply $\mathbf{A}$. Then pick one of the elements of the resultant vector (say, the largest), and normalize the vector such that the chosen element becomes 1.0. Once you pick an element, stick with it—always normalize such that this one becomes unity. Save the normalizing factor; it converges to the eigenvalue. Repeat the process until the change in the factor is negligible.

The method is so simple that an example probably will show it best. In the table, the 3X3 matrix, $\mathbf{A}$, is at left. The multiplying vectors are next (the first one being all ones). Next comes the result of the matrix multiplication; then the normalizing—with the factor shown first

Matrix			x Vector	Product	Factor	Normalized
9	1	7	1.0000000	17.0000000		1.0000000
3	3	3	1.0000000 =	9.0000000 =	17.0000000	0.5294117
−3	−1	−1	1.0000000	−5.0000000		−0.2941170
9	1	7	1.0000000	7.4705882		1.0000000
3	3	3	0.5294120 =	3.7058823 =	7.4705882	0.4960630
−3	−1	−1	−0.2941170	−3.2352941		−0.4330702
9	1	7	1.0000000	6.4645669		1.0000000
3	3	3	0.4960630 =	3.1889764 =	6.4645669	0.4933009
−3	−1	−1	−0.4330702	−3.0629921		−0.4738124

In the example shown, if the iterations were continued, the eigenvalue would emerge as 6.00 and the eigenvector converges to $\{1.0, 0.5, -0.5\}$.

Note: this matrix is ***not symmetric***. It could therefore have had complex eigenvalues and vectors. In that case, the convergence is quite different. Although iteration does work for the complex case, it will not be discussed here. Even if the eigenvalues and vectors are real, it is necessary to transpose $\mathbf{A}$ and iterate for the row eigenvector (the eigenvalue will be the same).

If it is desired to continue the procedure for the next largest eigenvalue—vector, then the new matrix is formed by subtracting out the results of the first iteration: $\mathbf{B} = \mathbf{A} - \lambda_1 \mathbf{v}_1 \mathbf{u}_1$. As in any iterative procedure, it is necessary to keep many significant figures. Even then, only the first "few" results will be within acceptable accuracy. Usually iteration is only done on large symmetric matrices, and only to derive the first one or two eigenvalues—vectors.

6.8   THE EIGENVALUE ANALYSIS OF SIMILAR MATRICES; DANILEVSKY'S METHOD

The eigenvalue analysis of a matrix has always been considered formidable, especially before the digital computer was available to do the messy calculations. It is no surprise, therefore, to find that methods have been developed to shorten and simplify the work. In the present day, the best of these methods are those that are coded easily on the computer. A method will be discussed which uses a "similarity transform" to develop a new matrix whose eigenvalue analysis is very simple to perform. And in this case, the matrix transformation, somewhat analogous to the Gauss-Jordan reduction method in determinants, is not a difficult one.

In Chapter 5, Section 5.4, the subject of "similar" matrices is introduced. In particular, two matrices, say $\mathbf{A}$ and $\mathbf{P}$, are defined as being *"similar"* if there exists a relation:

$$\mathbf{A} = \mathbf{SPS}^{-1} . \tag{6.66}$$

That is, the pre- and postmultiplying transform matrices are inverses of one another. Of special interest at present is the fact that similar matrices are possessed of the same eigenvalues. To show this we begin with the Cayley-Hamilton equation for $\mathbf{A}$:

$$(\mathbf{A} - \lambda_1 \mathbf{I})(\mathbf{A} - \lambda_2 \mathbf{I})(\mathbf{A} - \lambda_3 \mathbf{I}) \cdots (\mathbf{A} - \lambda_n \mathbf{I})\mathbf{v} = \mathbf{0} . \tag{6.67}$$

We can substitute $\mathbf{A}$ from (6.66) into (6.67)

$$(\mathbf{SPS}^{-1} - \lambda_1 \mathbf{I})(\mathbf{SPS}^{-1} - \lambda_2 \mathbf{I})(\mathbf{SPS}^{-1} - \lambda_3 \mathbf{I}) \cdots (\mathbf{SPS}^{-1} - \lambda_n \mathbf{I})\mathbf{v} = \mathbf{0}$$

and since $\mathbf{I} = \mathbf{SS}^{-1}$:

$$(\mathbf{SPS}^{-1} - \lambda_1 \mathbf{SS}^{-1})(\mathbf{SPS}^{-1} - \lambda_2 \mathbf{SS}^{-1})(\mathbf{SPS}^{-1} - \lambda_3 \mathbf{SS}^{-1}) \cdots (\mathbf{SPS}^{-1} - \lambda_n \mathbf{SS}^{-1})\mathbf{v} = \mathbf{0}$$

$$\mathbf{S}(\mathbf{P} - \lambda_1 \mathbf{I})(\mathbf{P} - \lambda_2 \mathbf{I})(\mathbf{P} - \lambda_3 \mathbf{I}) \cdots (\mathbf{P} - \lambda_n \mathbf{I})\mathbf{S}^{-1}\mathbf{v} = \mathbf{0}$$

$$(\mathbf{P} - \lambda_1 \mathbf{I})(\mathbf{P} - \lambda_2 \mathbf{I})(\mathbf{P} - \lambda_3 \mathbf{I}) \cdots (\mathbf{P} - \lambda_n \mathbf{I})\mathbf{x} = \mathbf{0}; \quad \mathbf{x} = \mathbf{S}^{-1}\mathbf{v} . \tag{6.68}$$

(6.68) clearly shows the same Cayley-Hamilton equation, with the same eigenvalues, as (6.67).

Of course, there are infinities of similarity transforms. The trick is to find one in which the analysis of the $\mathbf{P}$ matrix is easier to perform than the analysis of the original $\mathbf{A}$.

In Danilevsky's method this is definitely the case.

6.8.1   DANILEVSKY'S METHOD

The objective of this method is to derive, from the given input $\mathbf{A}$ matrix, the *similar* $\mathbf{P}$ matrix:

$$\mathbf{P} = \begin{bmatrix} p_{11} & p_{12} & p_{13} & p_{14} \\ 1 & 0 & 0 & 0 \\ 0 & 1 & 0 & 0 \\ 0 & 0 & 1 & 0 \end{bmatrix}.$$

In the above, and some of the displays that follow, a 4X4 will be shown, in preference to writing out a completely general case. The 4X4 will be clearer to follow (extension to nXn should be obvious).

Note that the unity elements are not on the main diagonal – but, are one diagonal down. All of the data in the original $\mathbf{A}$ matrix has been "squeezed" into the elements of the first row. In order to derive the characteristic equation, we subtract λ from the main diagonal and solve for the determinant. This is most easily done by expanding in minors of the first row. The result:

$$f(\lambda) = \lambda^n - p_{11}\lambda^{n-1} \ldots - p_{1,n-1}\lambda - p_{1n} = 0, \quad \text{in general} \tag{6.69}$$
$$f(\lambda) = \lambda^4 - p_{11}\lambda^3 - p_{12}\lambda^2 - p_{13}\lambda - p_{14} = 0, \quad \text{in the 4X4}.$$

That is, the first row elements of $\mathbf{P}$ are none other than the (negatives of) the characteristic equation coefficients. We are assured that the eigenvalues of $\mathbf{P}$ are the same as those of $\mathbf{A}$, by the argument above. We must therefore conclude that the characteristic equation is the same, and is given by (6.69). *Once the characteristic equation is derived, a separate method is used to determine the eigenvalues, the roots of the polynomial.*

Since $\mathbf{P}$ appears so very different from $\mathbf{A}$, it would seem that the transform would be a very complex one. But, that is not the case. The transform is affected sequentially by a series of very simple matrices, $\mathbf{M}_{k-1}^{-1}$ and $\mathbf{M}_{k-1}$, where k takes on the values n, $n-1$, ..., 2, (n being the order of $\mathbf{A}$). Note that the $\mathbf{M}$ matrices are required to be inverses.

Then, the first transform will be $\mathbf{A}_{n-1} = \mathbf{M}_{n-1}^{-1}\mathbf{A}\mathbf{M}_{n-1}$. Next, transform $\mathbf{A}_{n-1}, \mathbf{A}_{n-2}$, etc.:

$$\mathbf{A}_{n-2} = \mathbf{M}_{n-2}^{-1}\mathbf{A}_{n-1}\mathbf{M}_{n-2} = \mathbf{M}_{n-2}^{-1}(\mathbf{M}_{n-1}^{-1}\mathbf{A}\mathbf{M}_{n-1})\mathbf{M}_{n-2}$$
$$\mathbf{A}_{n-3} = \mathbf{M}_{n-3}^{-1}\mathbf{A}_{n-2}\mathbf{M}_{n-3} = \mathbf{M}_{n-3}^{-1}(\mathbf{M}_{n-2}^{-1}(\mathbf{M}_{n-1}^{-1}\mathbf{A}\mathbf{M}_{n-1})\mathbf{M}_{n-2})\mathbf{M}_{n-3}$$

until finally
$$\begin{cases} \mathbf{P} = \mathbf{S}^{-1}\mathbf{A}\mathbf{S} = \mathbf{M}_1^{-1}\mathbf{M}_2^{-1}\cdots\mathbf{M}_{n-1}^{-1}[\mathbf{A}]\mathbf{M}_{n-1}\cdots\mathbf{M}_2\mathbf{M}_1 \\ \mathbf{S}^{-1} = \prod_{k=1}^{n-1} \mathbf{M}_k^{-1}; \text{ and } \mathbf{S} = \prod_{k=n-1}^{1} \mathbf{M}_k^{-1}. \end{cases}$$

A picture of the matrices $\mathbf{M}$ (with $k = n$) is:

$$\mathbf{M}_{n-1}^{-1} = \begin{bmatrix} 1 & 0 & 0 & 0 \\ 0 & 1 & 0 & 0 \\ a_{n1} & a_{n2} & a_{n3} & a_{nn} \\ 0 & 0 & 0 & 1 \end{bmatrix}; \mathbf{M}_{n-1} = \begin{bmatrix} 1 & 0 & 0 & 0 \\ 0 & 1 & 0 & 0 \\ -\dfrac{a_{n1}}{a_{n,n-1}} & -\dfrac{a_{n2}}{a_{n,n-1}} & +\dfrac{1}{a_{n,n-1}} & -\dfrac{a_{nn}}{a_{n,n-1}} \\ 0 & 0 & 0 & 1 \end{bmatrix}.$$

$$(6.70)$$

A description of these matrices is: (for $k = n, n\text{-}1, \ldots 2$)

Matrix $\mathbf{M}_{k-1}^{-1}$	**Matrix $\mathbf{M}_{k-1}$**
This matrix is a unit matrix, with its $k - 1$ row replaced by the elements of the kth row (in (6.70), $k = n$).	This matrix is a unit matrix, with its $k - 1$ row replaced by the negatives of the kth row elements divided by the $k, k - 1$ element. However, the $k - 1, k - 1$ element is positive, and just the reciprocal of the $k, k - 1$ element.

A note: Equation (6.70) shows the character "n," because in that display the second to last row is shown. But, in the later transforms, it is not the n-1 row that is modified. So, the references to "n" in these equations *changes*, but of course the order of the matrix does not. At each step an index "k" (whose initial value was n) will decrease, causing the corresponding row in $\mathbf{M}$ to move up.

For example, in the second step (i.e., n-2), we define $\mathbf{M}_{n-2}$ and $\mathbf{M}_{n-2}^{-1}$. They are constructed from a unit matrix, with the n-2nd row taken from elements of the n-1st row of the newly defined $\mathbf{A}_{n-1}$ matrix. They are just like those of (6.70) – but with the modified row "moved up" one.

When k equals 2, then k-1 is equal to 1 (the 1st row), the final two $\mathbf{M}$ matrices are formed – from unit matrices, with their first rows taken from the elements of the 2nd row of the matrix defined in the previous step. When the transform of this step is completed, the $\mathbf{P}$ matrix is complete.

Equation (6.70) implies that a great many matrices must be kept around during the transform, but in fact none of the $\mathbf{M}$ matrices need actually be calculated or saved. Instead, each transform is done in two parts:

$$1)\mathbf{B} = [\mathbf{A}]\mathbf{M}_{n-1}; \text{ (this is again shown with } k = n)$$

and $$2)\mathbf{C} = \mathbf{M}_{n-1}^{-1}[\mathbf{B}]. \text{ (The result, } \mathbf{C}, \text{ is } \mathbf{A}_{n-1}) .$$

These are done using the following algorithms:

$$\text{For } k = n, n - 1, n - 2, \ldots, 2 \begin{cases} b_{ij} = a_{ij} - (a_{i,k-1})(\dfrac{a_{kj}}{a_{k,k-1}}), \text{ for } i < k, \text{ and } j \neq k - 1 \\ b_{i,k-1} = (a_{i,k-1})(\dfrac{1}{a_{k,k-1}}) \\ b_{kj} = 0, \text{ for all } j \neq k - 1; \ b_{k,k-1} = 1 . \end{cases}$$

$$(6.71)$$

Note especially, the last row of **B**. In a 4X4, that row will be $\{0, 0, 1, 0\}$. This is already the last row of **P**. Note also that the premultiplication of $\mathbf{M}_{n-1}^{-1}$ will not disturb the last row. In fact, the only row affected by this premultiplication is the k-1st row. That is, in $\mathbf{C} = \mathbf{M}_{k-1}^{-1}[\mathbf{B}]$:

$$\begin{cases} c_{ij} = b_{ij} \text{ for all } i \neq k-1, \text{ and all } j \\ c_{k-1,j} = \sum_{s=1}^{n} a_{ks}b_{sj}, \text{ for all } j . \end{cases} \tag{6.72}$$

In forming **B** and **C**, it is not necessary to actually multiply matrices. The relations shown in (6.71) and (6.72) are all that are needed. In carrying on to the next step, we simply set **A** equal to **C**, and then proceed with k decreased by one. That is, we "move up one row." And so it goes until $k = 2$.

Notice that the definition of the **M** matrix elements includes a division. For example, in the first step we divide by $a_{n,n-1}$. If any of these terms happens to be zero, then one must search upward along the n-1st column (or the "k-1st" column) to find a corresponding element that is not zero; and then interchange the two rows. This is the same as multiplying both **M** and $\mathbf{M}^{-1}$ by the unit matrix, with the same two rows interchanged. Then the transform remains a similar transform, and the development can proceed normally. In the event that no nonzero element can be found, the method fails.

Since the $\mathbf{M}_k$ and $\mathbf{M}_k^{-1}$ matrices (i.e., those defined in Equation (6.70)) are never actually calculated, the **S** and $\mathbf{S}^{-1}$ matrices will not be determined unless there is a reason to do so. If only the eigenvalues are required, **S** and $\mathbf{S}^{-1}$ are not needed. But, a complete eigenvalue analysis requires the vectors as well. Equation (6.68) already implies that these matrices will, then, be required.

Distinct eigenvalues. The following paragraphs outline the method for determining the eigenvectors. It will be noted that ***for each eigenvalue, just one pair of eigenvectors (row and column) is formed***. If the eigenvalues are not distinct, the method fails.

Defining the Eigenvectors

Returning to the eigenvalue analysis of **P**; for each root, we have the following equation to solve for the column eigenvectors: (the eigenvectors of **P**, are defined as **x**, column and **z**, row):

$$\begin{bmatrix} p_{11} - \lambda_i & p_{12} & p_{13} & p_{14} \\ 1 & -\lambda_i & 0 & 0 \\ 0 & 1 & -\lambda_i & 0 \\ 0 & 0 & 1 & -\lambda_i \end{bmatrix} \begin{Bmatrix} x_1 \\ x_2 \\ x_3 \\ x_4 \end{Bmatrix} = \mathbf{0} . \tag{6.73}$$

First, arbitrarily assign the value 1.0 to x_4 (x_n, in general). Then, using the last 3 equations (n-1 equations in general), the elements of the ith **x** vector are:

$$x_n = 1; \qquad x_k = \lambda x_{k+1}, \qquad \text{for } k = n-1, n-2, \dots, 1 . \tag{6.74}$$

For example, in the 4X4 case, $\mathbf{x}_i = \{\lambda_i^3, \lambda_i^2, \lambda_i, 1\}$.

For the row eigenvectors, we have $[\mathbf{z}_i][\mathbf{P}\text{-}\lambda_i\mathbf{I}] = \mathbf{0}$, a row equation:

$$
\begin{bmatrix} z_1 & z_2 & z_3 & z_4 \end{bmatrix}
\begin{bmatrix}
p_{11} - \lambda_i & p_{12} & p_{13} & p_{14} \\
1 & -\lambda_i & 0 & 0 \\
0 & 1 & -\lambda_i & 0 \\
0 & 0 & 1 & -\lambda_i
\end{bmatrix} = \mathbf{0}.
\tag{6.75}
$$

In this case, set $z_1 = 1$, and then

$$
z_k = \lambda_i z_{k-1} - p_{1,k-1}, \text{ for } k = 2, 3, \ldots n .
\tag{6.76}
$$

Because of the simplicity of $\mathbf{P}$, its eigenvectors are easily derived. But, although the eigenvalues of $\mathbf{P}$ are the same as those of $\mathbf{A}$, *the eigenvectors are different*. Starting with $[\mathbf{A}\text{-}\lambda_i\mathbf{I}]\mathbf{v}_i$:

$$
(\mathbf{A} - \lambda_i\mathbf{I})\mathbf{v}_i = (\mathbf{SPS}^{-1} - \lambda_i\mathbf{I})\mathbf{v}_i = \mathbf{S}(\mathbf{P} - \lambda_i\mathbf{I})\mathbf{S}^{-1}\mathbf{v}_i \rightarrow (\mathbf{P} - \lambda_i\mathbf{I})\mathbf{x}_i
$$

we see that we must transform $\mathbf{v}_i$ by $\mathbf{S}^{-1}$. That is $\mathbf{x}_i = \mathbf{S}^{-1}\mathbf{v}_i$, where $\mathbf{x}$ is the column vector in (6.73). Therefore, to obtain $\mathbf{v}$ from $\mathbf{x}$, we must premultiply by $\mathbf{S}$. In the row vector case, the logic is very similar, and the resulting transforms are:

$$
\begin{aligned}
\mathbf{v}_i &= \mathbf{S}\mathbf{x}_i \\
[\mathbf{u}_i] &= [\mathbf{z}_i]\mathbf{S}^{-1} .
\end{aligned}
\tag{6.77}
$$

In Equations (6.77), the square brackets are used just to emphasize that $\mathbf{u}_i$ and $\mathbf{z}_i$ are row vectors. Since both $\mathbf{S}$ and $\mathbf{S}^{-1}$ are used in the definition of the vectors, *then for a complete eigenvalue analysis these matrices must be retained as the similarity transform proceeds*.

Recall, from (6.71) and (6.72), that the original $\mathbf{A}$ matrix is updated via the intermediate matrices, $\mathbf{B}$, and $\mathbf{C}$ (only one of which has to be kept – i.e., $\mathbf{C}$ is the "in-place" update of $\mathbf{B}$). In the same sense, define a matrix, $\tilde{\mathbf{S}}$, which will be used to update $\mathbf{S}$. At the end of each update cycle, $\mathbf{S}$ will be set equal to $\tilde{\mathbf{S}}$, and the next cycle will again update $\tilde{\mathbf{S}}$. The emerging matrix $\mathbf{S}^{-1}$ will be updated in-place.

The relationships are very similar to those of (6.71) and (6.72):

($\mathbf{S}, \mathbf{S}^{-1}$, and $\tilde{\mathbf{S}}$ initialized to unit matrices)

$$
\left\{
\begin{aligned}
&\tilde{s}_{k-1,j} = -\frac{a_{k,j}}{a_{k,k-1}}, \text{ for all } j \neq k - 1; \ \tilde{s}_{k-1,k-1} = \frac{1}{a_{k,k-1}} \\
&\tilde{s}_{i,j} = s_{i,j} - s_{i,k-1} \bullet \frac{a_{k,j}}{a_{k,k-1}}, \text{ for } (k - 1) < i < n, \text{ and } j \neq k - 1 \\
&\tilde{s}_{i,k-1} = \frac{s_{i,k-1}}{a_{k,k-1}}, \quad \text{ for } (k - 1) < i < n, \text{ and } j = k - 1.
\end{aligned}
\right.
\tag{6.78}
$$

$$
s_{k-1,j}^{-1} = \sum_{p=1}^{n} a_{k,p} \bullet s_{k,j}^{-1}, \text{ for all } j .
\tag{6.79}
$$

In (6.79), the display, $s_{i,j}^{-1}$, means the i, jth element of $\mathbf{S}^{-1}$.

After the eigenvectors are determined, see (6.77), they still must be normalized, such that the product $\mathbf{u}_i \bullet \mathbf{v}_i = 1$. The task is trivial when the vectors are real; it is somewhat tricky when they are complex.

In the event that the original $\mathbf{A}$ matrix is symmetric, only the $\mathbf{x}$ (column) vectors are needed. They transform via the first Equation (6.77), and are normalized to unit length easily, since they are real.

6.8.2 EXAMPLE OF DANILEVSKY'S METHOD

The following $\mathbf{A}$ matrix will be discussed at some length in the next chapter

$$\mathbf{A} = \begin{bmatrix} 0 & 0 & 1 & 0 \\ 0 & 0 & 0 & 1 \\ -\dfrac{16}{9} & \dfrac{7}{9} & -\dfrac{2}{9} & \dfrac{1}{9} \\ 1 & -2 & \dfrac{1}{7} & -\dfrac{2}{7} \end{bmatrix} \tag{6.80}$$

Danilevsky's method will be used, here to determine the eigenvalues and eigenvectors of $\mathbf{A}$. The form of this matrix (whose upper half consists of a null matrix, and a unit matrix) arises in vibrations problems in which damping is present. Thus, physical considerations indicate that the eigenvalues will be complex (with negative real parts). In turn, the eigenvectors will also be complex. Since $\mathbf{A}$ is real, the 4 eigenvalues will be in 2 pairs of complex conjugates. The 4 eigenvectors will also come in 2 pairs of complex conjugates.

$$(\mathbf{A}) \begin{Vmatrix} 0.000000 & 0.000000 & 1.000000 & 0.000000 \\ 0.000000 & 0.000000 & 0.000000 & 1.000000 \\ -1.777778 & 0.777778 & -0.222222 & 0.111111 \\ 1.000000 & -2.000000 & 0.142857 & -0.285714 \end{Vmatrix}$$

The first display, (A), is simply the input. The next two, marked (1) and (2), are the $\mathbf{S}$ and $\mathbf{S}^{-1}$ matrices of the first transform step. Note that since they are the first step, then $\mathbf{S}$ will be equal to $\mathbf{M}_{n-1}^{-1}$, and $\mathbf{S}^{-1}$ will be equal to $\mathbf{M}_{n-1}$.

$$\begin{matrix} (1) \\ (\mathbf{S}) \end{matrix} \begin{Vmatrix} 1.000000 & 0.000000 & 0.000000 & 0.000000 \\ 0.000000 & 1.000000 & 0.000000 & 0.000000 \\ -7.000000 & 14.000000 & 7.000000 & 2.000000 \\ 0.000000 & 0.000000 & 0.000000 & 1.000000 \end{Vmatrix}$$

$$\begin{matrix} (2) \\ (\mathbf{S}^{-1}) \end{matrix} \begin{Vmatrix} 1.000000 & 0.000000 & 0.000000 & 0.000000 \\ 0.000000 & 1.000000 & 0.000000 & 0.000000 \\ 1.000000 & -2.000000 & 0.142857 & -0.285714 \\ 0.000000 & 0.000000 & 0.000000 & 1.000000 \end{Vmatrix}$$

Notice that in matrix (1), the (3,1) element is

$$(1)_{3,1} = -\frac{a_{41}}{a_{43}} = -7.0$$

Matrix (3) was not actually calculated as a matrix product, but instead, the relations (6.71) and (6.72) were used. Matrix (3) is now the new $\mathbf{A}$ matrix (in the text, it was labeled $\mathbf{A}_{n-1}$).

$$
\begin{matrix}
(3) \\
(A)
\end{matrix}
\begin{Vmatrix}
-7.000000 & 14.000000 & 7.000000 & 2.000000 \\
0.000000 & 0.000000 & 0.000000 & 1.000000 \\
-7.031748 & 13.666671 & 6.492066 & -0.047618 \\
0.000000 & 0.000000 & 1.000000 & 0.000000
\end{Vmatrix}
$$

Matrices (4), (5), and (6) are the results of the second transform. (4) is not the S matrix, yet. It is the product of $\mathbf{M}_{n-1}$ and $\mathbf{M}_{n-2}$. The matrix (6) has its last 2 rows transformed, on its way to becoming the P matrix.

$$
\begin{matrix}
(4) \\
(S)
\end{matrix}
\begin{Vmatrix}
1.000000 & 0.000000 & 0.000000 & 0.000000 \\
0.514518 & 0.073171 & -0.475029 & 0.003484 \\
0.203252 & 1.024390 & 0.349593 & 2.048780 \\
0.000000 & 0.000000 & 0.000000 & 1.000000
\end{Vmatrix}
$$

$$
\begin{matrix}
(5) \\
(S^{-1})
\end{matrix}
\begin{Vmatrix}
1.000000 & 0.000000 & 0.000000 & 0.000000 \\
-0.539682 & 0.682540 & 0.927438 & -1.902494 \\
1.000000 & -2.000000 & 0.142857 & -0.285714 \\
0.000000 & 0.000000 & 0.000000 & 1.000000
\end{Vmatrix}
$$

$$
\begin{matrix}
(6) \\
(A)
\end{matrix}
\begin{Vmatrix}
0.203252 & 1.024390 & 0.349593 & 2.048780 \\
-1.429217 & -0.711188 & -2.505872 & -0.739837 \\
0.000000 & 1.000000 & 0.000000 & 0.000000 \\
0.000000 & 0.000000 & 1.000000 & 0.000000
\end{Vmatrix}
$$

After the final transformation, all three of the matrices are fully formed. (9) now displays the P matrix, and (7) and (8) are S and S^{-1}, respectively.

$$
\begin{matrix}
(7) \\
(S)
\end{matrix}
\begin{Vmatrix}
-0.699684 & -0.497607 & -1.753318 & -0.517652 \\
-0.360000 & -0.182857 & -1.377143 & -0.262857 \\
-0.142212 & 0.923251 & -0.006772 & 1.943567 \\
0.000000 & 0.000000 & 0.000000 & 1.000000
\end{Vmatrix}
$$

$$
(8) \atop (S^{-1}) \quad \begin{Vmatrix} -3.551272 & 4.526329 & -1.017565 & 1.329158 \\ -0.539682 & 0.682540 & 0.927438 & -1.902494 \\ 1.000000 & -2.000000 & 0.142857 & -0.285714 \\ 0.000000 & 0.000000 & 0.000000 & 1.000000 \end{Vmatrix}
$$

$$
(9) \atop (P) \quad \begin{Vmatrix} -0.507937 & -3.825397 & -0.730159 & -2.777778 \\ 1.000000 & 0.000000 & 0.000000 & 1.000000 \\ 0.000000 & 1.000000 & 0.000000 & 0.000000 \\ 0.000000 & 0.000000 & 1.000000 & 0.000000 \end{Vmatrix}
$$

From (9), the characteristic equation is:

$$p^4 + 0.507937 p^3 + 3.825397 p^2 + 0.730159 p + 2.777778 = 0$$

to six decimal places. The calculations used "extended" type variables for high precision.

$$\lambda_1, \lambda_2 = -0.06250 \pm j0.99811$$
$$\lambda_3, \lambda_4 = -0.19147 \pm j1.16555 .$$

The matrices (7) and (8) are inverses, because each stage in their derivation used inverse matrices. Further, since A and P are *similar*, then the product SPS^{-1} produces the original A matrix. These two checking operations will be left to the reader.

It is notable that P, S, S^{-1} are all real. However, the eigenvalues are obviously complex, and so will be the eigenvectors. The development of the first eigenvector is shown in the accompanying table.

These vectors are determined by first calculating the x and z vectors (eigenvectors of P), using Equations (6.74) and (6.76). From there, the v and u vectors are found by using the transforms in (6.77). The first column in the table shows the first x vector (the x vectors are eigenvectors of P). Then $v_1 = Sx_1$. The middle column shows the result of this calculation.

First Column Eigenvector

x_1 vector	v_1	v_1(norm)
0.18655	-0.04481	0.01101
-0.98266	-1.00038	0.46837
-0.99233	-0.06249	0.01931
-0.12476	-0.99798	0.46742
-0.06250	1.00129	-0.46817
0.99811	0.01780	-0.01823
1.00000	1.00000	-0.46775
0.00000	0.00000	-0.00994

All the vectors, both row and column, are transformed similarly, defining the (complex) matrices $\mathbf{U}$ and $\mathbf{V}$. After that, these two matrices must be normalized such that their product is the unit matrix. The normalization can be accomplished in many ways (each might produce a different normalized $\mathbf{v}_1$ vector in the table). The choice made, here, was to divide both $\mathbf{u}_i$ and $\mathbf{v}_i$ by the square root of the dot product of $\mathbf{u}_i \bullet \mathbf{v}_i$.

6.8.3   DANILEVSKY'S METHOD—ZERO PIVOT

Each loop of Danilevsky's method uses the $(k, k-1)$ element as a divisor. If this elelment approximates zero, the method will fail unless an altering change can be made. Such a change is possible—which will be shown using the example of the 6X6 shown here.

$$
\begin{bmatrix}
a'_{11} & a'_{12} & a'_{13} & a'_{14} & a'_{15} & a'_{16} \\
a'_{21} & a'_{22} & a'_{23} & a'_{24} & a'_{25} & a'_{26} \\
a'_{31} & a'_{32} & a'_{33} & a'_{34} & a'_{35} & a'_{36} \\
a'_{41} & a'_{42} & 0 & a'_{44} & a'_{45} & a'_{46} \\
0 & 0 & 0 & 1 & 0 & 0 \\
0 & 0 & 0 & 0 & 1 & 0
\end{bmatrix}
$$

In the position shown, the value of k is 4 and the (4,3) element happens to be zero (Note that the elements are shown "primed" indicating that these are not the original a_{ij} values; eg., a_{43} was not necessarily zero at the beginning of the procedure).

At this point, the elements in row k (4, here) to the left of the zero element are tested for non zero. In this example, if either the 41 or 42 elements are non zero, the procedure can be continued by exchanging the column containing the nonzero element with the k-1 column. The column exchange can be viewed as *postmultiplying by a unit matrix with the same columns exchanged*, $\mathbf{I}_{j,k-1}$.

Recall that each stage of the Danilevsky reduction involves the calculation of the type:

$$\mathbf{A}'' = \mathbf{M}_{k-1}^{-1} \mathbf{A}' \mathbf{M}_{k-1} .$$

And that these $\mathbf{M}$ matrices are very carefully constructed to be inverses. Then, the postmultiplication by $\mathbf{I}_{j,k-1}$ must be accompanied ("balanced") by the *premultiplication of the inverse of* $\mathbf{I}_{j,k-1}$. But the inverse of $\mathbf{I}_{j,k-1}$ is simply $\mathbf{I}_{j,k-1}$. That is, the balancing operation to be performed is the *interchange of rows* j and $k-1$. In this example, assume that a'_{41} is non zero. In this case, columns 1 and 3 would be interchanged, and rows 1 and 3 interchanged. In this way, the method can be continued, and the "similarity" of the $\mathbf{A}$ and $\mathbf{P}$ matrices is maintained.

If all the elements in the kth row are zero, the row and column interchanges described above do not help. This case is illustrated with a 6X6 $\mathbf{A}'$ matrix below.

$$
\begin{bmatrix}
a'_{11} & a'_{12} & a'_{13} & a'_{14} & a'_{15} & a'_{16} \\
a'_{21} & a'_{22} & a'_{23} & a'_{24} & a'_{25} & a'_{26} \\
a'_{31} & a'_{32} & a'_{33} & a'_{34} & a'_{35} & a'_{36} \\
0 & 0 & 0 & a'_{44} & a'_{45} & a'_{46} \\
0 & 0 & 0 & 1 & 0 & 0 \\
0 & 0 & 0 & 0 & 1 & 0
\end{bmatrix}.
$$

Note that this matrix is "naturally" partitioned

$$
\mathbf{A}' = \begin{bmatrix} \mathbf{A}'_1 & \mathbf{A}'_3 \\ 0 & \mathbf{A}'_2 \end{bmatrix}.
$$

The matrix $\mathbf{A}'_2$ is already in the correct form, and the elements in its top row are the negative coefficients of (in this case) a 3rd degree polynomial. Further, the matrix $\mathbf{A}'_1$ can now be analyzed separately—which will result in another 3rd degree polynomial. The roots of these two polynomials are the eigenvalues of the original matrix.

When the original method fails (6X6, at the point shown in $\mathbf{A}'$) the development of $\mathbf{S}$ and $\mathbf{S}^{-1}$ ceases. To correct this, use the transforms for $\mathbf{A}'_1$. That is, the Danilevsky method produces

$$
\mathbf{P}' = \mathbf{Q}^{-1}\mathbf{A}'\mathbf{Q} \quad \text{(See Equation (6.70))}.
$$

This is a 3X3 transformation, in this case, with $\mathbf{Q}$ the transform matrix. Now, form $\mathbf{M}$ and $\mathbf{M}^{-1}$:

$$
\mathbf{M}^{-1} = \begin{bmatrix} \mathbf{Q}^{-1} & 0 \\ 0 & \mathbf{I} \end{bmatrix}, \text{ and } \mathbf{M} = \begin{bmatrix} \mathbf{Q} & 0 \\ 0 & \mathbf{I} \end{bmatrix}.
$$

Note that these are 6X6 inverse matrices. Pre- and post- multiply these onto the original 6X6 transform matrices, $\mathbf{S}$ and $\mathbf{S}^{-1}$. The result will be the overall 6X6 transform matrices.

$$
\mathbf{P} = [\mathbf{M}^{-1}\mathbf{S}^{-1}]\,\mathbf{A}\,[\mathbf{S}\mathbf{M}].
$$

Thus, even in this case, the complete Danilevsky similarity transform is available.

6.9   EXERCISES

6.1. Derive the general characteristic equation for a (3X3) by expanding $|\lambda\mathbf{I} - \mathbf{A}|$.

6.2. Using the expansion from exercise 1, find the characteristic equation and then the eigenvalues and vectors for the matrix, $\mathbf{A}$:

$$
\mathbf{A} = \begin{bmatrix}
23 & 4 & -6 \\
-18 & 1 & 6 \\
75 & 12 & -20
\end{bmatrix}
$$

6.3. Using the eigenvalue data from Problem 2, find the $\lambda_k\{v_k\}[u_k]$ matrices (for $k = 1, 2, 3$) and find the sum of these three matrices.

6.4. (a) Using the same data, find the matrix $B = \lambda_1\{v_1\}[u_1] + \lambda_2\{v_2\}[u_2]$ and find its characteristic equation.

(b) Show that the eigenvectors of B are the same as those for A.

(c) Given $B = \begin{bmatrix} 9 & -6 & -4 \\ -12 & 3 & 4 \\ 24 & -18 & -11 \end{bmatrix}$, and the A matrix from Problem 2, find BA and AB. Explain.

6.5. For the matrix $A = \frac{1}{2}\begin{bmatrix} 5 & -3 \\ -3 & 5 \end{bmatrix}$, find $\sqrt{A}$.

6.6. For the matrix $A = \begin{bmatrix} -0.7 & 2 \\ -0.6 & 1.5 \end{bmatrix}$, find $\sin(A)$.

6.7. (a) Given $Ax = \lambda x + c$, define the conditions under which a solution exists.

(b) Solve the equation assuming the necessary conditions.

(c) If the (2X2) A matrix is that from Problem 5, and the c vector is $c = \{1, -1\}$ solve the set in terms of the parameter λ. Does a solution exist when $\lambda = 1$?

6.8. In the polynomial $\prod_{j=1}^{6} (x - x_j) = x^6 + c_1 x^5 + \cdots + c_n$ find c_2 and c_3. Describe the formation of each of the coefficients.

6.9. Given $A = \begin{bmatrix} 6 & -3 & 0 \\ -3 & 6 & -3 \\ 0 & -3 & 4 \end{bmatrix}$

(a) Use Danilevsky's method to find the coefficients of its characteristic polynomial.

(b) Use matrix iteration (Section 6.7.2) to find the largest eigenvalue.

(c) "Divide-out" the root from (b) and solve the quadratic for the remaining eigenvalues of A.

CHAPTER 7

Matrix Analysis of Vibrating Systems

7.1   INTRODUCTION

The eigenvalue problem, the details of which were discussed in the previous chapter, has application in many important areas in engineering. Certainly one of the most interesting is in the study of (linearized) vibrating systems.[1] These systems are a perfect and direct example of the Characteristic Value problem. We will begin there, and add the non-homogeneous set as well:

$$\begin{cases} \mathbf{Ax} - \lambda\mathbf{x} = \mathbf{0} \\ \mathbf{Ax} - \lambda\mathbf{x} = \mathbf{c} \end{cases} \tag{7.1}$$

Given that $\mathbf{A}$ is "diagonalizable" (we omit the defective matrix case from discussion), there exist n values of λ ("eigenvalues") for which the *homogeneous* set has a solution. For each of these values the associated solutions are the "eigenvectors," $\mathbf{u}$ (row) and $\mathbf{v}$ (column). These two sets of solutions are orthogonal to one another: $\mathbf{u}_i \bullet \mathbf{v}_j = 0$ ($i \neq j$). In the event that $\mathbf{A}$ is symmetric, the $\mathbf{u}$ set is simply the transpose of the $\mathbf{v}$. In either event, the sets are normalized such that $\mathbf{u}_i \bullet \mathbf{v}_i = 1$.

For the *non-homogeneous* equation we assume a solution of the form $\mathbf{x} = \mathbf{Vy}$. Then:

$$\begin{cases} \mathbf{AVy} - \lambda\mathbf{Vy} = \mathbf{c}\,; \ \text{Premultiply by } \mathbf{U}: \\ \mathbf{UAVy} - \lambda\mathbf{y} = \mathbf{Uc}\,; \ (\mathbf{UAV} = \Lambda): \\ (\Lambda - \lambda\mathbf{I})\mathbf{y} = \mathbf{Uc} \\ \mathbf{y} = (\Lambda - \lambda\mathbf{I})^{-1}\mathbf{Uc} \end{cases} \tag{7.2}$$

therefore $\mathbf{x} = \mathbf{Vy} = \mathbf{V}(\Lambda - \lambda\mathbf{I})^{-1}\mathbf{Uc}\,.$ (7.3)

Apparently, the inverse of $(\mathbf{A} - \lambda\mathbf{I})$ is $\mathbf{V}(\Lambda - \lambda\mathbf{I})^{-1}\mathbf{U}$. Of course this inverse does not exist when λ is equal to one of the eigenvalues, λ_i. This fact is the more clear when the inverse is written as:

$$(\Lambda - \lambda I)^{-1} = \left[\delta_{ij} \bullet \frac{1}{(\lambda_j - \lambda)} \right]$$

a diagonal matrix (note the Kronecker delta, δ_{ij}). Then, in general the non-homogeneous set has no solution when λ equals one of the eigenvalues. If, however, the vector, $\mathbf{c}$, is orthogonal to $\mathbf{u}_i$, then the solution (7.3) holds: we maintain the orthogonality while allowing λ to approach λ_i. In the

---

[1]It is assumed that the reader is familiar with the differential equations which govern the motion of linear vibrating systems.

particular case in which one of the eigenvalues is zero, **A** is singular. It may be recalled (Chapter 4, Section 4.3) that **Ax** = **c** was shown to have no solution, when **A** is singular, unless **c** is orthogonal to all solutions of the transposed set, **A'z** = 0. In (7.3), above, the row vectors, $\mathbf{u}_i$, are solutions to the transposed set.

There are other displays and interpretations of (7.3). The most important of these will show that the Equation (7.3) can be written:

$$\mathbf{x} = \mathbf{V}(\Lambda - \lambda\mathbf{I})^{-1}\mathbf{Uc} = \sum_{i=1}^{n} \frac{\{\mathbf{v}_i\}[\mathbf{u}_i]}{(\lambda_i - \lambda)} \mathbf{c} = \sum_{i=1}^{n} \frac{\mathbf{u}_i \bullet \mathbf{c}}{(\lambda_i - \lambda)} \mathbf{v}_i \ . \tag{7.4}$$

The first summation shown in (7.4), is a summation of nXn matrices, $\{\mathbf{v}_i\}[\mathbf{u}_i]$, each of which is postmultiplied by **c**. The second summation shows the result of the multiplication, changing into a sum of the vectors $\mathbf{v}_i$, multiplied by the scalar dot products divided by the λ difference terms. This, final, form will be found to be most interesting, and will provide a direct solution to the differential equations of the vibration problem.

It will be found that much of this chapter deals with equations like (7.4). In particular, the non-homogeneous differential equations have a solution whose form is exactly the same. In that sense, we have already summarized much of this chapter.

7.2   SETTING UP EQUATIONS, LAGRANGE'S EQUATIONS

The systems that will be discussed herein are simple; their equations of motion will be almost trivial to set up. However, those that are found in practice are often anything but simple. It is therefore worthwhile to mention Lagrange's equations. His intentions were to simplify and formalize the derivation of equations — the force diagrams, and the (tricky) determination of the correct sign to attach to the forces.

Beginning at the most simple, a mass $m = \frac{W}{g}$ is suspended on a spring of spring constant, k, in Figure 7.1. Assume that motion is constrained to be "vertical" and in the plane of the paper. If the

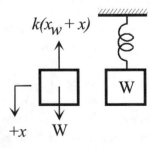

Figure 7.1:

mass is disturbed from equilibrium, the ensuing motion will be oscillatory in this one dimension.

The mathematical spring is defined to produce a restraining force on the mass proportional to a change of its length. The constant of proportionality is the parameter, k. In the force diagram to the left side of the figure, the upward force is $k(x_w + x)$. The force kx_w is exactly the amount necessary to statically balance the weight, W.

If the mass is disturbed from its static equilibrium position, Newtons Laws are used to equate the acceleration to the unbalanced force:

$$m\ddot{x} = -kx .$$ (7.5)

As vibration continues, energy is continually being transferred from kinetic to potential, and back again. No energy is lost from this theoretical system, since it has no energy dissipation terms. The kinetic (T) and potential (V) energies can be written as

$$T = \frac{m}{2}\dot{x}^2 \text{ and } V = \frac{k}{2}x^2 .$$

Note that $\frac{dT}{d\dot{x}} = m\dot{x}$ and $\frac{dV}{dx} = kx$, and therefore these terms could be introduced into Equation (7.5) as follows:

$$\frac{d}{dt}\left[\frac{dT}{d\dot{x}}\right] + \frac{dV}{dx} = 0 .$$

Then the original equations of motion can be written in this way, which is the Lagrange equation for this system.

In a more general system, there may be multiple coordinates required to describe the system motion. These may not all be rectilinear motion; the equations may describe torsion and angular motion, or charges/currents in electrical networks. Then, we must introduce the idea of "generalized coordinates," q, and presuppose that multiple coordinates are present, which turn the spatial derivatives into partial derivatives:

$$\frac{d}{dt}\left[\frac{\partial T}{\partial \dot{q}}\right] + \frac{\partial V}{\partial q} = 0$$ (7.6)

which is Lagrange's equation for conservative systems with no external forces present.

7.2.1   GENERALIZED FORM OF LAGRANGE'S EQUATIONS

One of the most useful forms of the equations is

$$\frac{d}{dt}\left[\frac{\partial T}{\partial \dot{q}_i}\right] - \frac{\partial T}{\partial q_i} + \frac{\partial V}{\partial q_i} + \frac{\partial D}{\partial \dot{q}_i} = f_i \text{ (component of external force)}$$ (7.7)

$$(1) \qquad (2) \qquad (3) \qquad (4) .$$

(1) Inertial forces, derived from kinetic energy.

(2) Gyroscopic and centrifugal forces. Derived from kinetic energy, from changes in direction.

(3) Potential forces.

(4) Viscous damping forces.

Derived from "Rayleigh's dissipation function." D.

Rayleigh's dissipation function is usually denoted by the letter "F." Here a "D" is used to avoid confusion with the external force, F. The term $\frac{\partial D}{\partial q_i}$ refers to the derivative of Rayleigh's dissipation function. It is introduced in order to account for the effects of dissipative, frictional effects. In this function, the forces are considered to be proportional to the velocity term $\dot{q}_i$. For a single particle the function is simply $D = \frac{1}{2}c\dot{x}^2$. The parameter, c is the proportionality between the dissipation force and the velocity which produces it. Its electrical analog describes the power loss in the electrical network, $\frac{1}{2}Ri^2$.

For the systems of interest here, the kinetic, potential, and dissipation functions are simple quadratic forms. For example, for a system of springs and masses:

$$ T = \tfrac{1}{2}\dot{\mathbf{q}}'\mathbf{M}\dot{\mathbf{q}}; \quad V = \tfrac{1}{2}\mathbf{q}'\mathbf{K}\mathbf{q}; \quad \text{and} \quad D = \tfrac{1}{2}\dot{\mathbf{q}}'\mathbf{C}\dot{\mathbf{q}} . $$

Then, we could define the vectors

$$ \nabla_q \equiv \left\{ \frac{\partial}{\partial q_i} \right\} \quad \text{and} \quad \nabla_{\dot{q}} = \left\{ \frac{\partial}{\partial \dot{q}_i} \right\} . $$

In this case, we can write Lagrange's equations as:

$$ \tfrac{d}{dt}\nabla_{\dot{q}}T + \nabla_{\dot{q}}D + \nabla_q V = \mathbf{f} $$
$$ \mathbf{M\ddot{x}} + \mathbf{C\dot{x}} + \mathbf{Kx} = \mathbf{f} $$

7.2.2 MECHANICAL / ELECTRICAL ANALOGIES

The following is the equation of motion for the simple spring-mass system, accompanied by the voltage equation for the R-L-C circuit – the diagrams for both are shown in Figure 7.2.

$$ \begin{cases} m\ddot{x} + c\dot{x} + kx = f_0 \sin \omega t \\[2mm] L\ddot{q} + R\dot{q} + \dfrac{q}{C} = e_0 \sin \omega t . \end{cases} $$

It is apparent that the mathematics is the same for both systems, and that the solutions will consist of damped sinusoids. These systems are, then, analogues of one another. From Figure 7.2, and the equations, the following analogues can be defined. The list, below, is adequate to compare and discuss the systems dealt with herein; but, it is not an exhaustive list.

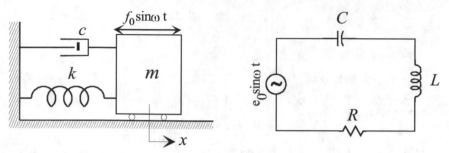

Figure 7.2: Mechanical and Electrical Analogues.

Mechanical		Electrical
Displacement, x	—	Charge, q
Velocity, $\dot{x}$	—	Current, $\dot{q}$ or i
Force, f	—	Voltage, e
Mass, m	—	Inductance, L
Spring Constant, k	—	Elastance, $S = 1/C$
Compliance, $\frac{1}{k}$	—	Capacitance, C
Damping Coefficient, c	—	Resistance, R

Most of the examples to be discussed in later paragraphs will be mechanical systems. It is important to note that the same solutions can be applied to their electrical analogies.

7.2.3   EXAMPLES USING THE LAGRANGE EQUATIONS

As an example of the method, consider the electrical network in Figure 7.3. By inspection

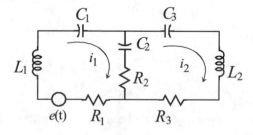

Figure 7.3: Electrical LRC Network.

$$2T = L_1 i_1^2 + L_2 i_2^2 = [i_1 \ i_2] \begin{bmatrix} L_1 & 0 \\ 0 & L_2 \end{bmatrix} \begin{Bmatrix} i_1 \\ i_2 \end{Bmatrix}$$

$$2V = S_1 q_1^2 + S_2 (q_1 - q_2)^2 + S_3 q_2^2; \quad \text{where}$$

$$S = \frac{1}{C} \quad \text{and} \quad q = \int i \, dt$$

$$2V = [q_1 \ q_2] \begin{bmatrix} S_1 + S_2 & -S_2 \\ -S_2 & S_2 + S_3 \end{bmatrix} \begin{Bmatrix} q_1 \\ q_2 \end{Bmatrix}$$

$$2D = R_1 i_1^2 + R_2 (i_1 - i_2)^2 + R_3 i_2^2 = [i_1 \ i_2] \begin{bmatrix} R_1 + R_2 & -R_2 \\ -R_2 & R_2 + R_3 \end{bmatrix} \begin{Bmatrix} i_1 \\ i_2 \end{Bmatrix}$$

then the equation set is: $\mathbf{L\ddot{q}} + \mathbf{R\dot{q}} + \mathbf{Sq} = \mathbf{e}$ where $\mathbf{L}$, $\mathbf{R}$, and $\mathbf{S}$ are the 2X2 matrices, above. This is a voltage equation, and could (perhaps more easily) have been determined using Kirchhoff's laws. Even in this case, though, note that there was no trouble or hesitation with the correct signs to use. For example, in both V and D the difference terms, e.g., $(i_1 - i_2)^2$ could have been written $(i_2 - i_1)^2$.

Further, it is often not that easy. Try this next example – a double pendulum. The use of Lagrange's equations comes in particularly handy.

Take, as the origin, the point of support of both pendulums, O. The inertial, rectangular coordinates x and y are to be measured from this point, and the generalized coordinates θ_1 and θ_2 will be referred to x, and y. The upper weight is at (x_1, y_1), the lower at (x_2, y_2). The kinetic energy

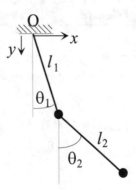

Figure 7.4: Double Pendulum.

is

$$T = \frac{m_1}{2}(\dot{x}_1^2 + \dot{y}_1^2) + \frac{m_2}{2}(\dot{x}_2^2 + \dot{y}_2^2) .$$

There are 4 relations between the generalized and the inertial coordinates. They are

$$\begin{cases} x_1 = l_1 \sin\theta_1; & x_2 = l_1 \sin\theta_1 + l_2 \sin\theta_2; \\ y_1 = l_1 \cos\theta_1; & y_2 = l_1 \cos\theta_1 + l_2 \cos\theta_2 . \end{cases}$$

These relations must be differentiated and plugged into the expression for T, to eliminate x and y in favor of the angular measurements. The result is

$$T = \frac{m_1}{2}l_1^2\dot{\theta}_1^2 + \frac{m_2}{2}[l_1^2\dot{\theta}_1^2 + l_2^2\dot{\theta}_2^2 + 2l_1 l_2\dot{\theta}_1\dot{\theta}_2\cos(\theta_1 - \theta_2)] \ .$$

The potential energy is solely due to vertical position within the gravitational field:

$$V = m_1 g l_1(1 - \cos\theta_1) + m_2 g l_1(1 - \cos\theta_1) + m_2 g l_2(1 - \cos\theta_2) + \text{ constant}.$$

The form of Lagrange's equation to use is:

$$\frac{d}{dt}\left[\frac{\partial T}{\partial \dot{\theta}_i}\right] - \frac{\partial T}{\partial \theta_i} + \frac{\partial V}{\partial \theta_i} = 0; \quad i = 1, 2$$

after some algebraic manipulation of the derivatives involved, the two nonlinear equations in θ_1 and θ_2 are:

$$(m_1 + m_2)l_1\ddot{\theta}_1 + (m_1 + m_2)g\sin\theta_1 + m_2 l_2\{\ddot{\theta}_2\cos(\theta_1 - \theta_2) + \dot{\theta}_1^2\sin(\theta_1 - \theta_2)\}$$
$$l_2\ddot{\theta}_2 + g\sin\theta_2 + l_1\{\ddot{\theta}_1\cos(\theta_1 - \theta_2) - \dot{\theta}_1^2\sin(\theta_1 - \theta_2)\} \ .$$

These equations can be linearized, for small amplitude vibrations, to:

$$\begin{bmatrix} (m_1 + m_2)l_1 & m_2 l_2 \\ l_1 & l_2 \end{bmatrix}\begin{Bmatrix} \ddot{\theta}_1 \\ \ddot{\theta}_2 \end{Bmatrix} + \begin{bmatrix} (m_1 + m_2)g & 0 \\ 0 & g \end{bmatrix}\begin{Bmatrix} \theta_1 \\ \theta_2 \end{Bmatrix}.$$

This problem, and especially its derivation, is a classic one found in many applied mathematics texts. The derivation is included here to show the power and comparative ease of the Lagrange equations. It is doubtful that any other approach would be successful. Fortunately, the other examples used in this chapter are very much simpler.

7.3   VIBRATION OF CONSERVATIVE SYSTEMS

Begin with an analysis of "conservative systems," which have no dissipative elements — no "dashpots" in the mechanical case, no resistance elements in the electrical network. The absence of such elements makes these networks "conservative" in that no energy escapes the system. Vibrations once started continue indefinitely.

The analysis of conservative systems is simpler, and moreover, will provide the method by which the more complex non-conservative networks are handled.

Both of the diagrams of Figure 7.5 depict conservative systems in which two dynamic variables are required to describe the complete vibration (e.g., 2 currents, i_1 and i_2, in Figure 7.5 (a)). The analysis will not be limited to two variables, since the development will be in terms of matrix elements.

The two networks of Figure 7.5 are analogues. As discussed in the previous section, the same equation type is used for both. There are two basic ways in which to derive these "equations of

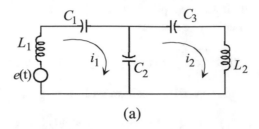

(a)

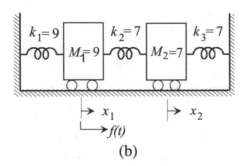

(b)

Figure 7.5: (a) Electrical LC Network, (b) Analogous Mechanical System.

motion." First, for the electrical network, we could use Kirchhoff's Laws, summing voltage drops around each loop:

$$L_1 \frac{di_1}{dt} + \frac{1}{C_1} \int i_1 dt + \frac{1}{C_2} \int (i_1 - i_2) dt = e(t)$$

$$L_2 \frac{di_2}{dt} + \frac{1}{C_3} \int i_2 dt + \frac{1}{C_2} \int (i_2 - i_1) dt = 0 \,.$$

Alternatively, by using the T and V from Figure 7.3 (the same as Figure 7.5 (a), just neglect the resistance elements), in terms of charge, q, and using elastance in place of capacitance:

$$\mathbf{L\ddot{q}} + \mathbf{Sq} = \mathbf{e}(t) = \begin{bmatrix} L_1 & 0 \\ 0 & L_2 \end{bmatrix} \mathbf{\ddot{q}} + \begin{bmatrix} S_1 + S_2 & -S_2 \\ -S_2 & S_2 + S_3 \end{bmatrix} \mathbf{q} = \begin{Bmatrix} e(t) \\ 0 \end{Bmatrix}. \tag{7.8}$$

The mechanical equivalent of using Kirchhoff's Laws would be to sum forces on each of the masses, m_1 and m_2, and (using Newton's Laws) equating to the acceleration force. However, since the systems of Figure 7.5 (a) and 7.5 (b) are analogues, and knowing that the analog of inductance is mass, m, the analog of q (charge) is displacement, x, and the analog of elastance is spring stiffness, k, the equations for the mechanical system can be written directly:

$$\mathbf{M\ddot{x}} + \mathbf{Kx} = \mathbf{d}f(t); \quad \mathbf{d} = \{1, \ 0\} \,. \tag{7.9}$$

The matrix elements can be taken directly from their analogues in (7.8):

$$\begin{bmatrix} m_1 & 0 \\ 0 & m_2 \end{bmatrix} \{\ddot{\mathbf{x}}\} + \begin{bmatrix} k_1 + k_2 & -k_2 \\ -k_2 & k_2 + k_3 \end{bmatrix} \{\mathbf{x}\} = \{\mathbf{f}(t)\} = \{\mathbf{d}\} \, f(t). \qquad (7.10)$$

Note that $f(t)$ is a scalar multiplier of the vector, $\mathbf{d}$. In the example, $\mathbf{d} = \{1, 0\}$, signifying that the driving function is applied to m_1 only. Of course this need not be the case — $f(t)$ might well be applied to all the masses, or a different force function might be applied to each. In this latter case (different drivers) solutions for each excitation are determined separately, then added together at the end. This strategy is successful when the subject systems are *linear*.

In an nXn case (e.g., n masses in Figure 7.5 (b)), the equations of motion are still written:

$$\begin{aligned} \mathbf{M}\ddot{\mathbf{x}} + \mathbf{K}\mathbf{x} &= \mathbf{d}f(t) \\ \mathbf{L}\ddot{\mathbf{q}} + \mathbf{S}\mathbf{q} &= \mathbf{d}e(t) \end{aligned} \qquad (7.11)$$

In paragraphs that follow, solutions for the first of Equations (7.11) will be discussed. It should be clear that the analysis holds equally for the electrical analog.

In (7.11), the matrix $\mathbf{M}$ is often diagonal, and always symmetric and positive definite. The matrix $\mathbf{K}$ is often tridiagonal (having non zero elements on only the main diagonal, and the adjoining "codiagonals"), always symmetric, and positive. It may not be positive definite, because it is sometimes singular. The result is that the eigenvalues and eigenvectors describing these networks will always be real (not complex). Further, the $\mathbf{M}$ and $\mathbf{K}$ matrices will *be diagonalized simultaneously by means of the eigenvectors*, as shown in following paragraphs.

7.3.1 CONSERVATIVE SYSTEMS – THE INITIAL VALUE PROBLEM

Beginning with (7.11), the driving vector is neglected and the resultant set solved to determine the "natural vibrations" which would occur if the system is disturbed from its static equilibrium state. At the instant of the disturbance, each mass may be given an initial displacement, x_0, and/or an initial velocity, $\dot{x}_0$. We will see that these two initial conditions will be just enough to determine the constants of integration in the solution. The homogeneous equations are:

$$\mathbf{M}\ddot{\mathbf{x}} + \mathbf{K}\mathbf{x} = \{\mathbf{0}\} \, .$$

Often, this set is written in terms of the "Dynamical Matrix," $\mathbf{D} = \mathbf{K}^{-1}\mathbf{M}$, or the inverse dynamical matrix ($\mathbf{D}^{-1}$). We will use the inverse dynamical matrix, and will call it "$\mathbf{A}$." That is, by premultiplying by $\mathbf{M}^{-1}$ the set becomes

$$\ddot{\mathbf{x}} + \mathbf{A}\mathbf{x} = \{\mathbf{0}\}, \text{ where } \mathbf{A} = \mathbf{M}^{-1}\mathbf{K} \, . \qquad (7.12)$$

Assume a solution set of the form $\mathbf{x} = \mathbf{v}e^{j\omega t}$, $\dot{\mathbf{x}} = j\omega \mathbf{v}e^{j\omega t}$, $\ddot{\mathbf{x}} = -\omega^2 \mathbf{v}e^{j\omega t}$ and (7.12) becomes:

$$\left\{ \begin{aligned} -\omega^2 \mathbf{v} + \mathbf{A}\mathbf{v} &= \{\mathbf{0}\} \\ \mathbf{A}\mathbf{v} &= \omega^2 \mathbf{v} \\ \mathbf{A}\mathbf{v} &= \lambda \mathbf{v}, \text{ with } \lambda = \omega^2 \, . \end{aligned} \right. \qquad (7.13)$$

The eigenvalue problem is discussed in Chapter 6 where it is shown that if $\mathbf{A}$ is nXn, there will be n solutions to (7.13), each associated with a separate eigenvalue (for now, this discussion is limited to "distinct" eigenvalues). The matrix $\mathbf{A}$ is generally not symmetric, although its eigenvalues will all be real. Then for each eigenvalue, λ_i:

$$\mathbf{A}\mathbf{v}_i = \lambda_i \mathbf{v}_i$$
$$\mathbf{u}_i \mathbf{A} = \lambda_i \mathbf{u}_i \ . \tag{7.14}$$

That is, the non-symmetric matrix $\mathbf{A}$ has both row eigenvectors, $\mathbf{u}_i$, and column eigenvectors, $\mathbf{v}_i$. The $\mathbf{u}_i$ vector associated with λ_j is orthogonal to the $\mathbf{v}_i$ vector associated with λ_i ($\mathbf{u}_i \bullet \mathbf{v}_j = 0, i \neq j$). The row vectors are brought together into matrix, $\mathbf{U}$, and the columns, respectively, numbered, into $\mathbf{V}$, usually normalize such that $\mathbf{UV} = \mathbf{I}$. This is all well known, from Chapter 6.

In this case, however, some additional orthogonality conditions exist. In (7.14), premultiply by $\mathbf{M}$, remembering that $\mathbf{A} = \mathbf{M}^{-1}\mathbf{K}$, then write, for two eigenvectors:

$$\mathbf{K}\mathbf{v}_i = \lambda_i \mathbf{M}\mathbf{v}_i$$
$$\mathbf{K}\mathbf{v}_j = \lambda_j \mathbf{M}\mathbf{v}_j \ . \tag{7.15}$$

Premultiply the first equation by $\mathbf{v}'_j$ and the second by $\mathbf{v}'_i$. Now, transpose the second equation. Since both $\mathbf{K}$ and $\mathbf{M}$ are symmetric:

$$\mathbf{v}'_j \mathbf{K}\mathbf{v}_i = \lambda_i \mathbf{v}'_j \mathbf{M}\mathbf{v}_i$$
$$\mathbf{v}'_j \mathbf{K}\mathbf{v}_i = \lambda_j \mathbf{v}'_j \mathbf{M}\mathbf{v}_i \ .$$

Now, when the second equation is subtracted from the first, the identical left sides cancel

$$(\lambda_i - \lambda_j)\mathbf{v}'_j \mathbf{M}\mathbf{v}_i = 0 \ .$$

Since the two eigenvalues are not equal (by hypothesis), then it must be concluded that

$$\mathbf{v}'_j \mathbf{M}\mathbf{v}_i = 0, \ \text{and thus } \mathbf{v}'_j \mathbf{K}\mathbf{v}_i = 0 \ ,$$

and this is an important and useful conclusion. The column vectors, $\mathbf{v}$, are said to be orthogonal "with respect to $\mathbf{M}$, or $\mathbf{K}$." The total equation set can be assembled as follows:

$$\mathbf{KV} = \mathbf{MV}\Lambda$$
$$(\mathbf{V'KV}) = (\mathbf{V'MV})\Lambda \ . \tag{7.16}$$

In (7.16), the $\mathbf{V}$ matrix is the ordered assemblage of the column eigenvectors. The Λ matrix is diagonal, with its ordered set of eigenvalues on the diagonal. "Ordered" means that the position of the eigenvalue on the diagonal of Λ must correspond with the position of its eigenvector in $\mathbf{V}$. The second Equation (7.16) is clearly all-diagonal. The eigenvector set diagonalizes both $\mathbf{M}$ and $\mathbf{K}$, *simultaneously*. If the eigenvectors are normalized to $\mathbf{V'MV} = \mathbf{I}$, then $\mathbf{V'KV}$ will be equal to Λ.

In addition to all this "new orthogonality," recall from Chapter 6:

$$\mathbf{AV} = \mathbf{V}\Lambda, \text{ and } \mathbf{UA} = \Lambda\mathbf{U} .$$

And if $\mathbf{U}$ and $\mathbf{V}$ are normalized such that $\mathbf{UV} = \mathbf{I}$(the usual case)

$$\mathbf{UAV} = \Lambda, \quad (\text{with } \mathbf{UV} = \mathbf{VU} = \mathbf{I}) .$$

The system shown in Figure 7.5(b) will be used to illustrate the analysis of conservative systems. Using equation (7.10), with the parameter values from the figure, a "by-hand" eigenvalue analysis is given below. In a more complex case this eigenvalue analysis would be done by computer

$$\mathbf{A} = \begin{bmatrix} \dfrac{k_1 + k_2}{m_1} & \dfrac{-k_2}{m_1} \\ \dfrac{-k_2}{m_2} & \dfrac{k_2 + k_3}{m_2} \end{bmatrix}; \ (\mathbf{A} - \lambda\mathbf{I}) = \mathbf{A}(\lambda) = \begin{bmatrix} \dfrac{k_1 + k_2}{m_1} - \lambda & \dfrac{-k_2}{m_1} \\ \dfrac{-k_2}{m_2} & \dfrac{k_2 + k_3}{m_2} - \lambda \end{bmatrix}$$

$$|\mathbf{A} - \lambda\mathbf{I}| = |\mathbf{A}(\lambda)| = \lambda^2 + (\frac{k_1 + k_2}{m_1} + \frac{k_2 + k_3}{m_2})\lambda + \frac{k_1 k_2 + k_1 k_3 + k_2 k_3}{m_1 m_2} .$$

Using the values $m_1 = 9$, $m_2 = 7$, $k_1 = 9$, $k_2 = 7$, $k_3 = 7$ from Figure 7.5 (b):

$$\mathbf{A} = \begin{bmatrix} \frac{16}{9} & -\frac{7}{9} \\ -1 & 2 \end{bmatrix} \quad |\mathbf{A}(\lambda)| = \lambda^2 - \frac{34}{9}\lambda + \frac{175}{63}; \quad \begin{cases} \lambda_1 = \omega_1^2 = \frac{7}{7} = 1.0 \\ \lambda_2 = \omega_2^2 = \frac{25}{9} \end{cases}$$

$$\mathbf{A} - \lambda_1\mathbf{I} = \begin{bmatrix} \frac{7}{9} & -\frac{7}{9} \\ -1 & 1 \end{bmatrix}; \ \Rightarrow \mathbf{v_1} = \begin{Bmatrix} 1 \\ 1 \end{Bmatrix} \text{ and } \mathbf{u_1} = \{1, \ \frac{7}{9}\}$$

$$\mathbf{A} - \lambda_2\mathbf{I} = \begin{bmatrix} -1 & -\frac{7}{9} \\ -1 & -\frac{7}{9} \end{bmatrix}; \ \Rightarrow \mathbf{v_2} = \begin{Bmatrix} \frac{7}{9} \\ -1 \end{Bmatrix} \text{ and } \mathbf{u_2} = \{1, \ -1\}$$

then $\mathbf{V} = \begin{bmatrix} 1 & \frac{7}{9} \\ 1 & -1 \end{bmatrix}$, and after normalizing for $\mathbf{UV} = \mathbf{I}$, $\mathbf{U} = \frac{1}{16}\begin{bmatrix} 9 & 7 \\ 9 & -9 \end{bmatrix}$.

The product $\mathbf{V}'M V$, (not normalized to equal $\mathbf{I}$, since $\mathbf{UV}$ has been normalized to equal $\mathbf{I}$) is $\begin{bmatrix} 16 & 0 \\ 0 & \frac{112}{9} \end{bmatrix}$ and therefore $\mathbf{V}'\mathbf{KV}$ will not be Λ. However, note that:

$$\mathbf{V}'\mathbf{KV} = \begin{bmatrix} 16 & 0 \\ 0 & \frac{25}{9} \bullet \frac{112}{9} \end{bmatrix}, \text{ which does equal } \mathbf{V}'\mathbf{MV}\Lambda \text{ (see (7.16), above)}.$$

It is worthwhile to show that $\mathbf{UAV} = \Lambda$. That is

$$\mathbf{UAV} = \frac{1}{16}\begin{bmatrix} 9 & 7 \\ 9 & -9 \end{bmatrix}\begin{bmatrix} \frac{16}{9} & -\frac{7}{9} \\ 1 & -1 \end{bmatrix}\begin{bmatrix} 1 & \frac{7}{9} \\ 1 & -1 \end{bmatrix} = \begin{bmatrix} 1 & 0 \\ 0 & \frac{25}{9} \end{bmatrix} . \tag{7.17}$$

With the eigenvalue analysis complete, and its results in hand, return to the initial value problem:

$$\mathbf{M\ddot{x}} + \mathbf{Kx} = \{0\} \quad \Rightarrow \mathbf{\ddot{x}} + \mathbf{M}^{-1}\mathbf{Kx} = \{0\} = \mathbf{\ddot{x}} + \mathbf{A}x = \{0\} \ .$$

With the knowledge that the eigenvector matrices diagonalize $\mathbf{A}$, the equation set can be "decoupled" by the vector transform $\mathbf{x} = \mathbf{Vy}$. Then:

$$
\begin{aligned}
&\mathbf{\ddot{x}} + \mathbf{Ax} = \{0\}; \quad \text{substitute } \mathbf{x} = \mathbf{Vy} \\
&\mathbf{V\ddot{y}} + \mathbf{AVy} = \{0\}; \quad \text{premultiply by } \mathbf{U} \\
&\mathbf{UV\ddot{y}} + \mathbf{UAVy} = \{0\}; \quad \text{where } \mathbf{UV} = \mathbf{I} \\
&\mathbf{\ddot{y}} + \mathbf{\Lambda y} = \{0\} \ .
\end{aligned}
\tag{7.18}
$$

This wonderful result produces a $\mathbf{y}$ equation set that is completely decoupled — each y_i can be solved for separately, from $\ddot{y}_i + \lambda_i y_i = 0$, a very simple differential equation. We find the solution

$$y_i = a_i \cos \omega_i t + b_i \sin \omega_i t; \quad \text{where } \omega_i^2 = \lambda_i \ .$$

Now, assemble the individual solutions together to form the vector solution to (7.18), In the 2X2 case, it is simple to write the expanded matrices:

$$
\left.
\begin{aligned}
y_1 &= a_1 \cos \omega_1 t + b_1 \sin \omega_1 t \\
y_2 &= a_2 \cos \omega_2 t + b_2 \sin \omega_2 t
\end{aligned}
\right\}
\Rightarrow
\mathbf{y} =
\begin{bmatrix} \cos \omega_1 t & 0 \\ 0 & \cos \omega_2 t \end{bmatrix}
\begin{bmatrix} a_1 \\ a_2 \end{bmatrix}
$$

$$
+
\begin{bmatrix} \sin \omega_1 t & 0 \\ 0 & \sin \omega_2 t \end{bmatrix}
\begin{bmatrix} b_1 \\ b_2 \end{bmatrix} .
\tag{7.19}
$$

In the general (i.e., nXn) case, the *form* of the solution set is the same. Then for the nXn case, define two diagonal matrices $[\underline{C}]$ and $[\underline{S}]$ such that:

$$\mathbf{y} = [\underline{C}]\mathbf{a} + [\underline{S}]\mathbf{b}; \quad \text{where } [\underline{C}] \equiv [\delta_{ij} \cos \omega_i t] \text{ and } [\underline{S}] \equiv [\delta_{ij} \sin \omega_i t] \ . \tag{7.20}$$

The symbol, δ_{ij}, is the "Kronecker delta:" $\delta_{ij} = \begin{cases} 0, & i \neq j \\ 1, & i = j \end{cases}$ which forces the diagonal matrix construct for the "cos matrix" and "sin matrix" as used, above. In (7.20) there are two columns (nX1) of undetermined coefficients (2 times n coefficients in all). But, we have 2 columns of initial conditions that must factor into the solution. These (2 times n) conditions will serve to determine the $\mathbf{a}$ and $\mathbf{b}$ coefficient vectors. Denote the condition vectors as $\mathbf{x_0}$ and $\mathbf{\dot{x}_0}$, whose elements represent the initial displacement, and initial velocity of the masses in the system. These vectors must be transformed via $\mathbf{y} = \mathbf{Ux}$ to obtain the initial values for the variables $\mathbf{y}$.

First notice that $[\underline{S}]_{t=0} = [0]$, and $[\underline{C}]_{t=0} = \mathbf{I}$. Then, from (7.20):

$$
\begin{aligned}
\mathbf{y} &= [\underline{C}]\mathbf{a} + [\underline{S}]\mathbf{b}; \quad \mathbf{y_0} = \mathbf{a} = \mathbf{Ux_0} \\
\mathbf{\dot{y}} &= -[\underline{\omega}][\underline{S}]\mathbf{a} + [\underline{\omega}][\underline{C}]\mathbf{b}; \quad \mathbf{\dot{y}_0} = [\underline{\omega}]\mathbf{b} = \mathbf{U\dot{x}_0} \ .
\end{aligned}
\tag{7.21}
$$

In the second of (7.21) the matrix $[\underline{\omega}]$ is diagonal $= [\delta_{ij}\omega_i]$. Then $\mathbf{a} = \mathbf{U}\mathbf{x}_0$ and $\mathbf{b} = [\underline{\omega}]^{-1}\mathbf{U}\dot{\mathbf{x}}_0$

$$\mathbf{y} = [\underline{C}]\mathbf{U}\mathbf{x}_0 + [\underline{S}][\underline{\omega}]^{-1}\mathbf{U}\dot{\mathbf{x}}_0 \qquad (7.22)$$

and since $\mathbf{x} = \mathbf{V}\mathbf{y}$, we premultiply (7.22) by $\mathbf{V}$ to return to the x variables.

$$\mathbf{x} = \mathbf{V}[\underline{C}]\mathbf{U}\mathbf{x}_0 + \mathbf{V}[\underline{S}][\underline{\omega}]^{-1}\mathbf{U}\dot{\mathbf{x}}_0 . \qquad (7.23)$$

At first, (7.23) appears to be very formidable, and not easily programmed. $[\underline{C}]$ and $[\underline{S}]$ are functions of time. A straightforward expansion would be very messy. Fortunately, there is an excellent interpretation of (7.23) which not only makes it clearer to "see," but, is also easily programmed.

7.3.2 INTERPRETATION OF EQUATION (7.23)

In the previous chapter there is a discussion of the synthesis of a matrix by its eigenvalues and eigenvectors. Equation (6.14) of that chapter reads:

$$\mathbf{A} = \sum_{i=1}^{n} \lambda_i \{\mathbf{v}_i\}\lceil\mathbf{u}_i\rceil . \qquad \text{(Chapter 6, (6.14))}$$

This result occurs through an interpretation of $\mathbf{A} = \mathbf{V}\Lambda\mathbf{U}$. The central idea is that Λ is a diagonal matrix. Its jth main diagonal element, i.e., λ_j, multiplies the jth column of $\mathbf{V}$ (or form the product $\Lambda\mathbf{U}$ first, in which case, the jth eigenvalue will be a multiplier on the jth row vector in $\mathbf{U}$).

The same logic is used here concerning the term $\mathbf{V}[\underline{C}]\mathbf{U}\mathbf{x}_0$ in (7.23). In this case, $[\underline{C}]$ is the diagonal matrix. Its jth term is $\cos\omega_j t$ and it multiplies the jth column of $\mathbf{V}$. Now, view the $\mathbf{V}[\underline{C}]$ matrix as *partitioned by columns*:

$$\mathbf{V}[\underline{C}] = \left[\left\{ \begin{array}{c} \cdots \\ \mathbf{v}_1\cos\omega_1 t \\ \cdots \end{array} \right\} \left\{ \begin{array}{c} \cdots \\ \mathbf{v}_2\cos\omega_2 t \\ \cdots \end{array} \right\} \cdots \left\{ \begin{array}{c} \cdots \\ \mathbf{v}_n\cos\omega_n t \\ \cdots \end{array} \right\} \right]$$

and the $\mathbf{U}$ matrix *partitioned by rows*, and the product is written:

$$\mathbf{V}[\underline{C}]\mathbf{U} = \mathbf{v}_1\mathbf{u}_1\cos\omega_1 t + \mathbf{v}_2\mathbf{u}_2\cos\omega_2 t + \cdots + \mathbf{v}_n\mathbf{u}_n\cos\omega_n t = \sum_{i=1}^{n}\mathbf{v}_i\mathbf{u}_i\cos\omega_i t$$

a summation of nXn's, each multiplied by the corresponding diagonal element of the center matrix. This is the same (desired) result as before (Chapter 6; (6.14)) – with the λ_i values replaced by the $\cos\omega_i t$ terms. Now, the term $\mathbf{V}[\underline{C}]\mathbf{U}$ does not look at all formidable, ***since the time varying terms appear as multipliers on an entire matrix entity***.

And, it gets better. Note that $\mathbf{V}[\underline{C}]\mathbf{U}$ involves summing n nXn's. But, when the $\mathbf{x}_0$ vector is post multiplied, it actually simplifies the sum – ***it is easier to operate with vectors than matrices***. Since the $\mathbf{u}_i$ terms are *rows*, they are "available" to dot into the $\mathbf{x}_0$ vector. Then;

$$\mathbf{V}\left[C\right]\mathbf{U}\mathbf{x}_0 = \sum_{i=1}^{n} \mathbf{v}_i(\mathbf{u}_i \bullet \mathbf{x}_0)\cos \omega_i t \ . \tag{7.24}$$

And the term is now composed of just n eigenvectors, weighted as shown in (7.24). This is easily visualized, and easily coded. The time dependent (cos) terms are straightforward scalar multipliers, as are the dot product $\mathbf{u}_i \bullet \mathbf{x}_0$ terms.

Returning to (7.23), its second term can now be written out by inspection. Note that the product $\left[S\right]\left[\omega\right]^{-1}$ is still just a diagonal, sandwiched between $\mathbf{V}$ and $\mathbf{U}$. The inverse $\left[\omega\right]^{-1}$ is just $\left[\delta_{ij}\frac{1}{\omega_i}\right]$.

$$\mathbf{V}\left[S\right]\left[\omega\right]^{-1}\mathbf{U}\dot{\mathbf{x}}_0 = \sum_{i=1}^{n} \mathbf{v}_i(\mathbf{u}_i \bullet \dot{\mathbf{x}}_0)\frac{1}{\omega_i}\sin \omega_i t \tag{7.25}$$

and now, putting (7.14) and (7.25) together:

$$\boxed{\mathbf{x} = \sum_{i=1}^{n} \mathbf{v}_i(\mathbf{u}_i \bullet \mathbf{x}_0)\cos \omega_i t + \sum_{i=1}^{n} \mathbf{v}_i(\mathbf{u}_i \bullet \dot{\mathbf{x}}_0)\frac{1}{\omega_i}\sin \omega_i t.} \tag{7.26}$$

This is a general result, the initial value problem solution applicable to nXn systems (networks). Notice that the eigenvalue analysis is "sum and substance" of the solution. Except for the given initial conditions, all terms are from that analysis (it is required that the eigenvalue analysis produces both sets of eigenvectors, *normalized such that* $\mathbf{UV} = \mathbf{I}$.)

In the particular 2X2 example from Figure 7.5 (b), with the initial conditions $\mathbf{x}_0$ {1, 0} and $\dot{\mathbf{x}}_0 = \{0, \ 1\}$, we find

$$\mathbf{u}_1 \bullet \mathbf{x}_0 = \tfrac{9}{16} \times 1 + \tfrac{7}{16} \times 0 = \tfrac{9}{16}$$

$$\mathbf{u}_2 \bullet \mathbf{x}_0 = \tfrac{9}{16} \times 1 + (-\tfrac{9}{16}) \times 0 = \tfrac{9}{16}$$

$$\mathbf{u}_1 \bullet \dot{\mathbf{x}}_0 = \tfrac{9}{16} \times 0 + \tfrac{7}{16} \times 1 = \tfrac{7}{16}$$

$$\mathbf{u}_2 \bullet \dot{\mathbf{x}}_0 = \tfrac{9}{16} \times 0 + (-\tfrac{9}{16}) \times 1 = -\tfrac{9}{16} \ .$$

Plugging these values into (7.26):

$$\mathbf{x} = \frac{9}{16}\begin{Bmatrix} 1 \\ 1 \end{Bmatrix}\cos 1t + \frac{9}{16}\begin{Bmatrix} \tfrac{7}{9} \\ -1 \end{Bmatrix}\cos \tfrac{5}{3}t + \frac{7}{16}\begin{Bmatrix} 1 \\ 1 \end{Bmatrix}\sin 1t - \frac{9}{16}\bullet\frac{3}{5}\begin{Bmatrix} \tfrac{7}{9} \\ -1 \end{Bmatrix}\sin \tfrac{5}{3}t. \tag{7.27}$$

From the display in (7.27), it is clear how the $\mathbf{v}$ vectors sum to form the total solution. These $\mathbf{v}$ eigenvectors are called the "normal modes" of the vibration. The absolute amplitudes of the vibration are of course strongly affected by the initial conditions. But, at each of the frequencies, *the ratios of the amplitudes remains always the same – in the proportions given in the eigenvectors.* The eigenvectors form the structure of the solution.

Figure 7.6, below, shows this pictorially. Notice that the 1 rad/sec vibration is in phase, and in the proportion of 1:1. The $\frac{5}{3}$ rad/sec vibration is out of phase, in the ratio of -7:9. The total motion for both masses is shown in the right-hand diagram of the figure.

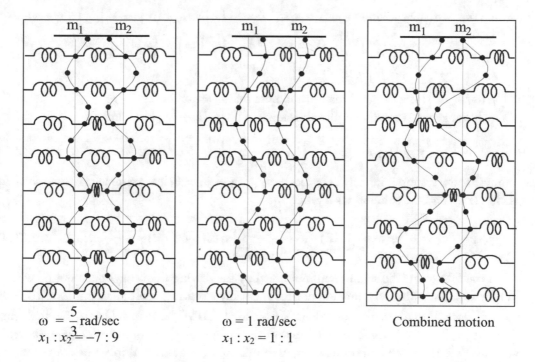

$\omega = \frac{5}{3}$ rad/sec $\omega = 1$ rad/sec   Combined motion
$x_1 : x_2 = -7 : 9$ $x_1 : x_2 = 1 : 1$

Figure 7.6: The Normal Modes of the system of Figure 7.5 (b), and how they sum together.

The figure shows several seconds of the solution of the initial value problem from Figure 7.5 (b), with the initial conditions { 1 0 }, as discussed, above.

Mathematically speaking, this motion would continue forever, without dying out, because this is a conservative system in which there are no elements to dissipate energy. Of course, such systems cannot be found in nature. There will always be some "damping" (resistance to motion), the simplest of which will be discussed below. Also, there are usually non-linearities, which we will not discuss.

7.3.3   CONSERVATIVE SYSTEMS - SINUSOIDAL RESPONSE

Consider, now, the same (conservative) system as before (Figure 7.5 (b)), but, now include the driving vector, as in Equations (7.11)

$$\mathbf{M}\ddot{\mathbf{x}} + \mathbf{K}\mathbf{x} = \mathbf{f}(t) = \{\mathbf{d}\}f(t) .\tag{7.28}$$

We will assume the function $f(t) = \cos \omega t$. Premultiply (7.28) by $\mathbf{M}^{-1}$ ($\mathbf{M}$ is nonsingular) and again make the vector transform $\mathbf{x} = \mathbf{V}\mathbf{y}$

$$\ddot{\mathbf{x}} + \mathbf{A}\mathbf{x} = \mathbf{M}^{-1}\mathbf{d} \cos t$$
$$\ddot{\mathbf{y}} + \Lambda\mathbf{y} = \mathbf{U}\mathbf{M}^{-1}\mathbf{d} \cos \omega t$$

We assume the particular solution $\mathbf{y} = \mathbf{Y}\cos \omega t$; $\dot{\mathbf{y}} = -[\omega]\mathbf{Y}\sin \omega t$; $\ddot{\mathbf{y}} = -[\omega^2]\mathbf{Y}\cos \omega t$:

$$-[\omega^2]\mathbf{Y} + \Lambda\mathbf{Y} = \mathbf{U}\mathbf{M}^{-1}\mathbf{d}; \quad [\omega^2] = [\omega][\omega]$$
$$\mathbf{Y} = [\delta_{ij}(\lambda_i - \omega^2)]^{-1}\mathbf{U}\mathbf{M}^{-1}\mathbf{d}$$

where $[\delta_{ij}(\omega_i^2 - \omega^2)]$ is a diagonal matrix; and note that $\lambda_i = \omega_i^2$.

The homogeneous solution is already known to be $[\underline{C}]\mathbf{a} + [\underline{S}]\mathbf{b}$. Then:

$$\mathbf{y} = [\underline{C}]\mathbf{a} + [\underline{S}]\mathbf{b} + [\delta_{ij}(\omega_i^2 - \omega^2)]^{-1}\mathbf{U}\mathbf{M}^{-1}\mathbf{d}\cos \omega t \ . \tag{7.29}$$

Assuming that the system is initially at rest $\mathbf{x}_0 = \mathbf{y}_0 = \dot{\mathbf{x}}_0 = \dot{\mathbf{y}}_0 = \mathbf{0}$, it is a simple matter to solve for $\mathbf{a}$ ($\mathbf{b}$ is clearly $\mathbf{0}$), and the solution, $\mathbf{x} = \mathbf{V}\mathbf{y}$, becomes

$$\mathbf{x} = \sum_{i=1}^{n} \mathbf{v}_i (\mathbf{u}_i \bullet \mathbf{M}^{-1}\mathbf{d}) \frac{\cos \omega t}{\omega_i^2 - \omega^2} - \sum_{i=1}^{n} \mathbf{v}_i (\mathbf{u}_i \bullet \mathbf{M}^{-1}\mathbf{d}) \frac{\cos \omega_i t}{\omega_i^2 - \omega^2} \ . \tag{7.30}$$

Note that the first term (the summation multiplied by the driving frequency) need not be written as a summation. Since the only function of time is already a separate multiplier, this term could be "interpreted-back" into the matrix operations: $\mathbf{V}[\Lambda - \omega^2]^{-1}\mathbf{U}\mathbf{M}^{-1}\mathbf{d}\cos \omega t$. That is, the single time function, $\cos \omega t$, (ω without subscript refers to the driving frequency) multiplies **all of the eigenvectors in its summation**. In the second summation, each multiplier, $\cos \omega_i t$, multiplies its corresponding vector, $\mathbf{v}_i$. Because of this (second) term, the vector summation is required – and it is the same sum as in the previous term. Therefore, the vector form of (7.30) is clearer, and the corresponding program simpler, written this way. In fact, (7.30) can be written:

$$\mathbf{x} = \sum_{i=1}^{n} \mathbf{v}_i (\mathbf{u}_i \bullet \mathbf{M}^{-1}\mathbf{d}) \frac{\cos \omega t - \cos \omega_i t}{\omega_i^2 - \omega^2} \ . \tag{7.31}$$

A conservative system should not be driven at a frequency equal, or very close to, one of the natural, "mode frequencies." Equation (7.30) clearly indicates why, with the difference frequencies in the denominator. However, note that if the corresponding dot product term $\mathbf{u}_i \bullet \mathbf{M}^{-1}\mathbf{d}$ is zero, then that eigenvector-term will not appear in the sum. The condition required for this to be true can be determined as follows:

It has already been established that $\mathbf{V}'\mathbf{M}\mathbf{V} = \mathbf{P}$ is a diagonal matrix. The values of the diagonal elements of $\mathbf{P}$ (they'll all be positive) depend on the normalization. But, the inverse of $\mathbf{P}$ is:

$$\mathbf{P}^{-1} = \mathbf{U}\mathbf{M}^{-1}\mathbf{U}'$$

which is clearly diagonal. Then, if the vector $\mathbf{d}$ is set equal to, say, $\mathbf{u}_1$, then the dot product of $\mathbf{u}_2 \bullet \mathbf{M}^{-1}\mathbf{u}_1$ will be zero – allowing the system to be driven at (or very close to) ω_2.

Now, if the initial conditions $\mathbf{x}_0$ and $\dot{\mathbf{x}}_0$ are not zero, then a bit of arithmetic gives:

$$\mathbf{x} = \sum_{i=1}^{n} \mathbf{v_i}(\mathbf{u}_i \bullet \mathbf{x}_0)\cos\omega_i t + \sum_{i=1}^{n} \mathbf{v_i}(\mathbf{u}_i \bullet \dot{\mathbf{x}}_0)\frac{1}{\omega_i}\sin\omega_i t + \sum_{i=1}^{n} \mathbf{v_i}(\mathbf{u}_i \bullet \mathbf{M}^{-1}\mathbf{d})\frac{\cos\omega t - \cos\omega_i t}{\omega_i^2 - \omega^2} .$$

(7.32)

Note that this is the sum of the initial value problem, plus the driven system solution with zero initial conditions (the sum of Equations (7.26) and (7.31)).

7.3.4   VIBRATIONS IN A CONTINUOUS MEDIUM

The vibrations in a continuous medium, like a beam, string, or reed, can be simulated in a matrix approach by "digitizing" the medium. This approach is used in Appendix C in the study of a vibrating string. Here, we consider a beam, or reed, using the same method.

Let it be required to find the lower natural frequencies and normal modes of a vibrating cantilever beam. Like the analysis of the vibrating string, the beam is to be "divided" into a (large) number of segments. The matrix that results is symmetric, large. Usually only a few natural frequencies are required, a situation that lends itself to the use of matrix iteration. See Section 6.7.2.

The diagram below shows a cantilevered beam. We visualize the mass of the beam to be concentrated at N points along its length — the remaining structure of the beam retains its bending properties. The mathematical model is no longer one of a continuous beam described by a partial

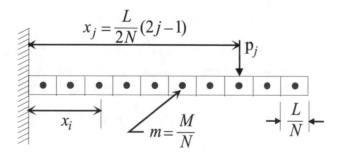

differential equation. Instead it resembles a spring mass system of order N. Distances are to be measured from the support (left) end. The length, L, is divided into N parts, at the center of each lies the mass, m, of that part. Let us number (index) the mass points from the left, starting from 1, and note that the dimension to the kth point is

$$x_k = \frac{L}{2N}(k-1) + \frac{L}{2N} = \frac{L}{2N}(2k-1) .$$

(7.33)

That is, the kth mass point lies at a distance x_k from the support end, where k and x_k are given by (7.33). The total mass of the beam is M, and each mass point has the mass $m = M/N$.

From the equations governing the bending of such a beam, the deflection, $y(x)$, at a point x, caused by a load, p, applied at a point s, is

$$y(x) = \frac{px^2}{EI}\left(\frac{s}{2} - \frac{x}{6}\right); \quad x < s$$

$$y(x) = \frac{ps^2}{EI}\left(\frac{x}{2} - \frac{s}{6}\right); \quad x > s$$

where E is Young's Modulus, the ratio of stress, psi, to strain, in./in.; and I is the second moment of the beam's cross sectional area, in.[4]. Note that the two equations are reciprocal in x and s.

The deflection at x, caused by the load at s, is the same as a deflection at s caused by the load at x. This "reciprocity" assures that the matrix to be defined below, will be symmetric.

Deflections are of interest only at the mass points. In particular, the deflection $y(x_i)$ is written y_i and we denote the positions of the loads, $p_{j'}$, as $x_{j'}$. Then the deflections are

$$y_i = \frac{1}{EI}\sum_j \frac{i'^2 L^2}{4N^2}\left(\frac{j'L}{4N} - \frac{i'L}{12N}\right)p_j; \quad \text{Note:} \quad \left(\begin{array}{c} i', j' = 1, 3, 5 \cdots (2N-1) \\ i' < j' \end{array}\right). \quad (7.34)$$

The deflection at point i' is the sum of the elemental deflections caused by all the loads at points $x_{j'}$. The term on the right side (a function of i' and j') defines the elements of an N by N matrix $\Phi_{i,j}$. The matrix equation can then be written

$$(y) = \frac{L^3}{EI}\Phi_{i,j}(p)$$

$$\Phi_{i,j} = \left[\frac{i'^2}{16N^3}\left(j' - \frac{i'}{3}\right)\right]; \quad \Phi_{j,i} = \Phi_{i,j}; \quad i' = 2i-1, \quad j' = 2j-1. \quad (7.35)$$

As explained, the vector $\mathbf{y} = \{y_i\}$ is the deflections at the mass points. The vector $\mathbf{p}$ consists of the loads at these same points. In the vibrating beam, these are the D'Alembert inertial loads and are seen to be in the negative direction – opposite to the direction of positive deflection:

$$p_j = -m\ddot{y}_j = -\frac{M}{N}\ddot{y}_j.$$

Now, Equation (7.35) can be updated:

$$\mathbf{y} = -\frac{ML^3}{EI}\Phi\ddot{\mathbf{y}}$$

$$\Phi_{i,j} = \frac{i'^2}{16N^4}\left(j' - \frac{i'}{3}\right); \quad \Phi_{j,i} = \Phi_{i,j}. \quad (7.36)$$

Assuming solutions of the form $\mathbf{y} = \mathbf{v}\cos\omega t$ ($\ddot{\mathbf{y}} = -\mathbf{v}\omega^2 \cos\omega t$):

$$\mathbf{v}\cos\omega t = -\frac{ML^3}{EI}\Phi(-\mathbf{v}\omega^2 \cos\omega t)$$

$$\lambda\mathbf{v} = \Phi\mathbf{v}; \quad \text{where } \lambda = \frac{EI}{\omega^2 ML^3}. \quad (7.37)$$

The last of Equations (7.37) is the eigenvalue problem, where the elements of Φ are defined in (7.36). From this point, iteration is used to derive the eigenvalues and eigenvectors as discussed in Chapter 6. Note that the iteration procedure converges to the largest eigenvalue, but to the smallest natural frequency — the one of most interest.

This very interesting approach, is also very simple. However, it is approximate by nature, and becomes more so as the matrix is deflated after each eigenvalue, eigenvector is found.

7.4 NONCONSERVATIVE SYSTEMS. VISCOUS DAMPING

The introduction of viscous damping terms is a very serious complication. There is no (eigenvector) matrix which simultaneously diagonalizes 3 matrices. Then, the equations:

$$\mathbf{L}\ddot{\mathbf{q}} + \mathbf{R}\dot{\mathbf{q}} + \mathbf{S}\mathbf{q} = \{\mathbf{d}\}e(t)$$
$$\mathbf{M}\ddot{\mathbf{x}} + \mathbf{C}\dot{\mathbf{x}} + \mathbf{K}\mathbf{x} = \{\mathbf{d}\}f(t)$$

(7.38)

cannot easily be attacked directly, at least in the same way that conservative systems were. For example, if $\mathbf{x} = \mathbf{v}e^{\lambda t}$ is substituted into (7.38), then, for the homogeneous system:

$$\left[\mathbf{M}\lambda^2 + \mathbf{C}\lambda + \mathbf{K}\right]\mathbf{v} = \mathbf{0}$$

wherein the nXn matrix elements can be written:

$$\left\| \begin{array}{cccc} m_{11}\lambda^2 + c_{11}\lambda + k_{11} & m_{12}\lambda^2 + c_{12}\lambda + k_{12} & \ldots\ldots & m_{1n}\lambda^2 + c_{1n}\lambda + k_{1n} \\ m_{21}\lambda^2 + c_{21}\lambda + k_{21} & \ldots\ldots & \ldots\ldots & \ldots\ldots \\ \ldots\ldots & \ldots\ldots & \ldots\ldots & \ldots\ldots \\ m_{n1}\lambda^2 + c_{n1}\lambda + k_{n1} & \ldots\ldots & \ldots\ldots & m_{nn}\lambda^2 + c_{nn}\lambda + k_{nn} \end{array} \right\|$$

This is the "Lambda Matrix" for the system. The characteristic equation can be found by hand-calculating the determinant of the matrix. The order of the polynomial will be $2n$, there will be $2n$ eigenvalues, and each will be associated with a row and column vector. But, there does not appear to be a more systematic way to attack the problem — one that uses the power of the computer, and/or one that provides the familiar $\mathbf{A}\mathbf{x} - \lambda\mathbf{x} = 0$, characteristic equation.

A systematic approach is available. ***The central point is to reduce the equation set to first order***. This can be done in the general case, where the original equations are order m. The result is a Lambda matrix of the recognizable type.

First, define the operator $p = \dfrac{d}{dt}$, and note that the original set is written:

$$[\mathbf{A}_0 p^n + \mathbf{A}_1 p^{n-1} + \cdots + \mathbf{A}_{n-1} p + \mathbf{A}_n]\mathbf{x} = \mathbf{f}$$

where $n > 1$, and the order of $\mathbf{A}_i$ is m. Assuming that the matrix $\mathbf{A}_0$ is not singular, the set can be reduced to first order, after premultiplying by the inverse of $\mathbf{A}_0$:

The reduced set is: $p\mathbf{y} = \mathbf{Ay} + \mathbf{g}$ where the vectors $\mathbf{y}$ and $\mathbf{g}$ are given by

$$\mathbf{y} = \{x_1 \ldots x_m, px_1 \ldots px_m, \ldots\ldots, p^{n-1}x_1 \ldots p^{n-1}x_m\}, \text{ and}$$
$$\mathbf{g} = \{0, 0, \ldots, \mathbf{A}_0^{-1}\mathbf{f}\}.$$

The new $\mathbf{A}$ matrix in $p\mathbf{y} = \mathbf{Ay} + \mathbf{g}$ is nmXnm. It is diagrammed, below. It is partitioned into square matrices of order m: each $\mathbf{0}$ represents an mXm null, and $\mathbf{I}_m$ represents the order m unit matrix: Note that the $\mathbf{I}_m$ submatrices are not on the main diagonal, but, the first upper codiagonal.

$$\mathbf{A} = \begin{bmatrix} 0 & \mathbf{I}_m & 0 & \cdots & 0 & 0 \\ 0 & 0 & \mathbf{I}_m & \cdots & 0 & 0 \\ \cdots & \cdots & & & & \cdots \\ 0 & 0 & 0 & \cdots & 0 & \mathbf{I}_m \\ -\mathbf{a}_n & -\mathbf{a}_{n-1} & & \cdots & & -\mathbf{a}_1 \end{bmatrix}, \text{ an nXn with each element an mXm.}$$

The mXm submatrices are $\mathbf{a}_i = \mathbf{A}_0^{-1}\mathbf{A}_i$.

In particular, when the equations are the Lagrangian equations for small vibrations of a linear system, in the generalized coordinates, $\mathbf{q}$, then

$$\mathbf{M\ddot{q}} + \mathbf{C\dot{q}} + \mathbf{Kq} = \mathbf{f}, \qquad |\mathbf{M}| \text{ not equal to zero}$$

can be reduced to the 2mX2m set

$$\mathbf{\dot{z}} - \mathbf{Az} = \mathbf{B}^{-1}\mathbf{h}, \quad \text{where} \quad \mathbf{A} = \begin{bmatrix} 0 & \mathbf{I} \\ -\mathbf{M}^{-1}\mathbf{K} & -\mathbf{M}^{-1}\mathbf{C} \end{bmatrix}; \quad \text{and } \mathbf{B} = \begin{bmatrix} \mathbf{C} & \mathbf{M} \\ \mathbf{M} & 0 \end{bmatrix}.$$

The very same results are achieved by a method attributed to K. A Foss,[2] although the development which follows is slightly different. It is worthwhile to follow from the beginning, because it very carefully preserves symmetric submatrices.

Considering the second Equation of (7.38), a trivial equation is added:

$$\mathbf{M\ddot{x}} + \mathbf{C\dot{x}} + \mathbf{Kx} = \{\mathbf{d}\}f(t)$$
$$\mathbf{M\dot{x}} - \mathbf{M\dot{x}} = 0 \tag{7.39}$$

Now, define $\mathbf{h} = \begin{Bmatrix} \mathbf{d} \\ 0 \end{Bmatrix}$ and $\mathbf{z} = \begin{Bmatrix} \mathbf{x} \\ \mathbf{\dot{x}} \end{Bmatrix}$, and write the set (7.39) as:

$$\begin{bmatrix} \mathbf{C} & \mathbf{M} \\ \mathbf{M} & 0 \end{bmatrix} \begin{Bmatrix} \mathbf{\dot{x}} \\ \mathbf{\ddot{x}} \end{Bmatrix} + \begin{bmatrix} \mathbf{K} & 0 \\ 0 & -\mathbf{M} \end{bmatrix} \begin{Bmatrix} \mathbf{x} \\ \mathbf{\dot{x}} \end{Bmatrix} = \begin{Bmatrix} \mathbf{f} \\ 0 \end{Bmatrix} = \mathbf{h}. \tag{7.40}$$

[2]"Coordinates Which Uncouple the Equations of Motion of Damped Linear Systems," submitted to the ASME Applied Mechanics Division, May, 1957, by K. A Foss, Massachusetts Institute of Technology.

This can be written:

$$\mathbf{B\dot{z}} + \mathbf{Gz} = \mathbf{h}; \quad \text{where} \quad \mathbf{B} = \begin{bmatrix} \mathbf{C} & \mathbf{M} \\ \mathbf{M} & \mathbf{0} \end{bmatrix}, \quad \text{and} \quad \mathbf{G} = \begin{bmatrix} \mathbf{K} & \mathbf{0} \\ \mathbf{0} & -\mathbf{M} \end{bmatrix}. \tag{7.41}$$

In (7.41), $\mathbf{0}$ is the nXn null matrix, and it should be recalled that $\mathbf{M}$, $\mathbf{C}$, and $\mathbf{K}$ are symmetric. Therefore, both $\mathbf{B}$ and $\mathbf{G}$ are symmetric! A good sign, it looks like the conservative case, except that the $\mathbf{z}$ vector appears in first derivative, rather than second.

The inverse of $\mathbf{B}$ is

$$\mathbf{B}^{-1} = \begin{bmatrix} \mathbf{0} & \mathbf{M}^{-1} \\ \mathbf{M}^{-1} & -\mathbf{M}^{-1}\mathbf{C}\mathbf{M}^{-1} \end{bmatrix}$$

and so premultiply (7.41) by $\mathbf{B}^{-1}$:

$$\mathbf{\dot{z}} - \mathbf{Az} = \mathbf{B}^{-1}\mathbf{h}, \quad \text{where} \quad \mathbf{A} = \begin{bmatrix} \mathbf{0} & \mathbf{I} \\ -\mathbf{M}^{-1}\mathbf{K} & -\mathbf{M}^{-1}\mathbf{C} \end{bmatrix}. \tag{7.42}$$

The 2nX2n $\mathbf{A}$ matrix is the same as given above. Now, (7.42) resembles the conservative case and will have similar orthogonality conditions.

7.4.1   THE INITIAL VALUE PROBLEM

As with the conservative case, this problem begins with (7.42), but with a zero driving vector:

$$\mathbf{\dot{z}} - \mathbf{Az} = \mathbf{0}. \tag{7.43}$$

In (7.43) the solution vector $\mathbf{z} = \mathbf{v}e^{\lambda t}$ is assumed, with the result:

$$\lambda \mathbf{v}e^{\lambda t} - \mathbf{A}\mathbf{v}e^{\lambda t} = \mathbf{0}, \quad \text{or} \quad \mathbf{A}\mathbf{v} - \lambda \mathbf{v} = \mathbf{0}$$

and the eigenvalue problem is evident.

This time, knowledge of the physical system predicts that the nonsymmetric matrix, $\mathbf{A}$, has complex eigenvalues and eigenvectors. The situation is summarized as:

- The original equation set is nXn. The symmetric matrices $\mathbf{M}$, $\mathbf{K}$, and $\mathbf{C}$ are real, not complex.

- Matrix $\mathbf{A}$ is 2nX2n. It is not symmetric, but, it is real.

- The eigenvalues are in complex conjugate pairs $\lambda_k = \sigma_k \pm j\omega_k$. There are n pairs of these. They will be found via the eigenvalue analysis. If the physical system is stable, the real parts of the eigenvalues will (must) be negative.

- As a matter of convenience, it will be assumed that the odd numbered eigenvalues are chosen as those with positive frequencies (e.g., $\lambda_1 = \sigma_1 + j\omega_1$, then $\lambda_2 = \sigma_1 - j\omega_1$).

- Each eigenvalue-pair is associated with a pair of column eigenvectors, and a pair of row eigenvectors. These are also complex conjugates. Even though they are "paired," each of these entities retains its own number. Then, for example, λ_1 and λ_2 are complex conjugate eigenvalues, and the associated vectors $(\mathbf{v}_1, \mathbf{v}_2)$ and $(\mathbf{u}_1, \mathbf{u}_2)$ are complex conjugates and are also available from the eigenvalue analysis. As before, it is assumed that the normalizing has been done such that $\mathbf{U}$ and $\mathbf{V}$ are reciprocal matrices: $\mathbf{u}_i \bullet \mathbf{v}_j = 0$, and $\mathbf{u}_i \bullet \mathbf{v}_i = 1$.

As in the conservative case, the eigenvectors, $\mathbf{v}$, diagonalize both $\mathbf{B}$ and $\mathbf{G}$, simultaneously. The proof of this is the same as before:

$$\begin{cases} \mathbf{G}\mathbf{v}_i = \mathbf{B}\mathbf{v}_i\lambda_i \\ \mathbf{G}\mathbf{v}_j = \mathbf{B}\mathbf{v}_j\lambda_j \end{cases}.$$

If the first of these equations is multiplied by $\mathbf{v}'_j$ and the second by $\mathbf{v}'_i$, and then the second is transposed (both $\mathbf{G}$ and $\mathbf{B}$ are symmetric), the left sides will again be identical, and will cancel, leaving:

$$\mathbf{v}'_j\mathbf{B}\mathbf{v}_i(\lambda_i - \lambda_j) = 0$$

which leads to the conclusion that $\mathbf{V}'\mathbf{B}\mathbf{V}$ and $\mathbf{V}'\mathbf{G}\mathbf{V}$ are diagonal, and that if $\mathbf{V}'\mathbf{B}\mathbf{V}$ were normalized to equal $\mathbf{I}$, then $\mathbf{V}'\mathbf{G}\mathbf{V}$ would equal Λ. However, for now it will be assumed that both row and column eigenvectors are available, and that they are normalized such that $\mathbf{U}\mathbf{V} = \mathbf{I}$. Then, in (7.43), the transform $\mathbf{z} = \mathbf{V}\mathbf{y}$ is made:

$$\begin{cases} \mathbf{V}\dot{\mathbf{y}} - \mathbf{A}\mathbf{V}\mathbf{y} = \mathbf{0} \\ (\mathbf{U}\mathbf{V})\dot{\mathbf{y}} - (\mathbf{U}\mathbf{A}\mathbf{V})\mathbf{y} = \mathbf{0} \\ \dot{\mathbf{y}} - \Lambda\mathbf{y} = \mathbf{0} \end{cases}. \qquad (7.44)$$

And the last of (7.44) shows that the equations are decoupled. Each equation $\dot{y}_i - \lambda_i y_i = 0$ can be solved separately. For each one the assumed solution is $y_i = y_0 e^{\lambda_i t}$ (where y_0 is the initial value, $y_i(0)$). Note that the vector $\mathbf{z}_0$ contains the initial conditions for both displacement and velocity, because $\mathbf{z} = \{\mathbf{x}, \dot{\mathbf{x}}\}$, and that $\mathbf{y}_0 = \mathbf{V}\mathbf{z}_0$. These solutions can be assembled into matrix form:

$$\mathbf{y} = [\delta_{ik} e^{\lambda_k t}]\mathbf{y}_0. \qquad (7.45)$$

The matrix, $[\delta_{ik} e^{\lambda_k t}]$, is diagonal. The return transform, back to the $\mathbf{z}$ Vector, is made by simply premultiplying by $\mathbf{V}$:

$$\mathbf{V}\mathbf{y} = \mathbf{z} = \mathbf{V}[\delta_{ik} e^{\lambda_k t}]\mathbf{U}\mathbf{z}_0.$$

This solution must be *interpreted* in exactly the same way that the conservative system solution was interpreted. Then:

$$\mathbf{z} = \mathbf{V}\mathbf{y} = \mathbf{V}[\delta_{ik} e^{\lambda_k t}]\mathbf{U}\mathbf{z}_0 = \sum_{k=1}^{2n} \mathbf{v}_k(\mathbf{u}_k \bullet \mathbf{z}_0)e^{\lambda_k t}$$

and since $\lambda_k = \sigma_k \pm j\omega_k$

$$\mathbf{z} = \sum_{k=1}^{2n} \mathbf{v}_k(\mathbf{u}_k \bullet \mathbf{z}_0)e^{\sigma_k t}e^{\pm j\omega_k t} = \sum_{k=1}^{2n} \mathbf{v}_k(\mathbf{u}_k \bullet \mathbf{z}_0)e^{\sigma_k t}(\cos \omega_k t \pm j \sin \omega_k t) \ . \tag{7.46}$$

The vectors $\mathbf{v}_k(\mathbf{u}_k \bullet \mathbf{z}_0)$ and $\mathbf{v}_{k+1}(\mathbf{u}_{k+1} \bullet \mathbf{z}_0)$ (where k is odd) are complex, and conjugate. Also, the numbering is such that when k is odd, ω is positive. Now, define the ith element of the vector $\mathbf{v}_k(\mathbf{u}_k \bullet \mathbf{z}_0) \equiv a_i + jb_i$, and write the pair of terms as:

$$(a_i + jb_i)(\cos \omega_k t + j \sin \omega_k t) + (a_i - jb_i)(\cos \omega_k t - j \sin \omega_k t)$$
$$2a_i \cos \omega_i t - 2b_i \sin \omega_i t$$

That is, the imaginary parts cancel. Therefore:

$$\boxed{\begin{aligned} x_i(1 \leq i \leq n) &= \sum_{k=1,3}^{2n-1} e^{\sigma_k t}(2a_{ki} \cos \omega_k t - 2b_{ki} \sin \omega_k t); \quad \text{where} \\ a_{ki} &= \text{Re} \ \{\mathbf{v}_{ki}(\mathbf{u}_k \bullet \mathbf{z}_0)\} \ ; \quad b_{ki} = \text{Im} \ \{\mathbf{v}_{ki}(\mathbf{u}_k \bullet \mathbf{z}_0)\} \ . \end{aligned}} \tag{7.47}$$

In (7.47), the notation Re{} reads "Real part of" whatever is in the brackets, and Im{} reads "The Imaginary part of" whatever is in the brackets.

Non-conservative System Example

To illustrate, take the earlier conservative system problem (Figure 7.5 (b)) and add "dashpots" as shown in the accompanying Figure 7.7. Give each of these the value of 1 unit of force per unit of

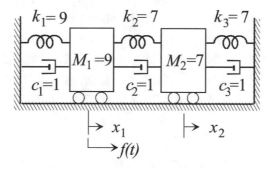

Figure 7.7: Nonconservative Mechanical System.

velocity. In this case, then, the $\mathbf{M}$ and $\mathbf{K}$ matrices will be the same as before, while the damping matrix will have the values:

$$\mathbf{C} = \begin{bmatrix} c_1 + c_2 & -c_2 \\ -c_2 & c_2 + c_3 \end{bmatrix} = \begin{bmatrix} 2 & -1 \\ -1 & 2 \end{bmatrix} \ .$$

Following the methods previously discussed, the defining equation is $\mathbf{M\ddot{x}} + \mathbf{C\dot{x}} + \mathbf{Kx} = \mathbf{0}$, for the initial value problem. The reduced (1st order) equation set is given by Equation (7.42) with the $\mathbf{A}$ matrix:

$$\mathbf{A} = \begin{bmatrix} \mathbf{0} & \mathbf{I} \\ \mathbf{M^{-1}K} & -\mathbf{M^{-1}C} \end{bmatrix} = \begin{bmatrix} 0 & 0 & 1 & 0 \\ 0 & 0 & 0 & 1 \\ -\frac{16}{9} & \frac{7}{9} & -\frac{2}{9} & \frac{1}{9} \\ 1 & -2 & \frac{1}{7} & -\frac{2}{7} \end{bmatrix} \tag{7.48}$$

The analysis of $\mathbf{A}$ yields the eigenvalues and eigenvectors shown below.

Eigenvalues: $\lambda_{1,2} = -0.06250 \pm j0.99811$; $\lambda_{3,4} = -0.19147 \pm j1.65552$

Row Eigenvectors, U   **Column Eigenvectors, V**

$$\begin{Vmatrix} -0.00587 & -0.02761 & -0.60194 & -0.46775 \\ -0.60314 & -0.46647 & -0.02350 & -0.00994 \\ -0.00587 & -0.02761 & -0.60194 & -0.46775 \\ 0.60314 & 0.46647 & 0.02350 & 0.00994 \\ 0.05629 & -0.04696 & 0.53062 & -0.53171 \\ 0.88447 & -0.88621 & 0.02174 & -0.03752 \\ 0.05629 & -0.04696 & 0.53062 & -0.53171 \\ -0.88447 & 0.88621 & -0.02174 & 0.03752 \end{Vmatrix} \quad \begin{Vmatrix} 0.01102 & 0.01102 & -0.01837 & -0.01837 \\ 0.46837 & -0.46837 & -0.24716 & 0.24716 \\ 0.01931 & 0.01931 & 0.01429 & 0.01429 \\ 0.46742 & -0.46742 & 0.31952 & -0.31952 \\ -0.46817 & -0.46817 & 0.41270 & 0.41270 \\ -0.01828 & 0.01828 & 0.01691 & -0.01691 \\ -0.46775 & -0.46775 & -0.53171 & -0.53171 \\ -0.00994 & 0.00994 & -0.03752 & 0.03752 \end{Vmatrix}$$

It appears that the damping is quite "light" inasmuch as the (negative) real parts of the eigenvalues are small. The frequencies of the undamped case were 1.0 and 5/3. *Note that damping lowers these frequencies* (to 0.998 and 1.655, respectively). *In the eigenvector display, the imaginary parts are shown below the reals.* For example, $u_{11} = -0.00587 - j0.60314$.

As a point of interest, note that the bottom 2 elements of the columns, v_i, are equal to λ_i times the top. The row vectors are those which will be dotted into the initial value vector, $\mathbf{z}_0$. They are displayed, here, in rows. They also in complex conjugate pairs, but unlike the columns, the last 2 elements are not λ times the first two elements.

Below are shown the $\mathbf{v}_k$ column eigenvectors, weighted (multiplied) by the scalar quantities $(\mathbf{u}_k \bullet \mathbf{z}_0)$. These determine the x variable coefficients. Note that the vectors are complex conjugate.

$$\mathbf{v}_k (\mathbf{u}_k \bullet \mathbf{z}_0) \ \textbf{Vectors}$$

$$\left\| \begin{matrix} 0.28192 & 0.28192 & 0.21807 & 0.21807 \\ -0.22857 & 0.22857 & 0.10194 & -0.10195 \\ 0.27741 & 0.27741 & -0.27742 & -0.27742 \\ -0.23321 & 0.23321 & -0.13980 & 0.13980 \\ 0.21052 & 0.21053 & -0.21053 & -0.21053 \\ 0.29568 & -0.29568 & 0.34150 & -0.34150 \\ 0.21544 & 0.21544 & 0.28456 & 0.28456 \\ 0.29147 & -0.29147 & -0.43250 & 0.43250 \end{matrix} \right\|$$

To derive the coefficients of the sin and cos terms, take the columns in pairs, and follow the interpretation (7.47). For example, 0.5638 is 0.28192 + 0.28192, while 0.4664 is the sum $0.2332 - (-0.2332)$. The total result:

$$\mathbf{x} = e^{-0.0625t} \left\{ \begin{bmatrix} 0.5638 \\ 0.5548 \end{bmatrix} \cos 0.9981t + \begin{bmatrix} 0.4572 \\ 0.4664 \end{bmatrix} \sin 0.9981t \right\}$$

$$+ e^{-0.1915t} \left\{ \begin{bmatrix} 0.4361 \\ -0.5548 \end{bmatrix} \cos 1.655t + \begin{bmatrix} -0.2039 \\ 0.2796 \end{bmatrix} \sin 1.655t \right\}. \qquad (7.49)$$

If the velocities, $\dot{\mathbf{x}}$, were required, the lower halves of the weighted eigenvectors might be handy ($\mathbf{z}$ defines both $\mathbf{x}$ and $\dot{\mathbf{x}}$), although it may be as easy to differentiate $\mathbf{x}$.

7.4.2   SINUSOIDAL RESPONSE

This is a more tedious problem, algebraically, than the initial value problem. Equation (7.42) provides the starting point, this time including the driving vector.

$$\dot{\mathbf{z}} - \mathbf{A}\mathbf{z} = \mathbf{B}^{-1}\mathbf{h}, \ \text{ where } \mathbf{A} = \begin{bmatrix} \mathbf{0} & \mathbf{I} \\ -\mathbf{M}^{-1}\mathbf{K} & -\mathbf{M}^{-1}\mathbf{C} \end{bmatrix}. \qquad (7.42) \text{ rewrite}$$

In this equation, the vector, $\mathbf{h}$, is $\{\mathbf{f}, 0\}$, where $\mathbf{f}$ is the driving vector. The case, $\mathbf{f} = \mathbf{d}\cos \omega t$, will be considered here. The product $\mathbf{B}^{-1}\mathbf{h}$ is much simpler than it looks.

$$\mathbf{B}^{-1}\mathbf{h} = \begin{bmatrix} \mathbf{0} & \mathbf{M}^{-1} \\ \mathbf{M}^{-1} & -\mathbf{M}^{-1}\mathbf{C}\mathbf{M}^{-1} \end{bmatrix} \left\{ \begin{matrix} \mathbf{f} \\ \mathbf{0} \end{matrix} \right\} = \left\{ \begin{matrix} \mathbf{0} \\ \mathbf{M}^{-1}\mathbf{d} \end{matrix} \right\} \cos \omega t \ .$$

In (7.42) the transform $\mathbf{z} = \mathbf{V}\mathbf{y}$ is made, knowing that the matrix $\mathbf{A}$ will then easily be diagonalized

$$\dot{\mathbf{z}} - \mathbf{A}\mathbf{z} = \mathbf{B}^{-1}\mathbf{h}, \text{ transform } \mathbf{z} = \mathbf{V}\mathbf{y} :$$
$$\mathbf{V}\dot{\mathbf{y}} - \mathbf{A}\mathbf{V}\mathbf{y} = \mathbf{B}^{-1}\mathbf{h}, \text{ Now, premultiply by } \mathbf{U} :$$
$$\dot{\mathbf{y}} - \Lambda\mathbf{y} = \mathbf{U}\mathbf{B}^{-1}\mathbf{h}$$
$$\dot{y}_i - \lambda_i y = \phi_i \cos \omega t, \text{ the typical equation in the set}$$
$$\phi_i = \left\{\mathbf{U}\mathbf{B}^{-1}\mathbf{h}\right\}_i , \text{ the ith component of } \mathbf{U}\mathbf{B}^{-1}\mathbf{h} .$$

The left side of $\dot{y}_i - \lambda_i y_i = \phi_i \cos \omega t$ will be a perfect differential if it is multiplied by $e^{-\lambda_i t}$. Then:

$$\frac{d}{dt}(y_i e^{-\lambda_i t}) = \phi_i e^{-\lambda_i t} \cos \omega t$$

$$y_i e^{-\lambda_i t} = \phi_i \int e^{-\lambda_i t} \cos \omega t = \frac{\phi_i e^{-\lambda_i t}}{\lambda_i^2 + \omega^2}(\omega \sin \omega t - \lambda_i \cos \omega t)$$

$$y_i = \frac{\phi_i}{\lambda_i^2 + \omega^2}(\omega \sin \omega t - \lambda_i \cos \omega t)$$

where the integrating factor, $e^{-\lambda_i t}$, is simply divided back out. The integral of the right side is taken from a table of integrals. The constants of integration are omitted, because only the "steady state" portion of the solution is required here. The transient portion, $ce^{\lambda_i t}$, has already been found (for the initial value problem). The complete solution is then:

$$y_i = c_i e^{\lambda_i t} + \frac{\phi_i}{\lambda_i^2 + \omega^2}(\omega \sin \omega t - \lambda_i \cos \omega t) . \tag{7.50}$$

In (7.50) the constant c no longer is the initial value, y_0. In order to calculate c, the initial conditions must be applied to this complete solution. The initial conditions assumed are $\mathbf{z}_0 = 0$. That is, both initial position and velocity are assumed zero. The result is

$$c = \frac{\phi_i \lambda_i}{\lambda_i^2 + \omega^2}, \text{ and } y_i = \frac{\phi_i \lambda_i}{\lambda_i^2 + \omega^2} e^{\lambda_i t} + \frac{\phi_i}{\lambda_i^2 + \omega^2}(\omega \sin \omega t - \lambda_i \cos \omega t) . \tag{7.51}$$

The assembly of solutions into matrix form yields:

$$\mathbf{z} = \mathbf{V}[\frac{\lambda_i e^{\lambda_i t}}{\lambda_i^2 + \omega^2}]\mathbf{U}\mathbf{B}^{-1}\mathbf{h} + \mathbf{V}[\frac{1}{\lambda_i^2 + \omega^2}]\mathbf{U}\mathbf{B}^{-1}\mathbf{h}\omega \sin \omega t - \mathbf{V}[\frac{\lambda_i}{\lambda_i^2 + \omega^2}]\mathbf{U}\mathbf{B}^{-1}\mathbf{h} \cos \omega t. \tag{7.52}$$

It is important to note that each of the terms within the square braces is a diagonal matrix. Just as before, the nature of the solutions is each term will be from the sum of vectors v, multiplied by scalar quantities taken from within the diagonal matrix. The transient part is:

$$\mathbf{z}_{tr} = \mathbf{V}[\frac{\lambda_i e^{\lambda_i t}}{\lambda_i^2 + \omega^2}]\mathbf{U}\mathbf{B}^{-1}\mathbf{h} = \sum_{k=1}^{2n} \mathbf{v}_k(\mathbf{u}_k \bullet \mathbf{B}^{-1}\mathbf{h})\frac{\lambda_k e^{\lambda_k t}}{\lambda_k^2 + \omega^2} .$$

In turn, this breaks into two terms, since $e^{\lambda_k t} = e^{\sigma_k t + j\omega_k t} = e^{\sigma_k t}(\cos \omega_k t + j \sin \omega_k t)$.

Since all of the terms will be handled in this same way, the first thing to do is to form the $2n\mathrm{X}2n$ matrix, $\mathbf{C}$, which begins as the matrix, $\mathbf{V}$, but then has each of its columns multiplied by the term $(\mathbf{u}_k \bullet \mathbf{B}^{-1}\mathbf{h})\dfrac{1}{\lambda_k^2 + \omega^2}$. Note that the multiplier is dependent on k, the column number, but all of the elements of the vector $\mathbf{v}_k$ are multiplied by the same (scalar) amount. From the elements of this matrix will come the coefficients of the *four* terms (four, since $\mathbf{z}_{tr}$ breaks into two).

7.4.3   DETERMINING THE VECTOR COEFFICIENTS FOR THE DRIVEN SYSTEM

Below is the diagram of a $4\mathrm{X}4$ (the elements, c_{ij}). Its columns are partitioned into upper and lower halves (c_u and c_l).

$k \mapsto$	1	2	3	4
c_u	c_{11}	c_{12}	c_{13}	c_{14}
c_u	c_{21}	c_{22}	c_{23}	c_{24}
c_l	c_{31}	c_{32}	c_{33}	c_{34}
c_l	c_{41}	c_{42}	c_{43}	c_{44}

Each column is a weighted eigenvector of the damped vibration problem:

$$\mathbf{c}_k = \left\{ \begin{array}{c} \mathbf{c}_{uk} \\ \mathbf{c}_{lk} \end{array} \right\} = \mathbf{v}_k \frac{(\mathbf{u}_k \bullet \mathbf{B}^{-1}\mathbf{h})}{\lambda_k^2 + \omega^2} . \tag{7.53}$$

Note that the entire multiplier on $\mathbf{v}_k$ in this expression is a scalar.

Each $2n\mathrm{X}1$ vector, $\mathbf{v}_k$, ($4\mathrm{X}1$ in the diagram) is an eigenvector of the reduced, first order equation $\mathbf{B}\dot{\mathbf{z}} + \mathbf{G}\mathbf{z} = \mathbf{h}$, where $\mathbf{z} = \left\{ \begin{array}{c} \mathbf{x} \\ \dot{\mathbf{x}} \end{array} \right\}$, and so $\mathbf{v}_k = \left\{ \begin{array}{c} \mathbf{e}_k \\ \lambda_k \mathbf{e}_k \end{array} \right\}$, where $\mathbf{e}_k$ is an $n\mathrm{X}1$ eigenvector of the original equation set. Then, in (7.53), $\mathbf{c}_l = \lambda_k \mathbf{c}_u$. Although the diagram shows a $4\mathrm{X}4$ (and so the original set is a $2\mathrm{X}2$, as in the example problem), the results given here are applicable to the $n\mathrm{X}n$ case in which the input is $\mathbf{d}\cos \omega t$. This is not a completely "general input case," but the method is the same for any linear input. For each, however, the multiplier in (7.53) will be different. Now, consider each solution term separately.

Transient Solution, $\mathbf{x}_{tr}$

In (1) and (2), below, consider k to be odd (i.e., 1, 3, 5, ...$2n-1$).

(1) $e^{\sigma_k t} \cos \omega_k t$: The $n\mathrm{X}1$ coefficient vector is $\mathbf{c}_{lk} + \mathbf{c}_{l,k+1}$. Note that the imaginary parts will cancel, the two identical real parts will add. Note also, that the sum is on $\mathbf{c}_l$, (lower).

(2) $e^{\sigma_k t} \sin \omega_k t$: The $n\mathrm{X}1$ coefficient is $j(\mathbf{c}_{lk} - \mathbf{c}_{l,k+1})$. In this case, the real parts will cancel, and the imaginaries will double. Multiplying then by j will make the vector real.

Steady State Solution, $\mathbf{x}_{ss}$

(1) $\omega \sin \omega t$. The nX1 coefficient will be $\sum\limits_{k=1}^{2n} \mathbf{c}_{uk}$. Sum the upper half vector. Note that the imaginary parts will cancel. Thus, the sum will actually be only the real parts. Also the sum is for all values of k. Especially note that the summation must be multiplied by ω.

(2) $\cos \omega t$. The term is negative, so the coefficient is $-\sum\limits_{k=1}^{2n} \mathbf{c}_{lk}$, the lower half summed and then negated. Again the imaginary parts will cancel.

Non-conservative System Example (Continued)

Returning to the damped system of Figure 7.7, with an input of $\mathbf{f} = \{18\cos 2t, 0\}$, with zero initial conditions. The eigenvalue analysis is the same; all that must be done is the construction of the $\mathbf{C}$ matrix from the row and column eigenvalue matrices, and build the solution.

$$\mathbf{c}_k = \left\{ \begin{array}{c} \mathbf{c}_{uk} \\ \mathbf{c}_{lk} \end{array} \right\} = \mathbf{v}_k \frac{(\mathbf{u}_k \bullet \mathbf{B}^{-1}\mathbf{h})}{\lambda_k^2 + \omega^2} = \left\| \begin{array}{cccc} 0.01068 & 0.01068 & 0.07469 & 0.07469 \\ -0.18720 & 0.18720 & -0.16648 & 0.16648 \\ 0.00734 & 0.00734 & -0.10268 & -0.10268 \\ -0.18709 & 0.18709 & 0.21191 & -0.21191 \\ \hline 0.18618 & 0.18618 & 0.26131 & 0.26131 \\ 0.02236 & -0.02236 & 0.15552 & -0.15552 \\ 0.18628 & 0.18628 & -0.33116 & -0.33116 \\ 0.01902 & -0.01902 & -0.21056 & 0.21056 \end{array} \right\|$$

From the data in the accompanying "C" Matrix, and using the rules given above, the complete solution to the driven system can be written:

$$\mathbf{x} = e^{-0.0625t} \left\{ \left\{ \begin{array}{c} 0.3724 \\ 0.3726 \end{array} \right\} \cos 0.99811t - \left\{ \begin{array}{c} 0.0447 \\ 0.0380 \end{array} \right\} \sin 0.99811t \right\} \tag{7.54}$$

$$+ e^{-0.1915t} \left\{ \left\{ \begin{array}{c} 0.5226 \\ -0.6623 \end{array} \right\} \cos 1.6555t + \left\{ \begin{array}{c} -0.3110 \\ 0.4211 \end{array} \right\} \sin 1.6555t \right\}$$

$$+ \left\{ \begin{array}{c} -0.8950 \\ 0.2898 \end{array} \right\} \cos 2t + \left\{ \begin{array}{c} 0.3414 \\ -0.3814 \end{array} \right\} \sin 2t$$

Sample Calculations:

$$-0.8950 = -(0.18618 + 0.18618 + 0.26131 + 0.26131)$$
$$0.4211 = j(-j0.21056 - j0.21056)$$

Since the initial positions are zero, the coefficients of the cos terms should sum to zero

$$0.3724 + 0.5226 - 0.8950 = 0$$
$$0.3726 - 0.6623 + 0.2898 = 0$$

A further check would be to differentiate for $\dot{x}$, and again check for a zero value at $t = 0$. This exercise will be left to the reader.

7.4.4   SINUSOIDAL RESPONSE – NONZERO INITIAL CONDITIONS

This case will be included here, because the result is very interesting. Equation (7.50) is the start:

$$y_i = c_i e^{\lambda_i t} + \frac{\phi_i}{\lambda_i^2 + \omega^2}(\omega \sin \omega t - \lambda_i \cos \omega t) . \qquad (7.50) \text{ rewrite}$$

The (now non-zero) initial conditions can be applied here, letting y_{0i} represent the initial value for y_i. Then:

$$y_{0i} = c_i - \frac{\phi_i \lambda_i}{\lambda_i^2 + \omega^2}; \quad \Rightarrow c_i = y_{0i} + \frac{\phi_i \lambda_i}{\lambda_i^2 + \omega^2}, \text{ and therefore}$$

$$y_i = y_{0i} e^{\lambda_i t} + \frac{\phi_i \lambda_i}{\lambda_i^2 + \omega^2} e^{\lambda_i t} + \frac{\phi_i}{\lambda_i^2 + \omega^2}(\omega \sin \omega t - \lambda_i \cos \omega t) . \qquad (7.55)$$

This is exactly like (7.51), but with the initial value solution ($[e^{\lambda_i t}]y_0$ in vector form) simply added in. Therefore, the interesting result is that, assuming the initial conditions to be the same as those chosen earlier in the solution to the initial value problem, then the solution (7.55) will be the sum of Equations (7.54) and (7.50). This same result occurred in the solution to the conservative system with sinusoidal driving function. See (7.32).

If the initial conditions are different than those resulting in (7.50) then, we would only need to re-solve the initial conditions problem, and add its solution to (7.54).

7.5   STEADY STATE SINUSOIDAL RESPONSE

Often in the solution of vibrating systems, only the steady state response from sinusoidal input is desired. This is a very significant reduction in effort compared to the complete solution since the eigenvalue analysis and transient solution are avoided.

A most simple example is the system in the previous Figure 7.5 whose equation is;

$$\mathbf{M}\ddot{\mathbf{x}} + \mathbf{K}\mathbf{x} = \mathbf{d} \cos \omega t = \left\{ \begin{array}{c} 18 \\ 0 \end{array} \right\} \cos 2t . \qquad (7.56)$$

A solution $\mathbf{x} = \mathbf{a} \cos 2t$ is assumed. Then $\ddot{\mathbf{x}} = -\mathbf{a}\omega^2 \cos \omega t = -4\mathbf{a} \cos 2t$, and

$$\mathbf{a} = [\mathbf{K} - 4\mathbf{M}]^{-1}\mathbf{d}, \text{ and then } \mathbf{x} = [\mathbf{K} - 4\mathbf{M}]^{-1}\mathbf{d} \cos 2t . \qquad (7.57)$$

Note that $[\mathbf{K} - 4\mathbf{M}]$ must not be singular. Also, in this idealized system the transient solution continues forever, which denies that this is the "steady state" solution. Rather, it is the vibration of the system at the driving frequency.

Using the parameter values from Figure 7.5, it will be found that (7.57) agrees with the sinusoidal response part of the complete solution found on page 187:

$$\mathbf{x} = \mathbf{V}[\Lambda - \omega^2\mathbf{I}]^{-1}\mathbf{U}\mathbf{M}^{-1}\mathbf{d}\cos 2t \ . \tag{7.58}$$

When the system includes dissipative elements, the situation is more complicated. For example, let it be required to find the steady state response of the mechanical system in Figure 7.7. the non conservative system whose complete solution is the example used in the text.

$$\mathbf{M}\ddot{\mathbf{x}} + \mathbf{C}\dot{\mathbf{x}} + \mathbf{K}\mathbf{x} = \mathbf{d}\cos 2t \ . \tag{7.59}$$

Now, an assumed solution of $\mathbf{a}\cos 2t$ will not do, because of the $\mathbf{C}\dot{\mathbf{x}}$ term. But, in this case, the driver can be changed to $\mathbf{d}e^{j\omega t} = \mathbf{d}(\cos\omega t + j\sin\omega t)$. This introduces the required imaginary part to the driver. With an assumed solution of $\mathbf{a}e^{j\omega t}$, real and imaginary parts satisfy the equations separately — so that the real part of the result comes from the cos input while the imaginary comes from the $j\sin$ input. When the actualinput is just $\cos\omega t$, the output is taken to be the real part of the result of the analysis:

$$(-\mathbf{M}\omega^2 + \mathbf{C}j\omega + \mathbf{K})\mathbf{a}e^{j\omega t} = \mathbf{d}e^{j\omega t}$$
$$\mathbf{a} = [\mathbf{K} - \omega^2\mathbf{M} + j\omega\mathbf{C}]^{-1}\mathbf{d} \tag{7.60}$$
$$\mathbf{x} = \mathrm{Re}\{\mathbf{a}e^{j\omega t}\} \ .$$

Where Re{ } reads "Real part of{ }." Equations (7.59) and (7.60) are general: the matrices are nXn, defined in the same way as previously discussed, although the specific example is a 2X2 system. The one complication is that the square nXn matrix to be inverted is now complex — and note that the vector, $\mathbf{a}$(nX1), will then, be complex. In this example, using the parameters of Figure 7.7:

$$\mathbf{a} = \left\{ \begin{array}{c} -0.8950 - j0.34145 \\ 0.28972 + j0.38137 \end{array} \right\} \ . \tag{7.61}$$

When this column for $\mathbf{a}$ is plugged back into $\mathbf{x} = \mathrm{Re}\{\mathbf{a}e^{j\omega t}\}$ the result checks with the values given in the steady state portion of the complete solution to the nonconservative system. (see Equation (7.55)).

Steady state response is often the requirement in electrical networks. For illustration, consider the electrical network shown here, which is the analog of the nonconservative mechanical system discussed above. The electrical network with these parameters is not very realistic with R in ohms, L in henries, and C in farads, but it will serve to illustrate the method. Then:

$$(\mathbf{L}p^2 + \mathbf{R}p + \mathbf{S})\mathbf{q} = \left\{ \begin{array}{c} 18 \\ 0 \end{array} \right\} \cos\omega t; \ \ S = \frac{1}{C}, \ \text{and} \ p \equiv \frac{d}{dt} \ . \tag{7.62}$$

$L_1 = 9, L_2 = 7, S_1 = 9, S_2 = 7, S_3 = 7,$ and $R_1 = R_2 = R_3 = 1,$ and with $e(t) = 18\cos\omega t.$

Let it be required to find the steady state output voltage, e_2, over a range of frequencies of the input. From the analysis of the mechanical system, the two resonant frequencies are 1 rps and 1.667 rps for

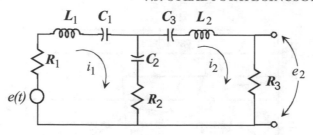

the conservative system and slightly less for the system with damping. A reasonable range, then would be from about $1/2$ rps to 2 rps. The example uses the same parameters as those in the mass-spring system.

For $\omega = 2$, the steady state solution to the set is identical to that given for the nonconservative mechanical system except it results in the charges q_1 and q_2, which will have to be differentiated (multiplied by $j\omega$) to obtain the current values, $\mathbf{i}$. In this example, only i_2 is of concern since $i_2 R_3$ is equal to the voltage output desired. Using numbers from (7.61), above:

$$\mathbf{i} = \text{Re}\{j\omega \mathbf{a} e^{j\omega t}\}$$
$$\mathbf{a} = [\mathbf{S} - \omega^2 \mathbf{L} + j\omega \mathbf{R}]^{-1} \mathbf{d} \tag{7.63}$$
$$e_2 = i_2 R_3 = (0.28975 + j0.38137) j\omega e^{j\omega t} = (-0.76274 + j0.5795) e^{j2t} .$$

It is common to write this result as an amplitude and phase angle. The amplitude is the sum of the squares of these numbers (0.91759), then

$$e_2 = 0.91759(\frac{-0.76274}{0.91759} + j\frac{0.5795}{0.91759}) e^{j\omega t} = 0.91759 e^{j\varphi} e^{j\omega t}$$
$$e_2 = 0.91759 e^{j(\omega t + \varphi)}; \quad \varphi = \tan^{-1}\frac{0.5795}{-0.76274} = 2.49\text{rad} = 142.7^\circ . \tag{7.64}$$

The objective, now, is to repeat this same solution, but using ω over the range $1/2$ rps to 2 rps.

The results are shown below. The greatest amplitude comes near the lower resonant frequency, and there is hardly any increase in magnitude at the upper resonant frequency.

The chart was created by stepping the driving frequency from 0.5 rps to 2.5 rps in 0.05 increments (fewer increments could have been used). At each frequency, the 2X2 complex matrix is inverted and premultiplied into the $\mathbf{d}$ vector, {18, 0}; in total there were a large number of operations — done by PC computer (in milliseconds).

In the following paragraphs, this same problem (i.e., finding amplitude and phase of the output over the same range of frequencies) will be accomplished a different way, and using a "special" type of determinant.

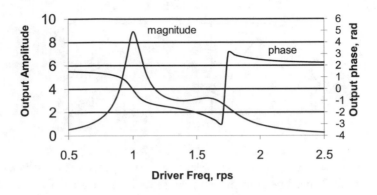

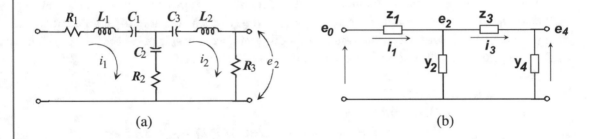

Figure 7.8: (a) Example Network. (b) Ladder Network.

7.5.1   ANALYSIS OF LADDER NETWORKS; THE CUMULANT[3]

When the network is a so-called "ladder" network, as shown in these diagrams, its analysis is again simplified. The "Example Network" shown is a ladder network; note the similarity of the two diagrams. In the ladder, the series elements are denoted by "z," impedance values, while the parallel elements are identified as "y," admittance values.

The numbering of these parameters follows the rule: if a series impedance comes first (as in the example), then impedances have odd numbers, beginning with one — along with the currents through them — while the admittances are even, with their associated voltages. If the leftmost immittance is an admittance (e.g., set $z_1 = 0$ in the diagram), then the admittances will have odd numbers (impedances will be even).

The ladder could have any number of rungs. The numbering scheme is continued in the manner described; and will run from 0 to $2n$ (or to $2n + 1$ when there is an impedance at the output end), where n is the number of rungs. Of interest here is the two rung example network shown.

---

[3] See "Synthesis of Filters" by Herrero and Willoner; Prentice-Hall EE Series, 1966. The methods of this reference are based upon the cumulant.

Comparing the two diagrams, z_1 will be equal to. $R_1 + L_1 j\omega + \dfrac{S}{j\omega}$ for steady state analysis. The other immittances are calculated in the same manner. The voltages across the top of the ladder are related, and the currents summed in the ladder rungs, by Kirchhoff's laws. This results in the following equation set:

$$
\begin{aligned}
e_0 &= i_1 z_1 + e_2 \\
i_1 - i_3 &= e_2 y_2 \\
e_2 &= i_3 z_3 + e_4 \\
i_3 &= e_4 y_4 .
\end{aligned}
\quad \text{in matrix format:} \quad
\begin{bmatrix}
z_1 & 1 & 0 & 0 \\
-1 & y_2 & 1 & 0 \\
0 & -1 & z_3 & 1 \\
0 & 0 & -1 & y_4
\end{bmatrix}
\cdot
\begin{Bmatrix}
i_1 \\
e_2 \\
i_3 \\
e_4
\end{Bmatrix}
=
\begin{Bmatrix}
e_0 \\
0 \\
0 \\
0
\end{Bmatrix} . \tag{7.65}
$$

These equations are written for the example problem, *but the point is the form of the matrix* in (7.65). It has the immittance values on the main diagonal (from 1 to 2n), with "1's" in the upper codiagonal, and "-1's" in the lower. The determinant, D, of this matrix is called a "*cumulant.*" With such a structured form, and so many zeros, it is not surprising that the cumulant is "special." For example, its numeric value, given the values of the diagonal elements, is easily calculated by the following algorithm:

Let the main diagonal elements be designated a_1, a_2, $\cdots$ a_{2n}. Then, the determinant value is found by:

```
a [0] := 1 + j0;
for k   := 1 to 2*n do
begin
  a[k] := {calculate the immittance value here}
    if k > 1 then a[k] := a[k]*a[k-1] + a[k-2];
end;
D := a[k]   {Note: D is the determinant value}
```

This algorithm assumes that a separate function (unique for every ladder network) is coded to calculate the a[k] values — i.e., the (complex) immittances that lie along the main diagonal of the matrix in (7.65). This code is executed for each frequency value. The separate function calculates an impedance value for k-odd, or an admittance value for k-even. For the example network, when k = 1, the function would return the complex number: $R_1 + j(L_1\omega - \dfrac{S_1}{\omega})$.

Following Cramer's Rule, the output voltage is found by replacing the 4th column in D with the column on the right side of the Equation (7.65). This determinant, call it D_4, expanded by the 4th column elements, has the simple value, e_0. Then:

$$
e_4 = \frac{D_4}{D} = \frac{e_0}{D} . \tag{7.66}
$$

In the example problem, $e_0 = 18$. It will be found that this method produces the same results that were obtained, above.

7.6   RUNGE-KUTTA INTEGRATION OF DIFFERENTIAL EQUATIONS

It would be of interest to verify the methods (results) of the previous section. To do so we will discuss the Runge-Kutta numerical method for integration of ordinary differential equations. We will reap a double benefit, since the method itself is a matrix application.

Numerical methods are approximate in nature, since they essentially extrapolate the equation set from its initial conditions. But, they are quick to set up and straightforward in implementation. Numerical solutions are of great value in the analysis of nonlinear problems. As in this case, such methods can also be valuable indicators of the validity of a direct approach.

Numerical methods are based on the following: Given a differential equation set that can be put into the form $\dot{\mathbf{x}} = f(\mathbf{x}, t)$, we divide the independent variable, t, into equal increments, τ, such that $t_n = n\tau$, and at the nth increment, $\dot{\mathbf{x}}_n = f(\mathbf{x}_n, t_n)$. The determination of $\mathbf{x}$ at the next step, t_{n+1}:

$$\mathbf{x}_{n+1} = \mathbf{x}_n + \tau \dot{f}(\mathbf{x}_n, t_n) .$$

That is, the $n+1$ step is estimated by adding to the previous values (at step n), the step size times a best estimate of the derivative of the function relating the $\mathbf{x}$ vector and the independent variable (time, t). The various methods differ in their estimation of the derivative. Notice that we must manipulate the functions into a first order derivative form. This step has already been done in the previous section. For a refresher, a given equation set:

$$[\mathbf{a}_0 p^n + \mathbf{a}_1 p^{n-1} + \cdots + \mathbf{a}_{n-1} p + \mathbf{a}_n]\mathbf{x} = \mathbf{d} \quad \text{where} \quad p \equiv \frac{d}{dt}, \quad p^2 \equiv \frac{d^2}{dt^2}, \quad \text{etc.} \quad (7.67)$$

and the $\mathbf{a}_i$ are mXm matrices. We define $\mathbf{x} = \mathbf{x}_1$, $p\mathbf{x} = \mathbf{x}_2$, $p^2\mathbf{x} = \mathbf{x}_3$, and so forth, up to $p^{n-1}\mathbf{x} = \mathbf{x}_n$. Then:

$$\mathbf{a}_0\dot{\mathbf{x}}_n + \mathbf{a}_1\mathbf{x}_n + \cdots + \mathbf{a}_{n-1}\mathbf{x}_2 + \mathbf{a}_n\mathbf{x}_1 = \mathbf{d} .$$

Now, premultiply by $\mathbf{a}_0^{-1}$ (assumed to be nonsingular), and it is easy to write, directly:

$$\begin{Bmatrix} \dot{\mathbf{x}}_1 \\ \dot{\mathbf{x}}_2 \\ \dot{\mathbf{x}}_3 \\ \cdots \\ \dot{\mathbf{x}}_n \end{Bmatrix} = \begin{bmatrix} 0 & \mathbf{I}_m & 0 & \cdots & 0 & 0 \\ 0 & 0 & \mathbf{I}_m & \cdots & 0 & 0 \\ \cdots & \cdots & & & & \cdots \\ 0 & 0 & 0 & \cdots & 0 & \mathbf{I}_m \\ -\mathbf{a}'_n & -\mathbf{a}'_{n-1} & & \cdots & & -\mathbf{a}'_1 \end{bmatrix} \begin{Bmatrix} \mathbf{x}_1 \\ \mathbf{x}_2 \\ \mathbf{x}_3 \\ \cdots \\ \mathbf{x}_n \end{Bmatrix} + \begin{Bmatrix} 0 \\ 0 \\ \cdots \\ 0 \\ \mathbf{a}_0^{-1}\mathbf{d} \end{Bmatrix} . \quad (7.68)$$

The matrices $\mathbf{a}'_i = \mathbf{a}_0^{-1}\mathbf{a}_i$. With the obvious definitions, $\dot{\mathbf{z}} = \mathbf{A}\mathbf{z} + \mathbf{h}$, and we are back to the equation set of Section 7.4. In general, if the order of the original $\mathbf{a}$ matrices is m, the order of (7.68) will be nm. For the systems discussed in this chapter, the degree $n = 2$ and the order is, therefore $2m$. Note that all the zeros shown in (7.68) are matrices. In the $\mathbf{A}$ matrix, they are mXm null matrices, and in the $\mathbf{h}$ vector they are mX1 columns.

The method to be described is the Runge-Kutta. It has reasonable accuracy, and it is a popular method, in current use. We will develop the method and run it against the problem, without a technical discussion of its relative merit, and/or its accuracy compared to other methods.

The initial value problem applies the initial conditions to $\mathbf{z}$, and to $\mathbf{h}$ at time t_0. Advancing the solution from step (time increment, τ) to step follows the Runge-Kutta algorithm below. Then to advance from the nth step to the $n + 1^{\text{st}}$ involves four intermediate steps (estimates), which are then put together in a manner like Simpsons rule to form the step:

$$\mathbf{y}_1 = \mathbf{A}\mathbf{z}_n + \mathbf{h}(t_n); \ t_n = n\tau$$
$$\mathbf{y}_2 = \mathbf{A}(\mathbf{z}_n + \tfrac{\tau}{2}\mathbf{y}_1) + \mathbf{h}(t_n + \tfrac{1}{2}\tau)$$
$$\mathbf{y}_3 = \mathbf{A}(\mathbf{z}_n + \tfrac{\tau}{2}\mathbf{y}_2) + \mathbf{h}(t_n + \tfrac{1}{2}\tau) \tag{7.69}$$
$$\mathbf{y}_4 = \mathbf{A}(\mathbf{z}_n + \tau\mathbf{y}_3) + \mathbf{h}(t_n + \tau), \ \text{then}$$
$$\mathbf{z}_{n+1} = \mathbf{z}_n + \tfrac{\tau}{6}(\mathbf{y}_1 + 2\mathbf{y}_2 + 2\mathbf{y}_3 + \mathbf{y}_4) .$$

In order to verify the direct approach to the non-conservative system, we return to Equation (7.43): $\dot{\mathbf{z}} - \mathbf{A}\mathbf{z} = \mathbf{0}$, or $= \mathbf{h}$, depending on whether or not the system is driven.

The non-conservative system of the previous section has

$$\mathbf{A} = \begin{bmatrix} \mathbf{0} & \mathbf{I} \\ -\mathbf{M}^{-1}\mathbf{K} & -\mathbf{M}^{-1}\mathbf{C} \end{bmatrix} = \begin{bmatrix} 0 & 0 & 1 & 0 \\ 0 & 0 & 0 & 1 \\ -\frac{16}{9} & \frac{7}{9} & -\frac{2}{9} & \frac{1}{9} \\ 1 & -2 & \frac{1}{7} & -\frac{2}{7} \end{bmatrix}; \text{ and } \mathbf{h} = \begin{Bmatrix} 0 \\ 0 \\ 2 \\ 0 \end{Bmatrix} \cos t . \tag{7.70}$$

In the first example, the (two-mass, three spring) system wasn't driven, but had the initial condition $\mathbf{z}_0 = \{ 1 \ \ 0 \ \ 0 \ \ 1 \ \}$, (i.e., $x_{10} = 1$ and $\dot{x}_{20} = 1$). Following the method (5,3), and noting the absence of a driving vector, the $\mathbf{y}_1$ vector (at $t_n = 0$) is simply $\mathbf{A}\mathbf{z}_0$.

If this method is continued, using a step size of 0.1 second, the curves shown below are the result. As an example, at time = 1.0 sec. The values for x_1 and x_2 are 0.44974 and 0.91952. These values agree with the true solution (see Equation (7.50)) to 4 decimal places. Making the step size smaller does not increase agreement, but the true solution numbers were only taken to 5 places. The point is that the Runge-Kutta produces an amazingly accurate replica of the true solution; certainly close enough to validate the "true solution."

The Runge-Kutta requires 4 matrix-vector multiplications, plus the same number of vector additions, per step. This problem ran to 120 steps. Thus, a lot of calculations were required to arrive at these curves. Imagine the task in the days before the computer! But, with a reasonably modern desk top, the tabular results are obtained almost instantly.

The second example of the previous section concerned the same mechanical system, but with a cosine input force directed at mass 1, in the amount of $18 \cos 2t$. The introduction of the driving vector just requires the addition of the $\mathbf{h}$ vector and a change to zero initial conditions (the two masses are initially at rest).

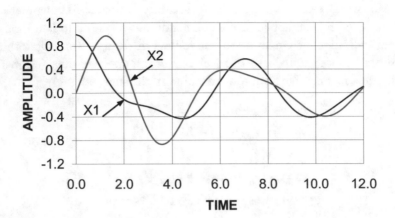

Damped System, Initial Value Problem

Over a 12 second time period, with a step size of 0.1 sec, the results again agree with the true solution in the first 4 decimals.

The graph below plots the motion of the two masses.

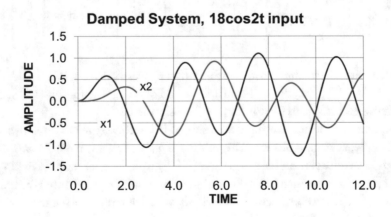

Damped System, 18cos2t input

7.7   EXERCISES

7.1.   Reduce the equation, $a\dddot{x} + b\ddot{x} + c\dot{x} + dx = f(t)$ to a set of 1st order equations.

7.2.   The single spring-mass system (a) shown below, has the natural frequency

$$\omega = \sqrt{\frac{k}{m}} \ \text{(system restricted to vertical motion)} .$$

If another spring and mass, with identical properties, is added as in (b), how many natural frequencies are there, what are they, and are any of them equal to ω ?

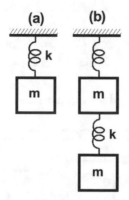

7.3. A machine weighing W_1 pounds is suspended on a foundation with spring constant, $k = 2k_1$. The machine is subjected to a vibrating force, $f_0 \cos \omega t$, frequencies very close to

$$\sqrt{\frac{k}{m_1}}; \quad \text{where } m_1 = \frac{W_1}{g}.$$

A vibration absorber is to be added, consisting of weight W_2 and spring constant k_2. Determine the relationship between k_2 and W_2 such that the motion of W_1 is minimized.

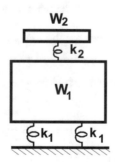

7.4. Given the text example non-conservative system of Figure 7.7, the matrix actually analyzed is the "reduced set" matrix, $\mathbf{A}$, given in Equation (7.48). This 4X4 has 4 eigenvalues, eigenvectors. If we define eigenvectors as $\mathbf{v}_i = \mathbf{q}_i, \lambda_i \mathbf{q}_i$, for $i = 1, 2, 3, 4$, show that each $\lambda_i, \mathbf{q}_i$ pair satisfy the given equation $(\mathbf{M}\lambda_i^2 + \mathbf{C}\lambda_i + \mathbf{K})\mathbf{q}_i$.

Show that, in general, if $\lambda_i, \mathbf{q}_i$ solve the given equation, then so do their complex conjugates.

7.5.   Calculate the eigenvalues and vectors for the system shown in the figure; then find the motion for all 3 masses, given the initial conditions:

$$x_{10} \ (W_1) = 2 \text{ inches, to the right}$$
$$x_{20} \ (W_2) = 1 \text{ inch, to the right}$$
$$x_{30} \ (W_3) = -1 \text{ inch, to the left}$$
$$\text{Initial velocities} = 0$$

The units for k are #/in, and for c, # per in/sec.

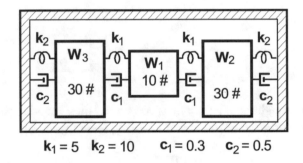

$$k_1 = 5 \quad k_2 = 10 \quad c_1 = 0.3 \quad c_2 = 0.5$$

7.6.   The system shown in Problem 5 is traveling to the right at 50 in/sec, when the container box is suddenly stopped (at time $t = 0$). Find the motion of W_1, and the unbalanced force on W_1 over the following second of time. What is the maximum force on W_1?

7.7.   If, in the system of Problem 5, W_2 and W_3 are "anchored together" (so that they must move together) would the results of Problem 6 change?

7.8.   Calculate the immittance values for the low pass filter in the diagram, at frequencies of 500 and 1000 cps. Solve the related cumulant to compare the ratio $\dfrac{e_0}{e_8}$ (input to output voltage — see text Equation (7.66) at the two frequencies.

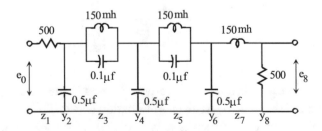

Capacitances are in microfarads, inductance is in millihenries, and resistance values in ohms. Note: the consistent set of parameters is ohms, henries, farads.

At zero frequency, the network is a simple voltage divider (shown here). In this case, the ratio of input to output voltage is 2.0 (6 db).

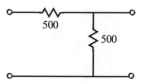

Partial Differentiation of Bilinear and Quadratic Forms

Begin with the definition of the partial differential operator, "del," ∇. This operator is defined as a *column vector*, an (mX1) matrix:

$$\nabla = \left\{ \begin{array}{c} \dfrac{\partial}{\partial x_1} \\[2mm] \dfrac{\partial}{\partial x_2} \\ \vdots \\ \dfrac{\partial}{\partial x_m} \end{array} \right\}.$$

(A.1)

The column length is determined by the number of independent variables, $(x_1, x_2, \cdots x_m)$, there are m of these, and del should be written ∇_x. indicating the independent variable. Herein, the subscript will be omitted unless there could be a confusion.

Note that ∇ is an operator; it is meaningless standing alone (A.1). But, given a function $q(x_1, x_2, \ldots x_m)$:

$$\nabla q = \nabla q(x_1, x_2, \cdots x_m) = \left\{ \begin{array}{c} \dfrac{\partial q}{\partial x_1} \\[2mm] \dfrac{\partial q}{\partial x_2} \\ \vdots \\ \dfrac{\partial q}{\partial x_m} \end{array} \right\}$$

(A.2)

certainly is meaningful, it being all the partials of q arranged into a column vector. As with all operators, one must be careful with the order in which the symbols are written. $q\nabla$ is as meaningless as ∇ itself.

Notation

In this appendix the superscript "t" will be used to indicate transposition. Example: $\mathbf{A}^t$.

This will allow the use of the "prime" to indicate differentitation

$$y_{ii}' = \frac{\partial y_j}{\partial x_i}$$

Given a set of functions $y_k(x_1, x_2 \cdots x_m)$; $k = 1 \ldots n$, they are arranged into an (nX1) column vector and denoted in boldface, $\mathbf{y}$. Although there is no matrix operation in which a column vector operates on another column vector, there is a way to display all of the partials of $\mathbf{y}$ with respect to the variables, x_i. It is written as $\nabla \mathbf{y}^t$. Note that $\mathbf{y}$ is *transposed to a row vector*.

$$\nabla \mathbf{y}^t = \left\{ \begin{array}{c} \dfrac{\partial}{\partial x_1} \\[2mm] \dfrac{\partial}{\partial x_2} \\[1mm] \vdots \\[1mm] \dfrac{\partial}{\partial x_m} \end{array} \right\} \begin{bmatrix} y_1 & y_2 & \cdots & y_n \end{bmatrix} = \begin{bmatrix} \dfrac{\partial y_1}{\partial x_1} & \dfrac{\partial y_2}{\partial x_1} & \cdots & \dfrac{\partial y_n}{\partial x_1} \\[2mm] \dfrac{\partial y_1}{\partial x_2} & \dfrac{\partial y_2}{\partial x_2} & \cdots & \dfrac{\partial y_n}{\partial x_2} \\[2mm] \vdots & \vdots & \ddots & \vdots \\[2mm] \dfrac{\partial y_1}{\partial x_m} & \cdots & \cdots & \dfrac{\partial y_n}{\partial x_m} \end{bmatrix} \quad \text{(mXn)} . \qquad \text{(A.3)}$$

Using the notation $y_{ij}' = \dfrac{\partial y_j}{\partial x_i}$, the Equation (A.3) reads:

$$\nabla \mathbf{y}^t = \begin{bmatrix} y_{11}' & y_{12}' & \cdots & y_{1n}' \\ y_{21}' & y_{22}' & & y_{2n}' \\ & & \ddots & \\ y_{m1}' & y_{m2}' & & y_{mn}' \end{bmatrix} \quad \text{where} \ \ y_{ij}' = \frac{\partial y_j}{\partial x_i} . \qquad \text{(A.4)}$$

The column-by-row matrix product is unusual, but it is conformable. Note that the ith row has all partials with respect to x_i, and the jth column contains the partials of y_j. Also, notice that $\nabla \mathbf{y}^t$ cannot be transposed: the result would again have ∇ operating on nothing.

A "bilinear form" is the dot product of two vectors, say $\mathbf{y} \bullet \mathbf{w}$, whose elements are functions of m independent x-variables, i.e., $y_k(x_1, x_2 \cdots x_m)$ and $w_k(x_1, x_2 \cdots x_m)$. Arranged into column vectors, both $\mathbf{y}$ and $\mathbf{w}$ must have the same number of elements, i.e., $k = 1, , n$, and, for example $\mathbf{y} = \{y_1, y_2, \cdots y_n\}$. Then

$$\mathbf{y} \bullet \mathbf{w} = \mathbf{y}^t \mathbf{w} = y_1 w_1 + y_2 w_2 + \cdots + y_n w_n . \qquad \text{(A.5)}$$

Note that each term contains a yw product with each variable in the first power — thus, "bilinear." If the functions $w_i = y_i$, the form would be a "quadratic form," containing only y_i^2 terms

The dot product in (A.5) is a scalar, and so (from Equation (A.2)) we expect that $\nabla(\mathbf{y^t w})$ is a vector, with each row, i, containing the partials $\mathbf{y \bullet w}$ with respect to x_i. In performing row by row differentiation, we choose to take all the partials of y first, then those of w. The general (kth) row is:

$$\begin{bmatrix} y'_{k1} & y'_{k2} & \cdots & y'_{kn} \end{bmatrix} \mathbf{w} + \begin{bmatrix} w'_{k1} & w'_{k2} & \cdots & y'_{kn-} \end{bmatrix} \mathbf{y} . \tag{A.6}$$

Now, when all the m rows are written, ($i = 1,, m$), the result will be

$$\nabla(\mathbf{y^t w}) = \begin{bmatrix} y'_{11} & y'_{12} & \cdots & y'_{1n} \\ y'_{21} & y'_{22} & & y'_{2n} \\ & & \ddots & \\ y'_{m1} & y'_{m2} & & y'_{mn} \end{bmatrix} \{w\} + \begin{bmatrix} w'_{11} & w'_{12} & \cdots & w'_{1n} \\ w'_{21} & w'_{22} & & w'_{2n} \\ & & \ddots & \\ w'_{m1} & w'_{m2} & & w'_{mn} \end{bmatrix} \{y\} = (\nabla\mathbf{y^t w}) + (\nabla\mathbf{w^t})\mathbf{y}. \tag{A.7}$$

As expected, the result is a (1Xn) vector.

In (A.5), if $\mathbf{w} = \mathbf{Az}$ the dot product would be $\mathbf{y \bullet w} = \mathbf{y^t Az}$. The $\mathbf{A}$ matrix (which "transforms" $\mathbf{z}$ to $\mathbf{w}$) is necessarily nXn. Since the elements of $\mathbf{A}$ *are not functions of the* x_j *variables*, the inclusion of $\mathbf{A}$ does not add complication:

$$\nabla(\mathbf{y^t w}) = \nabla(\mathbf{y^t Az}) = (\nabla\mathbf{y^t})\mathbf{Az} + (\nabla\mathbf{z^t A^t})\mathbf{y} = (\nabla\mathbf{y^t})\mathbf{Az} + (\nabla\mathbf{z^t})\mathbf{A^t y} . \tag{A.8}$$

In the case $\mathbf{z} = \mathbf{y}$ and $\mathbf{A}$ is a symmetric matrix, the dot product is a quadratic form $\mathbf{y^t Ay}$ and

$$\nabla(\mathbf{y^t Ay}) = 2(\nabla\mathbf{y^t})\mathbf{Ay} . \tag{A.9}$$

In Chapter 4 the quadratic form in the regression problem is $\mathbf{e^t e} = \mathbf{x^t A^t Ax} - 2\mathbf{x^t A^t b} + \mathbf{b^t b}$, in which the $\{y\}$ variables are, respectively, $y_k = x_k$. In this case, $y'_{ij} = x'_{ij} = \delta_{ij}$ (the Kronecker delta, whose value is zero except when $i = j$ when it is unity). Then:

$$\nabla\mathbf{e^t e} = \nabla(\mathbf{x^t A^t Ax}) - 2(\nabla\mathbf{x^t})\mathbf{A^t b} = 2(\nabla\mathbf{x^t})\mathbf{A^t Ax} - 2(\nabla\mathbf{x^t})\mathbf{A' b} . \tag{A.10}$$

Since the $\mathbf{b}$ vector is not a function of the x variables, its derivative is zero. Now, $(\nabla\mathbf{x^t})$ is $[\delta_{ij}]$, the unit matrix, so:

$$\nabla\mathbf{e^t e} = 2\mathbf{A^t Ax} - 2\mathbf{A' b} . \tag{A.11}$$

The Interest in Chapter 6 is just the differentiation of the quadratic form. Beginning with Equation (A.9), note that again $y_k = x_k$, and $\nabla\mathbf{y^t} = y'_{ij} = x'_{ij} = \delta_{ij}$. Then:

$$\nabla(\mathbf{x^t Ax}) = 2(\nabla\mathbf{x^t})\mathbf{Ax} = 2\mathbf{Ax} . \tag{A.12}$$

APPENDIX B

Polynomials

Associated with every square (nXn) matrix is a characteristic polynomial equation:

$$f(\lambda) = c_0\lambda^n + c_1\lambda^{n-1} + \cdots c_{n-1}\lambda + c_n = 0$$

whose degree is n. The roots of this polynomial are the eigenvalues of the matrix. Our interest in polynomials is fueled by the requirement to find these eigenvalues. Toward that end, some of the basic arithmetic algorithms are discussed here, with a display of "Pascal-like" code. Then, an outline of a recommended method for determining polynomial roots is given.

B.1 POLYNOMIAL BASICS

In this appendix the polynomial will be written as:

$$p(x) = c_0x^n + c_1x^{n-1} + \cdots + c_{n-1}x + c_n = 0 \,. \tag{B.1}$$

The coefficient, c_0, the coefficient of the highest power term, can always be made to equal unity. Only c_0 and c_n are required to be non-zero and both c_0 and c_n can be made unity by the transformation $x = kz$ with $k = \sqrt[n]{\dfrac{c_n}{c_0}}$. Note that in the representation, (B.1), the sum of the subscript on each term with its corresponding power always equals n.

The Equation (B.1) is not an identity; there are exactly n (generally complex) "roots," x_j, that cause $p(x)$ to vanish. ***In this discussion, we consider only polynomials with real coefficients. As a consequence, if a root is complex, its complex conjugate must also be a root.*** If the degree, n, is odd, there must be at least one real root.

It is sometimes desirable to define a related polynomial, defined by the transform $x = 1/z$:

$$p(z) = c_nz^n + c_{n-1}z^{n-1} + \cdots + c_1z + c_0 = 0 \tag{B.2}$$

which is the same as (B.1), but with the coefficients taken in reverse order, and possessing the inverses of the roots of (B.1). A root $x > 1$ is transformed to a root $z < 1$. As an example, if the root extraction method converges to the smallest (absolute value) root first, then (B.2) might be used to obtain roots in the reverse order.

Denoting the roots of (B.1) as x_j, $j = 1..n$, the polynomial can be written as in (B.3):

$$p(x) = \prod_{j=1}^{n}(x - x_j) = (x - x_1)(x - x_2)\cdots(x - x_n) \,. \tag{B.3}$$

By performing the indicated multiplication of the factors in (B.3), the relationships between the coefficients and the roots can be derived:

$$
\begin{cases}
c_1 = -\sum_{j}^{n} x_j = -\text{Sum of all roots} \\[2mm]
c_2 = \sum_{i<j}^{n} x_i x_j = \text{Sum of root products taken 2 at a time} \\[2mm]
c_3 = -\sum_{i<j<k}^{n} x_i x_j x_k = -\text{Sum of root products taken 3 at a time} \\[2mm]
\cdots \cdots \cdots \text{ etc.} \\[2mm]
c_n = (-1)^n \prod_{j}^{n} x_j = \text{Product of all roots.}
\end{cases}
\tag{B.4}
$$

The notation of (B.4) is unusual. The coefficient c_3 is the negative sum of the products of the roots, taken 3 at a time. If n were 4, then c_3 would be $x_1 x_2 x_3 + x_1 x_2 x_4 + x_1 x_3 x_4 + x_2 x_3 x_4$. The coefficient c_4 is the sum of all products of roots, taken four at a time. And so on, until there is only the single term product of all the roots, equal to c_n. Note the alternating signs in (B.4).

At first, it may seem unlikely that the operations in (B.4) will always produce real coefficients. However, in the simple example given ($n = 4$), if x_3 and x_4 are complex conjugates and x_1 and x_2 are conjugates, it is easily seen that the terms forming c's will be conjugates — their sums, real.

Equations (B.4) provide insight into the character of the roots, but if a set of roots is given, these relationships are not useful in calculating the coefficients. The recommended algorithm to generate the coefficients from the roots is surprisingly simple. The "Pascal-like" code is given below. Since the roots are generally complex, the routine must be executed using complex variable data types and using complex arithmetic. Note: c [0] must be 1.0 and real.

```
c[0]:=(1+j0);  {Calculate the coefficients c from roots, x}
for k:=1 to N do
begin
  c[k]:=0+j0;  { j = sqrt of -1 }
  for i:=k downto 1 do if (i = 1) then c[i]:=c[i]-x[k] else
    c[i]:=c[i]-x[k]*c[i-1];
end;
```

Pascal does not have a complex type, nor does it support complex arithmetic directly. Thus, the complex type must be defined in the program, and complex arithmetic must be done in separate procedures.

B.2   POLYNOMIAL ARITHMETIC

Polynomials are added/subtracted by adding/subtracting the coefficients of "like" terms (those having the same degree in the variable, x). Polynomial multiplication:

$$A^n(x) = a_0 x^n + \cdots + a_n = B^m(x) C^l(x) \tag{B.5}$$

is affected by multiplying every term in B by every term in C, and collecting "like terms." Note, in (B.5), the superscript on the capital letters indicates the degree of the polynomial. The result, A, will clearly be a polynomial of degree $n = l + m$.

It is instructive to indicate the terms to be collected by means of a diagram, Table B.1, showing an example whose product is to be $A^7 = B^3 C^4$. From the first row of the table, the coefficient a_0 will be just the product of b_0 and c_0. The succeeding rows indicate the terms to be multiplied and summed:

$$\begin{cases} a_0 = b_0 c_0 \\ a_1 = b_0 c_1 + b_1 c_0 \\ a_2 = b_0 c_2 + b_1 c_1 + b_2 c_0 \\ a_3 = b_0 c_3 + b_1 c_2 + b_2 c_1 + b_3 c_0 \\ \cdots = \text{etc} \end{cases} \tag{B.6}$$

Table B.1:

	$c_0 x^4$	$c_1 x^3$	$c_2 x^2$	$c_3 x$	c_4
$a_0 x^7$	$b_0 x^3$				
$a_1 x^6$	$b_1 x^2$	$b_0 x^3$			
$a_2 x^5$	$b_2 x$	$b_1 x^2$	$b_0 x^3$		
$a_3 x^4$	b_3	$b_2 x$	$b_1 x^2$	$b_0 x^3$	
$a_4 x^3$		b_3	$b_2 x$	$b_1 x^2$	$b_0 x^3$
$a_5 x^2$			b_3	$b_2 x$	$b_1 x^2$
$a_6 x$				b_3	$b_2 x$
a_7					b_3

The table was constructed by writing the A coefficients down the left, and putting the higher order (i.e., C) coefficients along the top. At the intersection of row/column, the B term is chosen such that its exponent, plus the top row exponent add to the A power at the left. Note that the terms of $B(x)$ appear to "slide" across the table from left toward right. In fact, if the b coefficients are written on a "sliding strip" the algorithm is shown clearly in Table B.2, below.

In this scheme, the "a" coefficient that lies just below b_0 is determined by the products of the b and c row terms. For example, when b_0 slides under c_3 the a_3 coefficient is calculated by multiplying

Fixed Strip					c_0	c_1	c_2	c_3	c_4	0	0	0
Sliding Strip	b_3	b_2	b_1	b_0	$\rightarrow$							
Results					a_0	a_1	a_2	a_3	a_4	a_5	a_6	a_7

Table B.2: **Polynomial multiplication**, by a "Sliding strip" method.

the adjacent b and c coefficients in the same column, going from right to left:

$$a_3 = b_0c_3 + b_1c_2 + b_2c_1 + b_3c_0 .$$

The computer method just "automates" Table B.1 and Equations (B.5) and (B.6).

The computer routine for multiplying polynomials B and C is written directly from the sliding strip display. It uses index 'k' to slide the b coefficients along, 'j' to choose a b coefficient, and (k-j) to choose the c. Nb and Nc are the respective degrees of the polynomials B and C. This routine is done in real arithmetic — we consider only polynomials with real coefficients. The coefficient variables are given in the lower case corresponding to the upper case polynomial designation.

```
{Polynomial Multiply: A = B times C}
  for k:=0 to Nb+Nc do
  begin
    a[k]:=0; for j:=0 to Nb do
    if((k-j)<=Nc) and ((k-j)>=0)then a[k]:=a[k]+b[j]*c[k-j];
  end;
```

To develop polynomial division, Equations (B.5) can be solved for the coefficients, c_j. Note that the coefficient b_0 *must equal one*:

$$\left\{\begin{array}{l} c_0 = a_0 \\ c_1 = a_1 - b_1c_0 \\ c_2 = a_2 - b_1c_1 - b_2c_0 \\ c_3 = a_3 - b_1c_2 - b_2c_1 - b_3c_0 \\ \cdots = \text{etc} \end{array}\right. \qquad \text{where } b_0 = 1.0 . \qquad (B.7)$$

In **polynomial division,** ($C = A$ divided by B, Equations (B.7)), the graphic scheme is similar, but different in that the divisor (b) coefficients must be reversed in sign, except for b_0, required to be 1.0. Most importantly, the b coefficients multiply previously determined c coefficients. In this case, it is clearer if the rows for Results, and Sliding Strip are interchanged. In Table B.3, the quotient, C polynomial coefficients, are to be calculated.

The only b coefficient that multiplies the a row is $b_0 = 1$. All the rest of the b coefficients multiply previously determined c coefficients ("feedback"). For example, slide the b strip until the "1" is under c_4. (c_4 has not yet been determined, but the c terms to its left have been). Thus:

$$c_4 = a_4 - b_1c_3 - b_2c_2 - b_3c_1 .$$

Table B.3:											
Fixed Strip					a_0	a_1	a_2	a_3	a_4	a_5	a_6	a_7
Results					c_0	c_1	c_2	c_3	c_4	0	0	0
Sliding Strip	$-b_3$	$-b_2$	$-b_1$	1	$\rightarrow$						

Note, again, that only the "+1" (i.e., b_0) on the b strip "reaches over" to multiply the a strip, a_4. This sliding strip scheme is one method of performing "**synthetic division**,"

Notice that there can be no further c coefficients after c_4. These locations are zero. However, the synthetic division process continues until the 1 slides under a_7. The terms (three, in this example) so calculated are the remainder, R, coefficients, whose degree is one less than the divisor. In this example

$$r_0 = a_5 - b_1 c_4 - b_2 c_3 - b_3 c_2$$
$$r_1 = a_6 - b_1 \times 0 - b_2 c_4 - b_3 c_3 = a_6 - b_2 c_4 - b_3 c_3$$
$$r_2 = a_7 - b_1 \times 0 - b_2 \times 0 - b_3 c_4 = a_7 - b_3 c_4 .$$

As an example, divide $A^7(x) = x^7 + 2x^6 + 5x^5 + 17x^4 - 49x^3 + 20x^2 + 54x - 18$ by

$B^3(x) = x^3 + 10x - 3$. The result will be $C^4(x)$ with remainder, $R^2(x)$.

Table B.4:								
	0	1	2	3	4	5	6	7
Fixed Strip	1	2	5	17	-49	20	54	-18
Results	1	2	-5	0	7	0	0	0
Sliding Strip				3	-10	0	1	$\rightarrow$
					R	5	-16	

The A coefficients are entered into the fixed strip, and the sliding strip is prepared with the B coefficients reversed in sign (except for the "1"). **Note that zero coefficients are included**. The results strip show the calculated c coefficients for the 4th degree polynomial, C. The remainder coefficients are shown in the bottom row. The position of the sliding strip is at the point that the coefficient r_0 is to be calculated ($20 + 0 \times 7 - 10 \times 0 - 5 \times 3 = 5$).

If calculations were to be made by hand, there is really no reason to prefer the sliding strip way to do synthetic division. However, calculations are to be made by computer — and the sliding strip method clearly shows the algorithm.

The Pascal code to accomplish this division is shown below. The dividend is A, the divisor, B, the quotient is Q, and the remainder is REM. Order(A) = N, order(B) = m. The index k "pushes the strip along," and the index j is used to gather up the product terms:

```
Begin {Begin Synthetic Division----}
```

```
   q [0]:=a[0];
 for k:=1 to N do if k <= (N-m) then
 begin
   q[k]:=a[k];
   for j:=1 to m do if k-j > -1 then q[k]:=q[k]-b[j]*q[k-j];
 end else
 begin    {when k > N-m the remainder coeffs are calculate'd)
   rem[k-N+m-1]:=a[k];
   for j:=1 to m do if (k-j) <= N-m then
   rem[k-N+m-1]:=rem[k-N+m-1]-b[j]*q[k-j];
 end;
end;
```

Again, no complex arithmetic is involved.

B.2.1   EVALUATING A POLYNOMIAL AT A AIVEN VALUE

Given $P(x)$, find $P(x_0)$. This is equivalent to dividing P by $(x - x_0)$. Back to synthetic division. And there is a "bonus," Upon repeated applications of synthetic division, $P(x_0)$ first appears, then $P'(x_0)$ (the apostrophe is used here to denote differentiation), then $\frac{1}{2}P''(x_0)$! The reason that this is a real bonus is that the Newton method for determining roots use these values.

As an example, take the polynomial, C, and determine these values at x = −2.

$$C(x) = x^4 + 2x^3 - 5x^2 + 7 \quad \text{at} \ x = -2 .$$

The following Table B.5 repeatedly divides by $(x - x_0) = (x + 2)$. The sign of "b_1" must be reversed, so that the sliding tab is $(-2\ 1)$. Also note that the degree of each row is one less than the one above (e.g., the degree of $P'(x)$ is one less than $P(x)$).

Table B.5:					
	4	3	2	1	0
P coeffs	1	2	-5	0	7
P(-2)	1	0	-5	10	-13
P'(-2)	1	-2	-1	12	
½P''(-2)	1	-4	7		
Slide	-2	1	→		

The numeric results are shown in the double outlined boxes (example, $P'(-2) = 12$). The slide strip that produced all these results is shown in the bottom row. It slides one less column in evaluating each derivative. Note that the "fixed strip" moves down one row. Example: the fixed strip for the calculation of $P'(-2)$ is the row just above.

In the above table, the rows can be filled in simultaneously. The table can be filled column by column, rather than row by row. This makes the computer coding all the easier. The code given below takes advantage of this. The polynomial is of degree N and its coefficients are c[k]. The value of x is given in xx. The value of the polynomial at x is in f, the derivative at x in f1, and $1/2$ the second derivative in f2. In the code, below, the polynomial coefficients are in array c[k].

```
Begin {Evaluate the polynomial and its derivatives at xx}
  f:=c [0]; f1:=c[0]; f2:=c[0];
  for k:=1 to N do
  begin
    f:=c[k]+xx*f;
    if k < N then f1:=f+xx*f1;
    if k < N-1 then f2:=f1+xx*f2;
  end;
end;
```

A very simple routine. However, xx may be complex, and so the routine must be done in complex arithmetic.

B.3   EVALUATING POLYNOMIAL ROOTS

Deriving the polynomial roots from its coefficients is certainly not a simple problem. For polynomials of degree greater than 4 there is no direct approach — only iterative methods are available. A recommended approach is summarized as follows:

- Laguerre's method (below) is recommended to provide an initial estimate of a root.

- When a close estimate is found (Laguerre), then Newton's method produces excellent results.

- Using the result from Newton's method, the original polynomial can be "deflated" by dividing the polynomial by the newly found root factor(s), to a product of the root factor times a polynomial of lesser degree. In this way, the degree of each "stage" becomes reduced until a quadratic, cubic or quartic polynomial remains. At that point, the remaining roots can be calculated directly.

The equations for the Laguerre and Newton methods will be given here, without derivation.

B.3.1   THE LAGUERRE METHOD

Given a polynomial, $P(x)$, of degree N, and an initial estimate of a root at x_0, define:

$$G \equiv \frac{P'(x_0)}{P(x_0)} \quad \text{and} \quad H \equiv G^2 - \frac{P''(x_0)}{P(x_0)} .$$

(Note: The apostrophe is used here to denote differentiation.)

Then $h = \dfrac{-N}{G \pm \sqrt{(N-1)(NH - G^2)}}$, and the next estimate of the root is $x_0 + h$.

B.3.2   THE NEWTON METHOD

When a root is located closely enough to ensure convergence of the Newton method:

$$\frac{1}{h} = -\frac{P'(x_0)}{P(x_0)} + \frac{\frac{1}{2}P''(x_0)}{P'(x_0)} \; .$$

Note that all the factors (P and its derivatives) are obtained directly via the above "Pascal-like" routine. Just put them together to find h. The next estimate of the root is $x_0 + h$.

B.3.3   AN EXAMPLE

An example will provide a general overview of the methods involved:

$$P(x) = 1.0x^5 + 2.0x^4 + 3.0x^3 + 4.0x^2 + 5.0x + 6.0 \; .$$

Using an initial value of $0 + j0$, the Laguerre method quickly finds roots near $(-0.806 \pm j1.223)$. Then Newton refines the roots to $(-0.8057865 \pm j1.22290471)$.

The deflation process reduces the polynomial to third degree (note that two roots have been found so that the deflation is from degree 5 to degree 3), with coefficients:

$$1.000000 \quad 0.388427 \quad 0.229234 \quad 2.797480$$

This cubic polynomial can be solved directly. However, for the example we continue the iterative methods.

The Laguerre method works with this new polynomial to find $(0.552 \pm j1.253)$. Now, the Newton method is used — but with this difference: ***Newton will always use the original polynomial,*** because inaccuracies in the first roots will effect the accuracies of the deflated polynomial. Newton finds $(0.551685 \pm j1.2533352)$.

Now a polynomial x + 1.491798, whose root is obviously -1.491798. And this completes the process.

It should be noted that these calculations must be done with high precision. This example problem was done using the "extended" variable type in Turbo Pascal (Delphi). The first root was actually determined to be:

$$-8.05786469389031\mathrm{E} - 0001 \quad 1.22290471337441E + 0000$$

Root evaluation is very sensitive to small changes in the coefficients. It may be that this sensitivity can be reduced somewhat by the transformation given at the beginning of the appendix, which results in both c_0 and c_n being set to 1.0. For the problems used in preparing this appendix, the accuracy was so great that the transformation appeared to make no difference. But in more realistic problems, it might.

APPENDIX C

The Vibrating String

This appendix analyzes the vibrations in a stretched string, first using a digitized matrix approach, then comparing it with the analysis of the continuous string. Refer to Chapter 6, where the interest is in the "normal modes" and how they sum together. In Chapter 7, the interest is in applying the same kind of digitized approach to the vibration analysis of a beam.

C.1 THE DIGITIZED – MATRIX SOLUTION

The figure shows the string anchored at each end and pulled tight. The string length, L, is imagined to be divided into n equal segments. Figure C.1(A) shows $n = 8$.

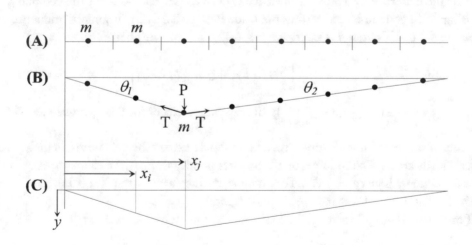

Figure C.1: Vibrating String.

The mass, m, of each segment is concentrated at the center of the segment. The string itself is then considered weightless. The mass, m, is equal to the total mass, M, divided by n.

When a vertical load, P, is applied to the j th mass, the string is deformed as in (B), and is resisted by the tension in the string. The static weights, mg, are very small compared to tension forces. Neglecting these weights, then, $P = T(\sin\theta_1 + \sin\theta_2)$. The deformation is small enough to consider $\sin\theta = \tan\theta$, and $\cos\theta = 1$ (note the chosen positive x and y directions, in (C)). Then

$$\sin\theta_1 \approx \frac{y(x_j)}{x_j} \quad \text{and} \quad \sin\theta_2 \approx \frac{y(x_j)}{L - x_j}, \quad \text{and}$$

$$P = y(x_j)T\left[\frac{1}{x_j} + \frac{1}{L - x_j}\right] = TL\frac{y(x_j)}{x_j(L - x_j)} \,.$$

Then the deflection at x_j due to the load, P, at x_j, is

$$y(x_j) = \frac{Px_j(L - x_j)}{TL} \,.$$

The diagram, (C), similar triangles, shows that, for $x_i < x_j$,

$$\frac{y(x_i)}{y(x_j)} = \frac{x_i}{x_j} \quad \Rightarrow \quad y(x_i) = \frac{Px_i}{TL}(L - x_j), \quad \text{for } x_i < x_j \,.$$

For $x_i > x_j$ just exchange x_j and x_i, $\Rightarrow y(x_i) = \dfrac{Px_j}{TL}(L - x_i)$, for $x_i > x_j$.

This is by virtue of the "reciprocity theorem" i.e., the deflection at x_i due to the load at x_j is the same as the deflection at x_j due to the same load at x_i. Or, just work out the geometry.

Using these equations, and setting the load, P, to unity, a dimensionless "influence matrix," $[w_{ij}]$, can be defined. First, note that $x_k = \frac{L}{n}(k - \frac{1}{2})$, for $k = 1 \, .. n$. Then, for $x_i < x_j$:

$$w_{ij} = \frac{x_i}{TL}(L - x_j) = \frac{1}{TL}\left[\frac{L}{n}(i - \tfrac{1}{2})\right]\left[L - \frac{L}{n}(j - \tfrac{1}{2})\right]$$

$$w_{ij} = \frac{L}{Tn^2}(i - \tfrac{1}{2})(n - j + \tfrac{1}{2}); \quad \text{dimensionless except for the } \frac{L}{T} \text{ multiplier} \,.$$

The $[w]$ matrix will be symmetric (prove it), and we should extract the L/T term as a multiplier. Now, if (vertical) loads are applied to some, or all the mass points, given as the elements of vector, $\mathbf{p}$, the resultant deflection $\mathbf{y}$ is just $\mathbf{y} = \frac{L}{T}\mathbf{Wp}$. For example, the displacement y_1 is equal to its displacement due to the load at $j = 1$, plus that at $j = 2$, plus ... etc.

If the string is vibrating freely, the loads are just the inertial forces, $-m_i\ddot{y}_i = -\frac{M}{n}\ddot{y}_i$

$$\mathbf{y}(t) = -\frac{LM}{T}\mathbf{W\ddot{y}}(t); \quad \text{with } w_{ij} = \frac{1}{n^3}(i - \tfrac{1}{2})(n - j + \tfrac{1}{2}) \,.$$

Note that an additional "n" factor is absorbed into $\mathbf{W}$, which has the eigenvalues, λ_i, and the eigenvectors, $\mathbf{v}_i$. Given the eigenvalue analysis of $\mathbf{W}$, the solution to this vector differential equation is constructed as a linear summation of its eigenvectors (or "**normal modes**"):

$$\mathbf{y}(t) = \sum_{r=1}^{n} \mathbf{v}_r(a_r \cos\omega_r t + b_r \sin\omega_r t); \quad \text{with } \omega_r = \sqrt{\frac{T}{LM\lambda_r}} \,.$$

The coefficients a_r and b_r can be determined from the initial conditions by using the orthogonality property of the eigenvectors: $\mathbf{v}_i \bullet \mathbf{v}_j = \delta_{ij}$ (i.e., the eigenvectors are normalized). That is, by setting $t = 0$, the a_r coefficients are found from the initial positions of the masses, and the b_r coefficients are determined from the initial velocities of the masses.

$$\mathbf{y}(0) = \sum_r \mathbf{v}_r a_r; \ \text{ Now dot through with the vector } \mathbf{v}_k : \ \Rightarrow \mathbf{v}_k \bullet \mathbf{y}(0) = \sum_r \mathbf{v}_k \bullet \mathbf{v}_r a_r$$

$$a_r = \mathbf{v}_r \bullet \mathbf{y}(0)$$

$$\dot{\mathbf{y}}(0) = \sum_r \mathbf{v}_r \omega_r b_r; \ \text{ again dot through by } \mathbf{v}_k : \mathbf{v}_k \bullet \dot{\mathbf{y}}(0) = \sum_r \mathbf{v}_k \bullet \mathbf{v}_r \omega_r b_r, \ \text{ and}$$

$$b_r = \frac{\mathbf{v}_r \bullet \dot{\mathbf{y}}(0)}{\omega_r} \ .$$

Summary: In this analysis, the eigenvalues determine the frequencies (in terms of length, mass, and tension) and for each of these there is a corresponding eigenvector, or normal mode. A normal mode is a spatial description of the string. A sum of these modes builds the solution along the x-dimension (the $\mathbf{y}$ vector). Each mode brings along a constant and sinusoidal time function. There are just enough initial conditions to determine the constants, since each mass point starts (time $t = 0$) with an initial position and velocity.

The determination of the constants depends on the orthogonality of the "normal modes,"

C.2   THE CONTINUOUS FUNCTION SOLUTION

Our objective is to compare the continuous solution to the one above; therefore, its derivation will be very brief. The highlights presented here follow the derivation in [4, p. 431].

The string is now viewed as continuous, and a solution $y(x, t)$ is sought. As above, the initial conditions are given, say, as $y_0(x)$, and $\dot{y}_0(x)$.

The governing equation for the vibration of the string is the one dimensional wave equation:

$$\frac{TL}{M} \frac{\partial^2 y}{\partial x^2} = \frac{\partial^2 y}{\partial t^2}$$

where T, L, and M are defined as in the matrix solution.

We assume a solution of the form $y(x, t) = e^{j\omega t} f(x)$.

$$\frac{\partial^2 y}{\partial x^2} = e^{j\omega t} f'', \ \text{and} \ \frac{\partial^2 y}{\partial t^2} = -\omega^2 e^{j\omega t} f \ .$$

Substituting these into the wave equation, results in the ordinary differential equation:

$$f'' + \frac{M\omega^2}{TL} f = 0, \ \text{whose solution is}$$

$$f(x) = c_1 \sin \sqrt{\frac{M}{TL}} \omega x + c_2 \cos \sqrt{\frac{M}{TL}} \omega x \ .$$

Since $f(0) = 0$, c_2 must be zero. However, $f(L) = 0$ also, and this cannot mean that $c_1 = 0$, else there would be only the trivial solution. Nevertheless

$$\sin \sqrt{\frac{M}{TL}} \omega L = \sin \sqrt{\frac{ML}{T}} \omega = 0 \,.$$

This condition can be met if $\sqrt{\frac{ML}{T}} \omega = n\pi$, or $\omega_n = \sqrt{\frac{T}{ML}} n\pi$, since sin = 0 at these values.

That is, there are an infinity of values, ω_n, where n can be any integer > 0. For each of these values, there are corresponding functions, $f_n(x)$ and $y_n(x, t)$

$$f_n(x) = c_1 \sin \frac{n\pi x}{L}; \text{ and } y_n(x, t) = c_1 e^{j\omega_n t} \sin \frac{n\pi x}{L} \,.$$

Note that both the real (cos) and imaginary (sin) parts of $y(x, t)$ solve the wave equation, their sum is also a solution:

$$y_n = \sin \frac{n\pi x}{L} \left(a_n \cos \sqrt{\frac{T}{ML}} n\pi t + b_n \sin \sqrt{\frac{T}{ML}} n\pi t \right) \,.$$

And the final solution is the infinite sum of all the y_n functions, $y(x, t) = \sum_{n=1}^{\infty} y_n$.

At time $t = 0$ the initial position of the string is $y_0(x)$:

$$y(x, 0) = y_0(x) = \sum_n a_n \sin \frac{n\pi x}{L}$$

and this shows clearly that the string position is a sum of sin functions, in the same way that the matrix solution string position is a sum of eigenvectors. Furthermore, a group of sin functions can form an orthogonal set — as shown in work with Fourier series. From a mathematical handbook

$$\int_0^{\pi} \sin ax \sin bx \, dx = 0, \quad (b \neq a); \text{ and } \int_0^{\pi} \sin^2 ax \, dx = \frac{\pi}{2} \,.$$

Then with the change of variable $z = \frac{L}{\pi} x$, the second integral value is $\frac{L}{2}$.

$$\text{Then} \qquad \int_0^L y_0(x) \sin \frac{k\pi x}{L} dx = \sum_n a_n \int_0^L \sin \frac{k\pi x}{L} \sin \frac{n\pi x}{L} dx = \frac{L}{2} a_k \,.$$

The coefficients b_k are calculated in the same way, resulting in the final solution:

$$y(x, t) = \sum_{n=1}^{\infty} \sin \frac{n\pi x}{L} \left(a_n \cos \sqrt{\frac{T}{ML}} n\pi t + b_n \sin \sqrt{\frac{T}{ML}} n\pi t \right)$$

$$a_n = \frac{2}{L} \int_0^L y_0(x) \sin \frac{n\pi x}{L} dx, \text{ and } b_n = \frac{2}{n\pi L} \int_0^L \dot{y}_0(x) \sin \frac{n\pi x}{L} dx \,.$$

It is truly remarkable how similar the two solutions are.

C.3 EXERCISES

C.1. Show that the $[w]$ matrix is symmetric.

C.2. Show that $\mathbf{y}(t) = \sum_{r}^{n} a_r \mathbf{v}_r \cos \omega_r t$ is a solution to $\mathbf{y}(t) = -\frac{LM}{T} \mathbf{W} \ddot{\mathbf{y}}(t)$.

APPENDIX D

Solar Energy Geometry

The distribution of solar energy, weather patterns, and the seasons, all depend on the geometry between the earth and the sun. Apparently, we are just far enough from the sun to benefit greatly from its heat without having been cooked into some other kind of life form(s). The tilt of the earth's axis provides our seasonal variations, and the geometry problem of solar energy — its variation, both daily and seasonally.

Although the orbit of the earth is not quite circular, it is assumed to be so in this picture.

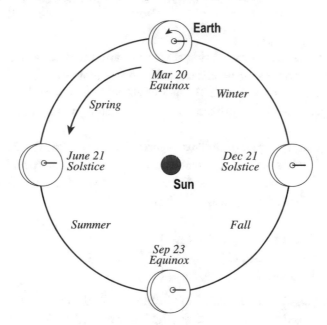

The view is the earth orbit from above the North Pole. The earth is shown in its 4 most important positions in this orbit. Note the direction of earth motion in the orbit, and also the direction of earth's rotation about its axis (straight line emerging from the small circle). **This axis always points to the right** and is tilted at an angle of 23.5° relative to the plane of the orbit. Note the short arc on the left sides of the 4 earths shown; this is the equator, visible because of the tilt.

All points on earth's surface at a given latitude experience the same variations of radiant energy. The earth's rotation produces the most frequent variation — from dawn 'til dusk. But, every day is different because of the earth's travel in its orbit and that 23.5° axis tilt. For example, summer in the northern hemisphere occurs when the north pole is tilted toward the sun. It is then winter in the southern hemisphere.

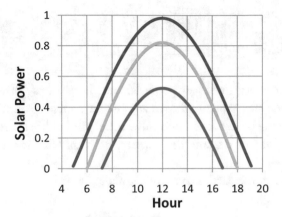

The equations necessary for calculating these variations are developed in Chapter 5 — Solar Angles. An example variation is shown here; the point chosen is at latitude 35° (north), its longitude is arbitrary. The top curve represents June 21, the middle curve represents both March 21 and Sept. 21, the bottom one December 21 (approximate dates). Similar curves could be drawn for all days in the year.

The above diagram is the plot of the "***cosine factor***," $C_f = \mathbf{s}_x \cdot \mathbf{p}_x$ for the days given, and over the time span shown. The results were calculated as if there were a solar panel lying on the ground at the given latitude. The "panel vector" has the coordinates $\mathbf{p}_x = \{1, 0, 0\}$. The sun vector, $\mathbf{s}_x$, has the coordinates given in Chapter 5, (5.26):

$$\mathbf{s}_x = \left\{ \begin{array}{c} C\varphi\, C\theta_s C\varphi_s + S\varphi\, S\varphi_s \\ S\theta_s\, C\varphi_s \\ C\varphi\, S\varphi_s - S\varphi\, C\theta_s\, C\varphi_s \end{array} \right\} (C \equiv \cos,\ S \equiv \sin) . \qquad \text{(Chapter 5 (5.26))}$$

The Greek characters (angles) are as defined in Chapter 5, and the capitals C and S are cos and sin, respectively. The angle φ is the panel latitude, φ_s, the "sun latitude," and θ_s measures the sweep angle of the sun vector as the earth rotates (producing the daylight hours).

If enough solar panels were laid on the ground whose total power output is 1 kw (kilowatt) when subjected to the direct rays of the sun ($C_f = 1$), the vertical axis of the diagram could be read as kw. The abscissa is in hours (24 hour clock), 12 being noon. The area(s) beneath the curves are the kw-hr energy(s) received during the respective days.

Of course the panel(s) are never just laid horizontal. On a flat roof the panel is oriented north-south with the north edge elevated as shown here. The optimum elevation angle is equal to the latitude of the panel. For example, if the installation is at 35° latitude, the best elevation angle is 35°.

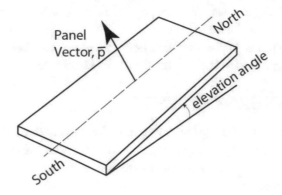

The Chapter 5 panel vector equation is:

$$\mathbf{p}_x = \{S\varphi_p, \quad S\theta_p C\varphi_p, \quad -C\varphi_p C\theta_p\}. \qquad \text{(Chapter 5 (5.27))}$$

In terms of this equation, φ_p is set at 55° (90 - 35)°, θ_p is 0° (i.e., the panel center line is directly north-south). The results are significantly different, as shown in this "power profile," The solar power curves for all 365 days of the year fall between the top and bottom curves shown — however, in a rather complicated way. The top curve represents both day 0 and day 180. Days 92 and 274 follow the lower curve, but days after the fall equinox become shorter. The day 274 daytime is between the vertical lines — from about 7.2 to .16.8 (9.6 hours).

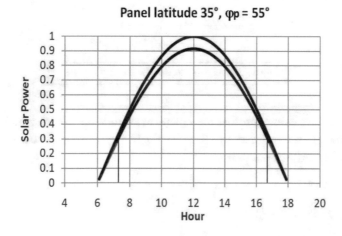

Almost the same results are found for installations at other latitudes. For example, if the panel is located farther north, say 45° (near the latitude of Fargo, ND), and the elevation angle is set at 45°, the results are about the same as those shown here for 35° (there is some difference: the fall and winter days are shorter). Solar energy is more popular in the southwestern US states because they receive more sunny days than states farther north.

Of course, installations in the southern hemisphere are the "reverse" of these. When the Northern Hemisphere tips toward the sun, the Southern Hemisphere is tipped away.

D.1 YEARLY ENERGY OUTPUT

As mentioned, the abscissa of these power curves has the units of time. Therefore, the area under the power curve for any given day determines the total energy output during that day. These daily energy figures can be plotted to show seasonal variation, and when all 365 days are summed the result is the total yearly energy.

Because the installation is usually on a pitched roof where the panel elevation is given by roof orientation and pitch, it is of some interest to compare results over a range of elevation angles.

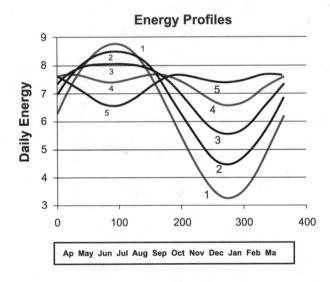

This diagram shows energy profiles for a panel at latitude 35°, given panel elevations of 0, 10, 20, 30, and 40 degrees, in the order of 1 to 5, respectively. For all 5 curves $\theta_p = 0$.

The total energy outputs per year are;

1. 0° elevation, 2238

2. 10° elevation, 2442 (-7.5%)

3. 20° elevation, 2578 (-2.4%)

4. 30° elevation, 2639 ($-<1\%$)

5. 40° elevation, 2624 ($-<1\%$)

The numbers in parentheses are the comparisons to 2641 — the yearly energy when the panel is at 35° elevation. The 1st result has the panel horizontal (an unrealistic choice); all the rest show less than a 10% loss. A solar panel on a house whose roof has a 4 inch per foot pitch would have an elevation of 18° (unless, of course, the panels are elevated further).

The total energy profile from the installation at 45° latitude, whose panel is elevated at 45°, is very similar to curve 4, above. Its total is slightly lower, at 2607.

D.2   AN EXAMPLE

The following compares panel(s) placed on a roof sloping at 20° ($\varphi_p = 70°$), whose axis points 45° off south toward the east ($\theta_p = +45°$), with the same panel(s) placed optimally — on a flat roof, oriented directly south, and whose elevation is 35°. Both sets of panels are at 35° latitude.

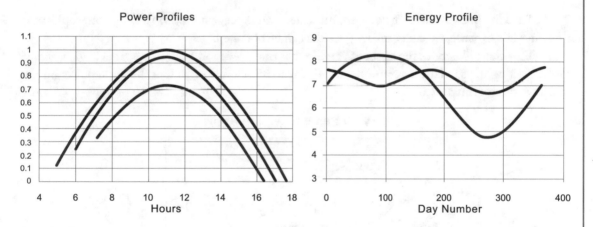

The comparative power profiles (curve 1) for the "35° optimum case" have already been shown. The energy profile comparison is interesting — especially because the "70—45"case (curve 2, above) does better than "optimum" in the early months of the year. It does lose overall — its total energy per year is 2476 compared to 2641, a loss of only 6+%.

These energy comparisons suggest that the elevation angle should be increased at the fall equinox, or even more often. This leads to the possibility of the panel "***tracking***" the sun position — changing the panel angles to point more directly toward the sun.

D.3   TRACKING THE SUN

Continuous tracking systems — used in large, industrial solar energy collectors — include sensors to determine the exact position of the sun, then point the collector toward it. But, the sun's position is repeated every year (with corrections) as shown in this chart. Tracking Angles: curve 1 is azimuth (from south), 2 is elevation above the horizon. Day 0 is the spring equinox.

Tracking Angles--Day 0

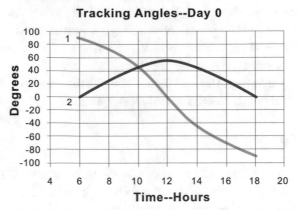

Tracking both of these angles is sophisticated and expensive. But, with almost no sophistication, and less expense, excellent tracking results can be obtained.

This chart illustrates the effect of varying the panel elevation angle. The panel angle is varied by subtracting the sun latitude from the initial (day 0) 35° elevation, so that the noon power output is always the maximum 1.0.

Varying Panel Elevation Angle

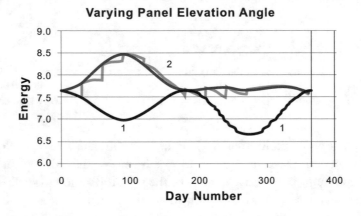

Curve 1 elevation is the fixed "optimum" of 35° shown in the previous chart. There are two Curves-2. The smooth one shows the results of varying the panel angle every day. The discontinuous one shows varying it only every 30 days. All results shown are at panel latitude 35°, and the axes of the panels are directly north-south ($\theta_p = 0$). At the spring and fall equinox 35° is truly optimum, and all curves are the same on these days.

The curves 2 yearly energy output is 2866, and 2852 — very nearly the same, and show an 8% increase over that of curve 1 — 2641 (the smooth curve is +8.5%).

It is assumed that the monthly panel angle changes can be done manually, at the same time the panels are cleaned. The support structure must be marked with at least 12 positions.

In order to make any further improvement in output, the panel will have to be moved during each sun day — and this cannot be done manually. Further, the best way to do this is to **rotate the panel about its north-south** axis, not swing the panel to an azimuth angle.

If the sun panel surface faces the sun directly, every minute of every day in the year, the total output energy number is 4377 — the theoretical best.

The data in the following chart is obtained by changing the panel elevation angle monthly, as above, and then rotating the panel about its north-south axis during daylight hours. An initial panel rotation angle of +60° (toward the east) is set before dawn. From 8am until 4pm the panel is rotated at 15° per hour (the same rate that the earth turns), leaving it at -60° at 4pm, when rotation stops. The 8 to 4 rotation hours are used during all 365 days.

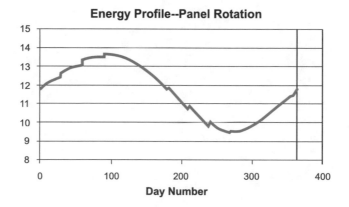

The total yearly energy output is 4259; very nearly the same output that could be obtained by a sophisticated tracking system.

But, it is likely that even this relatively simple tracking system is not economical for residential systems, due to the low energy output of a single solar panel. The power required to rotate it may be more than it is worth.

Solar panels on rooftops are like wings in the wind. They must be firmly anchored. When the panel has to move (rotate) the additional mechanical problems could be significant.

Nevertheless, these methods may be economic when adapted to larger systems.

This appendix has demonstrated the use of equations developed in Chapter 5. The numbers calculated are concerned with only the geometry, not with the physics and mechanical problems associated with real solar panels. For this reason comparative data has been emphasized. For example, changing the panel elevation angle monthly will certainly not produce energy equal to 2852, but it probably will increase the actual solar panel output by 7%.

The development algorithms might need some alterations. For example, the earth orbit is not circular, but elliptical. Its orbital speed, assumed to be constant, actually varies.

To implement any algorithm for tracking, the controller (computer) would have to know the clock time that the sun will be directly overhead (i.e., solar noon).

APPENDIX E

Answers to Selected Exercises

E.1 CHAPTER 1

2. There are m row vectors and n column vectors.

3. If **A** and B are square and conformable, BA is conformable. If **A** is mXn and B is nXm, BA is conformable

4. $\mathbf{u}_1 \cdot \mathbf{v}_1 = 0.32192$

5. The product $\{\mathbf{v}_1\}[\mathbf{u}_1]$ is a 4X4 matrix:

$$
\begin{bmatrix}
15.833 & 7.4538 & 1.7180 & 20.9373 \\
-4.1953 & -2.0197 & -0.4655 & -5.6731 \\
7.3584 & 3.5424 & 0.8165 & 9.9504 \\
-10.3222 & -4.9692 & -1.14534 & -13.9582
\end{bmatrix}
$$

6. $\mathbf{c} = x_1\mathbf{a}_1 + x_2\mathbf{a}_2 + \cdots + x_n\mathbf{a}_n$

7. $\mathbf{v} = \{0, 0, 0\}$. $\mathbf{u} = [\, 21, 6, 3\,]$

8. It is more efficient in all cases to calculate vectors. For example, in determining **ABCv** multiply **Cv** first. In (2) $\mathbf{v}_1\mathbf{u}_1\mathbf{v}_2\mathbf{u}_2$, calculate the dot product $\mathbf{u}_1\mathbf{v}_2$ first

14. $\mathbf{T}_1\mathbf{T}_2 = \begin{bmatrix} \cos\theta\cos\varphi & -\sin\theta & \cos\theta\sin\varphi \\ \sin\theta\cos\varphi & \cos\theta & \sin\theta\sin\varphi \\ -\sin\varphi & 0 & \cos\varphi \end{bmatrix}$, and note that the product of orthogonal matrices is always orthogonal. That is: $\mathbf{T}_1\mathbf{T}_2(\mathbf{T}'_2\mathbf{T}'_1) = \mathbf{I}$

17. The transformation matrix is formed by replacing the 3,1 element (a zero) of a unit matrix by the ratio $-\dfrac{a_{31}}{a_{11}}$.

E.2 CHAPTER 2

1. 5741326, s = 13; 35421, s = 8; 123465, s = 1; 654321, s = 15

2. $b_{44}b_{12}b_{31}b_{23} = b_{12}b_{23}b_{31}b_{44}$ $s = 2$, *plus* $c_{43}c_{22}c_{14}c_{51}c_{35}$, $s = 9$, *minus*

3. $|\mathbf{A}| = 9$; $|\mathbf{B}| = -14$

7. The rank of $\mathbf{A}_1$ is 1. It has just one independent vector.

11. The rank of $\mathbf{A}_2$ is 2.

13. Each factor of the expansion of the 3X3 will have 3 terms. They can all be arranged into column order. Now, just differentiate. The result will be three expansions in exactly the form to be shown.

E.3   CHAPTER 3

1. $\mathbf{Q} = \begin{bmatrix} 1 & 0 & 0 \\ -3 & 1 & 0 \\ -2 & 0 & 1 \end{bmatrix}$, whose determinant = 1;

$|\mathbf{A}| = 2$; $\mathbf{QA} = \begin{bmatrix} 1 & 3 & -1 \\ 0 & 2 & 4 \\ 0 & 0 & 1 \end{bmatrix}$; $\mathbf{Qc} = \left\{ \begin{matrix} 4 \\ 0 \\ -1 \end{matrix} \right\}$

2. From $\mathbf{QAx} = \mathbf{Qc}$, $x_3 = -1$, $x_2 = 2$, and then $x_1 = -3$, solved in that order.

3. In this book, complex matrices are often written with imaginary parts above the reals. The inverse is written here in this notation. Note also, the double bars (instead of bracket), denoting a matrix.

$$A^{-1} = \left\| \begin{matrix} 8.00 & -2.00 & -3.00 \\ 7.00 & 6.00 & -2.00 \\ \\ 6.00 & -1.00 & -2.00 \\ 5.00 & 6.00 & -2.00 \\ \\ -3.00 & -1.00 & 1.00 \\ 0.00 & --\,2.00 & 0.00 \end{matrix} \right\|$$

5. Given $\mathbf{Ax} = \mathbf{c}$, and $\mathbf{B} = \mathbf{A}^{-1}$, if two columns of $\mathbf{A}$ are interchanged, corresponding rows of both $\mathbf{B}$ and $\mathbf{x}$ will be interchanged. Compare the following to problems 1 & 2, above:

Given $\mathbf{Ax} = \mathbf{c}$, with $\mathbf{A} = \begin{bmatrix} 3 & 1 & -1 \\ 11 & 3 & 1 \\ 6 & 2 & -1 \end{bmatrix}$, then $\mathbf{A}^{-1} = \begin{bmatrix} 2.5 & 0.5 & -2 \\ -8.5 & -1.5 & 7 \\ -2 & 0 & 1 \end{bmatrix}$, and $\mathbf{x} = \left\{ \begin{array}{c} 2 \\ -3 \\ -1 \end{array} \right\}$

E.4   CHAPTER 4

2. $\mathbf{a}_2 \times \mathbf{a}_3 = -\{3,\ 6,\ 1\};$ $\mathbf{a}_2 \times \mathbf{a}_3 \bullet \mathbf{a}_1 = -14;$ $\mathbf{a}_2 \times \mathbf{a}_3 \bullet \mathbf{c} = -4.$ Then $x_1 = \frac{2}{7}$

4. The rank of $\mathbf{M}$ is just 2

5(a). Neither the columns nor the rows of $\mathbf{M}$ are independent.

5(c). Since the rank is 2, then $x_3, x_4,$ and x_5 can be chosen arbitrarily, and

$$\left\{ \begin{array}{c} x_1 \\ x_2 \end{array} \right\} = \left\{ \begin{array}{c} \frac{1}{3} \\ \frac{7}{3} \end{array} \right\} x_3 - \left\{ \begin{array}{c} \frac{2}{3} \\ \frac{2}{3} \end{array} \right\} x_4 - \left\{ \begin{array}{c} \frac{5}{3} \\ \frac{2}{3} \end{array} \right\} x_5$$

5(d). Given $\mathbf{M}\mathbf{x} = \mathbf{y}$, the set will be compatible iff $\mathbf{y}$ is orthogonal to $\{-1, 2, 1\}$.

6(a). The columns of $\mathbf{A}$ are independent.

6(b). The rows of $\mathbf{A}$ are dependent.

6(c). $\mathbf{z} = \{-14, -1, 20, 39, 1\}$ is orthogonal to the columns of $\mathbf{M}$.

6(d). $\mathbf{x} = \{-3, 1, 2, -1\}$

6(e). With the given $\mathbf{y}$ vector the set is incompatible.

12. Begin with the $\mathbf{X}$ matrix shown at right. The desired determinant is obtained by striking out row 2 and column 2 of $\mathbf{X}$. However, this determinant is just the minor of the 2,2 element in $\mathbf{X}$. $|\mathbf{X}| = (x_4 - x_3)(x_4 - x_2)(x_4 - x_2)(x_3 - x_2)(x_3 - x_1)(x_2 - x_1)$

$$\mathbf{X} = \begin{bmatrix} 1 & x_1 & x_1^2 & x_1^3 \\ 1 & x_2 & x_2^2 & x_2^3 \\ 1 & x_3 & x_3^2 & x_3^3 \\ 1 & x_4 & x_4^2 & x_4^3 \end{bmatrix}$$

In the text discussing Lagrange polynomials it is shown that $\mathbf{XA} = \mathbf{I}$, that is, the two matrices are inverses. Then the minor of the x_{22} element is a_{22} multiplied by $|\mathbf{X}|$.

$$a_{22} = \frac{x_1 x_3 + x_1 x_4 + x_3 x_4}{(x_2 - x_1)(x_2 - x_3)(x_2 - x_4)}$$

$|\mathbf{X}|a_{22}$ is equal to the determinant of the reduced matrix given in the exercise.

E.5   CHAPTER 5

1(a). $\mathbf{x} = \mathbf{Ty}$; $\quad \mathbf{T} = \begin{bmatrix} \cos\theta\cos\varphi & -\sin\theta & -\sin\varphi\cos\theta \\ \cos\varphi\sin\theta & \cos\theta & -\sin\varphi\sin\theta \\ \sin\varphi & 0 & \cos\varphi \end{bmatrix}$; $\quad \left\{ \begin{array}{l} \theta = \text{longitude of point A} \\ \varphi = \text{latitude of point A} \\ \theta = 286°; \;\; \varphi = 41° \end{array} \right\}$

1(b). The great circle distance is 2496 mi.

1(c). The distance along the 41° latitude line is 2529 mi.

1(d). Looking down at point A (down the y_1 axis) there is an angle α between the negative y_2 axis and the edge of the great circle path.

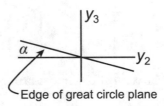

A unit vector along the great circle edge has the dimensions $\{0, -\cos\alpha, \sin\alpha\}$. The dot product of this vector and the normal to the great circle plane must be zero. Using this information, the angle is determined to be 16.28°. The heading is, then, 286.28°.

3. $\mathbf{R}^2 \varpi = -\mathbf{RWr} = -\mathbf{r} \times (\varpi \times \mathbf{r}) = -(\mathbf{r} \times \varpi) \times \mathbf{r}$, and note that $\omega \times \omega = 0$

4. The inertia matrix has no cross product terms because of symmetry:

$$\mathbf{J}_x = \tfrac{Ma^2}{3} \begin{bmatrix} 1 & 0 & 0 \\ 0 & 4 & 0 \\ 0 & 0 & I_{33} \end{bmatrix}.$$ The required torque is $\tfrac{Ma^2\omega^2}{2}$, and most important, its direction

is perpendicular to the surface of the plate as it turns.

E.6   CHAPTER 6

1. The determinant $\mathbf{A}(\lambda) = \begin{vmatrix} a_{11} - \lambda & a_{12} & a_{13} \\ a_{21} & a_{22} - \lambda & a_{23} \\ a_{31} & a_{32} & a_{33} - \lambda \end{vmatrix}$ can be expanded by the use of the

7th property of determinants (see chapter 2). As an example

$$\begin{vmatrix} a_{11} - \lambda & a_{12} & a_{13} \\ a_{21} & a_{22} - \lambda & a_{23} \\ a_{31} & a_{32} & a_{33} - \lambda \end{vmatrix} = \begin{vmatrix} a_{11} & a_{12} & a_{13} \\ a_{21} & a_{22} - \lambda & a_{23} \\ a_{31} & a_{32} & a_{33} - \lambda \end{vmatrix} - \begin{vmatrix} \lambda & a_{12} & a_{13} \\ 0 & a_{22} - \lambda & a_{23} \\ 0 & a_{32} & a_{33} - \lambda \end{vmatrix}.$$

With continued use of this property, the following results are obtained:

$$|\mathbf{A}(\lambda)| = c_0\lambda^3 + c_1\lambda^2 + c_2\lambda + c_3$$
$$c_0 = -1$$
$$c_1 = a_{11} + a_{22} + a_{33}$$
$$c_2 = -\begin{vmatrix} a_{11} & a_{13} \\ a_{31} & a_{33} \end{vmatrix} - \begin{vmatrix} a_{11} & a_{12} \\ a_{21} & a_{22} \end{vmatrix} - \begin{vmatrix} a_{22} & a_{23} \\ a_{32} & a_{33} \end{vmatrix}$$
$$c_3 = |\mathbf{A}|$$

This expansion is easier than it appears, and can be generalized to work for characteristic determinants of higher order.

2. Using the above expansion, the characteristic polynomial for the given matrix is:

$$P(\lambda) = c_0\lambda^3 + c_1\lambda^2 + c_2\lambda + c_3 = -\lambda^3 + 4\lambda^2 + 7\lambda - 10 = \lambda^3 - 4\lambda^2 - 7\lambda + 10$$
$$= (\lambda - 1)(\lambda + 2)(\lambda - 5)$$

The adjoint of $[\mathbf{A} - \mathbf{I}]$ is $[\mathbf{A} + 2\mathbf{I}][\mathbf{A} - 5\mathbf{I}] = \mathbf{A}^2 - 3\mathbf{A} - 10\mathbf{I} = \begin{bmatrix} 72 & 12 & 24 \\ 72 & -12 & -24 \\ -216 & 36 & 72 \end{bmatrix}$ which

yields the two vectors $v_1 = \{1, -1, 3\}$ and $u_1 = [\ 6, 1, -2\ ]$, and note that $u_1 \cdot v_1 = 1$.

In the same manner, the other two vector pairs are determined such that:

$$\mathbf{U} = \begin{bmatrix} 6 & -1 & -2 \\ 3 & 0 & -1 \\ 1 & 1 & 0 \end{bmatrix}; \quad \mathbf{V} = \begin{bmatrix} 1 & -2 & 1 \\ -1 & 2 & 0 \\ 3 & -7 & 3 \end{bmatrix}, \text{normalized such that UV = VU = I}$$

4. $\mathbf{B} = \begin{bmatrix} 18 & -1 & -6 \\ -18 & 1 & 6 \\ 60 & -3 & -20 \end{bmatrix}$, and $|\mathbf{B}(\lambda)| = \lambda^3 + \lambda^2 - 2\lambda = 0 = \lambda(\lambda - 1)(\lambda + 2)$.

Note that the third eigenvalue is zero (and $\mathbf{B}$ is singular). However,

$$\text{adj}\{\mathbf{B}(\lambda = 0)\} = -2\begin{bmatrix} 1 & 1 & 0 \\ 0 & 0 & 0 \\ 3 & 3 & 0 \end{bmatrix}, \text{which yields the same eigenvectors that were found for}$$

the original $\mathbf{A}$ matrix.

5. The (symmetric) $\mathbf{A}$ matrix has the eigenvalues, 1 and 4. Its eigenvectors are $\mathbf{V} = \frac{1}{\sqrt{2}}\begin{bmatrix} 1 & 1 \\ -1 & 1 \end{bmatrix}$

The square root matrix will have the same eigenvectors as $\mathbf{A}$. When this matrix is multiplied by itself the result is $\mathbf{V}^t \Lambda \mathbf{V} \mathbf{V}^t \Lambda \mathbf{V} = \mathbf{V}^t \Lambda^2 \mathbf{V}$. Therefore, the square root matrix must be:

$$\frac{1}{2}\begin{bmatrix} 1 & 1 \\ -1 & 1 \end{bmatrix}\begin{bmatrix} 2 & 0 \\ 0 & 1 \end{bmatrix}\begin{bmatrix} 1 & -1 \\ 1 & 1 \end{bmatrix} = \frac{1}{2}\begin{bmatrix} 3 & -1 \\ -1 & 3 \end{bmatrix}.$$

6. The characteristic equation of $\begin{bmatrix} -0.7 & 2 \\ -0.6 & 1.5 \end{bmatrix}$ is $\lambda^2 - 0.8\lambda + 0.15 = (\lambda - 0.3)(\lambda - 0.5)$

$\text{Adj}\{\mathbf{A}(\lambda=0.3)\}$ $=\mathbf{A}^a(.3)$ $= [\mathbf{A} - 0.5\mathbf{I}] = \begin{bmatrix} -1.2 & 2 \\ -0.6 & 1 \end{bmatrix}$;$\text{Adj}\{\mathbf{A}(\lambda=0.5)\}$ $= [\mathbf{A} - 0.3\mathbf{I}] =$

$\begin{bmatrix} -1 & 2 \\ -0.6 & 1.2 \end{bmatrix}$

$Z_1(\lambda_1 = .3) = \dfrac{[\mathbf{A} - 0.3\mathbf{I}]}{(0.3 - 0.5)} = \begin{bmatrix} 6 & -10 \\ 3 & -5 \end{bmatrix}$; $Z_2(\lambda_2 = .5) = \dfrac{[\mathbf{A} - 0.5\mathbf{I}]}{(0.5 - 0.3)} = \begin{bmatrix} -5 & 10 \\ -3 & 6 \end{bmatrix}$

$\text{Sin}(\mathbf{A}) = \sin(\lambda_1)Z_1 + \sin(\lambda_2)Z_2 = \begin{bmatrix} -0.624006 & 1.839053 \\ -0.551716 & 1.398952 \end{bmatrix}$

six decimals are given in the event that you would like to show that $\sin^2(\mathbf{A}) + \cos^2(\mathbf{A}) = \mathbf{I}$.

7(a). The solution includes the diagonal matrix $[\delta_{ij}\frac{1}{\lambda_j - \lambda}]$ which insists that there is no solution for $\lambda = \lambda_j$.

7(c). Yet, when the $\mathbf{c}$ vector is orthogonal to the j th normal mode (as in this case), the term $\lambda - \lambda_j$ does not appear in the solution; thus, a solution exists at that critical e-value. Though mathematically correct, it would be very dangerous to depend on this, physically.

9. $\mathbf{A}$ reduces to the $\mathbf{P}$ matrix $\begin{bmatrix} 16 & -66 & 54 \\ 1 & 0 & 0 \\ 0 & 1 & 0 \end{bmatrix}$ which yields the required coefficients. The column vectors are the same as the row vectors, since the given matrix is symmetric.

The roots are 1.08352, 9.86399, and 5.05249

Also used in this calculation are the following $\mathbf{S}$ and $\mathbf{S}^{-1}$ matrices:

$$\mathbf{S} = \begin{bmatrix} \frac{1}{9} & -1\frac{1}{9} & 1\frac{2}{3} \\ 0 & -\frac{1}{3} & 1\frac{1}{3} \\ 0 & 0 & 1 \end{bmatrix}; \quad \mathbf{S}^{-1} = \begin{bmatrix} 9 & -30 & 25 \\ 0 & -3 & 4 \\ 0 & 0 & 1 \end{bmatrix}$$

Note that the matrix product $[\mathbf{S}][\mathbf{S}^{-1}] = \mathbf{I}$.

E.7   CHAPTER 7

Problems 2 & 3. With x_1 and x_2 measured from fixed locations, the equations of motion of both are $\mathbf{M\ddot{x}} + \mathbf{Kx} = \mathbf{d}f\cos\omega t$ where:

$$\mathbf{M} = \begin{bmatrix} m_1 & 0 \\ 0 & m_2 \end{bmatrix};\ \mathbf{K} = \begin{bmatrix} k_1 + k_2 & -k_2 \\ -k_2 & k_2 \end{bmatrix};\ \mathbf{A} = \mathbf{M}^{-1}\mathbf{K} = \begin{bmatrix} \frac{k_1+k_2}{m_1} & \frac{-k_2}{m_1} \\ \frac{-k_2}{m_2} & \frac{k_2}{m_2} \end{bmatrix}$$

2. In this case, $m_1 = m_2$ and $k_1 = k_2$, and there is no driving force. Now, set $\omega = \sqrt{\frac{k}{m}}$ and

$$[\mathbf{I}\lambda - \mathbf{A}] = \begin{bmatrix} \lambda - 2\omega^2 & \omega^2 \\ \omega^2 & \lambda - \omega^2 \end{bmatrix};\ \text{Det} = \lambda^2 - 3\omega^2\lambda + \omega^4;\ \text{Then } \lambda = \frac{3 \pm \sqrt{5}}{2}\omega^2$$

And the 2 resonant frequencies (rad/sec), are 0.618ω and 1.618ω.

3. This time, the k and m values are not the same, and there is a driving force as shown above. In this equation we assume a solution $\mathbf{x} = \mathbf{a}\cos\omega t$. This reduces the above equation to an algebraic one by which we can solve for the maximum amplitudes (a_1 and a_2) of x_1 and x_2.

$$-\omega^2\mathbf{Ma} + \mathbf{Ka} = \mathbf{f};\ \text{where } \mathbf{f} = \{f_0,\ 0\}$$
$$\mathbf{a} = [\mathbf{K} - \omega^2\mathbf{M}]^{-1}\mathbf{f}$$
$$[\mathbf{K} - \omega^2\mathbf{M}] = \begin{bmatrix} k_1 + k_2 - \omega^2 m_1 & -k_2 \\ -k_2 & k_2 - \omega^2 m_2 \end{bmatrix}$$

Set D equal to the determinant of $[\mathbf{K} - \omega^2\mathbf{M}]$. Then:

$$\mathbf{a} = \frac{1}{D}\begin{bmatrix} k_2 - \omega^2 m_2 & k_2 \\ k_2 & k_1 + k_2 - \omega^2 m_1 \end{bmatrix}\begin{Bmatrix} f_0 \\ 0 \end{Bmatrix}$$

Note that if the ratio of k_2 divided by m_2 is made equal to ω^2, the term $k_2 - \omega^2 m_2$ goes to zero. In both problems 2 and 3, the mass, m_1 will be motionless, if m_1 is driven at the resonant frequency of the single mass/spring system—*unless* the determinant also goes to zero. Does it?

5. The matrices are: (It is easier to work with the inverse of $\mathbf{M}$)

$$\mathbf{M}^{-1} = \begin{bmatrix} 38.64 & 0.0 & 0.0 \\ 0.0 & 12.88 & 0.0 \\ 0.0 & 0.0 & 12.88 \end{bmatrix};\ \mathbf{K} = \begin{bmatrix} 10 & -5 & -5 \\ -5 & 15 & 0 \\ -5 & 0 & 15 \end{bmatrix};$$

$$\mathbf{C} = \begin{bmatrix} 0.6 & -0.3 & -0.3 \\ -0.3 & 0.8 & 0 \\ -0.3 & 0 & 0.8 \end{bmatrix}$$

The **A** matrix is: $\mathbf{A} = \begin{bmatrix} \mathbf{0} & \mathbf{I} \\ -\mathbf{M}^{-1}\mathbf{K} & -\mathbf{M}^{-1}\mathbf{C} \end{bmatrix}$, a 6X6 matrix.

An eigenvalue analysis of this **A** matrix yields the characteristic numbers:

−2.65415	−5.1520	−14.08986
± $j9.89244$	± $j12.91$	± $j16061$

The "solution vectors" are given in the table, below. These vectors are the coefficients of the cosine and sine terms in the solution. See equation (4.12) in chapter 7. All these decimal places are not necessary, but it will make it easier for the student to check answers.

λ_1		λ_2		λ_3	
0.434494	0.184987	0.000	0.000	1.565506	1.287241
0.324751	0.126032	1.000	0.399084	−0.324751	−0.298648
0.324751	0.126032	−1.000	−0.399084	−0.324751	−0.298648

Note how the initial conditions are met by the sums of the cosine term coefficients. For example, $x_{10} = 2 = 0.434494 + 1.560556$.

6. This problem differs from 5 by only the initial conditions. The motion of the masses, and the unbalanced force on W_1 are plotted here:

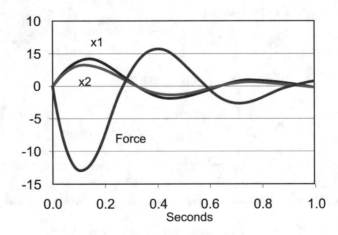

The force is in lbs., its maximum is 12.8#, at time 0.11sec. The graph shows the displacement plot for x_1 and x_2. The motion of W_3 is the same as x_2.

7. Symmetry suggests that the motion of W_1 and W_2 will be the same as in the previous problem. Then, a tie between them would make no difference, given the same initial conditions, as in the problem.

8. At 500 cps the immitances are:

z_1	500	z_5	$j5.531E2$
y_2	$j1.571E\text{-}3$	y_6	$j1.571E\text{-}3$
z_3	$j5.531E2$	z_7	$j4.712E2$
y_4	$j1.571E\text{-}3$	y_8	$2.0E\text{-}3$

At this frequency the voltage ratio $\frac{e_0}{e_8}$ is $-1.91 + j0.668 = 2.028$ @ 2.81 rad. This is very nearly the same as the zero frequency ratio (it's within the pass of the filter). But, the ratio at 1kcps is $-15.11 + j77.77 = 79.22$ @ 1.763 rad, or 38 db.

Bibliography

[1] Hildebrand, F. B., *Methods of Applied Mathematics*. Prentice-Hall, Englewood Cliffs, New Jersey, 1958.
Chapter 1 of this book is an excellent introduction to: "Matrices, Determinants, and Linear Equations."

[2] Frazer, R. A., Duncan, W. J., Collar, M.A., *Elementary Matrices, and Some Applications to Dynamics and Differential Equations*. Cambridge Press.
This is a very complete treatise, very concise. It isn't all that "elementary." Whatever the question, the answer is probably in this book.

[3] Lanczos, C., *Applied Analysis*. Prentice-Hall, Englewood Cliffs, New Jersey, 1958. 160
A wonderful writer and "explainer." My favorite.

[4] Pipes, L. A., *Applied Mathematics for Engineers and Physicists*. McGraw-Hill Book Co., 1958. 166, 239

[5] Pipes, L. A., *Matrix Methods for Engineering*. Prentice-Hall, Englewood Cliffs, New Jersey, 1963.
Both of these books are gems. *Matrix Methods* is a must for any engineer.

[6] Wylie, C. R., *Advanced Engineering Mathematics*. McGraw Hill, 1960.
A fine undergraduate book, written by a great teacher at University of Utah.

[7] Faddeeva, V. N., *Computational Methods of Linear Algebra*. Dover Publications, Inc. New York.
This reference has influenced this book through the notes of a dear friend and boss, Dr. Lee. I. Wilkinson at the General Electric Co., and later, Honeywell. The method of Danilevsky described herein follows Lee's notes — which references this book.

[8] Press, W., Flannery, B., Teukolsky, S, and Vetterling, W. *Numerical Recipes in Pascal*. Cambridge University Press, 1989.
Laguerre's method for determining the initial estimate of a polynomial root, described in Appendix B, was taken from this book, page 296.
This reference describes Crout's method, which is more efficient in the solution of simultaneous equations than that found in this work.

Author's Biography

MARVIN J. TOBIAS

A native Utahn (Salt Lake), **Marvin J. Tobias** graduated in Mechanical Engineering (ME) from the University of Utah.

He joined the General Electric Company and began a series of rotating assignments through various Departments in the East, initially in the General Engineering Laboratory, then in the Jet Engine Department. In the fall, Marv successfully applied for a three year intensive training program—the Advanced Engineering Program. During the latter years of the program his assignments were in radar signal and data processing.

Assignments in both ME and EE proved the need for further education and training in mathematics, so the middle year (B course—applied mathematics) was particularly interesting.

Over the years, Marv became a lecturer for the B course, developing many notes on matrix algebra and calculus. As the PC became more powerful, with sophisticated word processing and graphics software, those notes became the content of this book.

Index

Printed in the United States
by Baker & Taylor Publisher Services